AF477203

Principles in
Seed Vigour Testing

Principles in
Seed Vigour Testing

Dr. Satyanarayana Rao Vallabhaneni

BSP **BS Publications**

A unit of **BSP Books Pvt. Ltd.**

4-4-309/316, Giriraj Lane, Sultan Bazar,

Hyderabad - 500 095.

Phone : 040 - 23445605, 23445688

Principles in Seed Vigour Testing *by Dr. Satyanarayana Rao Vallabhaneni*

© 2021, *by Publisher*

All rights reserved

Published by:

BS Publications
A unit of **BSP Books Pvt., Ltd.**

4-4-309/316, Giriraj Lane, Sultan Bazar,
Hyderabad - 500 095
Phone : 040 - 23445600, 23445688
e-mail : info@bspbooks.net
www.bspbooks.net

ISBN : 978-93-88305-91-4

With love on our illustrious parents and fountains of inspiration

(Late) Vallabhaneni Chandra Mouli,
B.Sc., B.Ed.,

Smt. Babi Sarojini , V

Gudlavalleru, Krishna (Dist)A.P

Dr. Yanamadala Radha Venkata Krishna Rao
M.B.B.S.
Smt. Basaveswara Lakshmi

Avanigadda, Krishna (Dist.) A.P.

ACHARYA N.G. RANGA AGRICULTURAL UNIVERSITY
Administrative Office : LAM, Guntur - 522 034, A.P., India
http://www.angrau.ac.in
Tel : 91-0863-2347011 (O)
email : vicechancellorangrau@gmail.com

Dr. V. DAMODARA NAIDU
Vice-Chancellor

Foreword

Seed is the utmost important primary biological input, the master key to success in Agriculture. The best initial plant stand establishment in the field, status of the crop, efficiency of other agricultural inputs and finally the crop productivity/production depends on seed vigor. This clearly emphasizes the necessity of timely availability of quality seeds at affordable prices to farmers. Seed vigor is associated with aspects of seed performance i.e., during the storage, rate and uniformity of seed germination, seedling growth and emergence ability to germinate after storage. Henceforth, the evaluation of germination and the identification of seed lots with high vigor and outstanding performance is important for successful crop production and making agriculture an attractive and highly lucrative business.

In the changing global agricultural scenario exposed to climate change recently, the seed vigor counts much to adopt the crop, its establishment and remain robust across diverse environmental conditions. Indian agriculture is of no exception for this, but for diversified agro-climatic regions, a drastic change is required to produce seeds with good vigor and with other desirable traits required both now and in the future. In India, both public and private sectors produce the seeds, but private sector has the edge in their efficient supply to the farmers. But, it is disheartening that, complaints are often being received from farming community regarding the loss of crop or missing the crop season due to poor vigor of seeds. In this context, this book entitled" Principles in Seed Vigor Testing" aims at providing necessary information in the form of basic principles and testing methodologies of seed vigor and these are explained in a lucid manner so that, teachers, students, farmers, researchers and other stakeholders can judge the quality of seeds on their own.

I compliment Dr. Satyanarayana Rao Vallabhaneni for bringing out this book timely that practically aims for testing one of the most important quality parameters for seeds, i.e., the seed vigor before commencing the crop production. The contents in this book are also dealt more scientifically and addressed the issues like seed production, processing, testing, and marketing.

వల్లభనేని దామోదర నాయుడు

(V. DAMODARA NAIDU)

ACHARYA N.G. RANGA AGRICULTURAL UNIVERSITY
Advanced Post Graduate Centre – Lam – Guntur 522 034
Andhra Pradesh – India

Dr. K.L. Narasimha Rao	E-mail: apgc.angrau@gmail.com
Special Officer	Phone: +91-9989557444, +91-7207911999

Foreword

This book covers the basic principles and also the advanced latest technologies related to seed. Hence, this book will be highly useful tool for the post graduate students, researchers, and also field workers in seed industry.

(K. L. NARASIMHA RAO)

Preface

The book consists of ten chapters to cover a wide range of topics basically devoted to understanding seed vigour.

The Chapter 1 provides information on general aspects like introduction, scope and out line of the subject of seed vigour.

Chapter 2 deals with the origin and meaning of seed vigour; relationship between different seed quality parameters like germination, viability and vigour, and components of seed vigour.

Chapter 3 deals with the need, history, definitions and concepts of seed vigour testing.

Chapter 4 covers key elements, significance and importance of seed vigour;

Chapter 5 deals with various factors responsible for differences in seed vigour, and manifestation of differences in seed vigour.

Chapter 6 deals with the relationship of seed vigour with crop growth and seed yield.

Chapter 7 deals with applications and uses of seed vigour.

Chapter 8 deals with the mechanism of seed vigour.

Chapter 9 deals with seed vigour testing requirements & procedures, applications, usefulness, limitations and precautions during seed vigour testing.

Chapter 10 deals with precision, reporting and interpretation of vigour test results.

Acknowledgements

With reverence, I owe my gratitude to my parents Smt. V. Baby Sarojini and Late Sri Chandra Mouli, Vallabhaneni., B.Sc. B.Ed. of Gudlavallueru, in shaping my career with their affection, love and encouragement. I am also grateful to my in-laws, Smt. Y.B. Lakshmi and Dr, and Dr. Radha Venkata Krishna Rao, Yanamadala, M.B; B.S; of Avanigadda, Krishna District for their blessings showered on me. I have no words left to express my sincere thanks to my wife, Dr. Radha, Yanamadala, currently working as Professor and Head, Department of Agricultural Economics, at Agricultural College, Bapatla for her constant support

I am grateful to Dr. M M K Durga Prasad, Principal Scientist (Oilseeds) who taught me basics in writing Research papers and, Dr. C. Panduranga Rao, Professor and Head Department of Genetics and Plant Breeding at Agricultural College Bapatla, who entrusted me the responsibility of teaching UG Seed Technology course.

I express my thanks to all the authors and publishers online and off line journals, whose books I have frequently consulted and referred in this book. I am also grateful to the teaching staff in department of Genetics and Plant Breeding, Agricultural College, Bapatla for providing me with ample time to prepare the script of this book. My special thanks are due to Miss. Kavya Pati and Miss. Sruthi Koradi, Research Scholars in Genetics and Plant Breeding Department, Agricultural College, Bapatla for their help in proof reading and uploading the script files to the publishers. I am thankful to Mr. Naresh, Manager and M/s. Nikhil Shah and Anil Shah of BS Publications in bringing out this book.

The author is thankful to the Vice –Chancellor, Acharya N G ranga Agricultural university, Lam, Guntur for granting permission to publish the book.

Contents

Foreword ....(vii & ix)

Preface .... (xi)

Acknowledgements .... (xiii)

CHAPTER 1

Seed Vigour *(Quality knows no border)*

1.1 Introduction 1

1.2 Seed Quality 2

    1.2.1 Components of Seed Quality 5

    1.2.2 Major Reasons for Seed Deterioration in Storage Level 6

    1.2.3 At Physiological Level 6

    1.2.4 At Biochemical Level 6

    1.2.5 At Molecular Level 7

1.3 Seed Vigour 7

    1.3.1 What it is? 7

    1.3.2 A Critical Factor 8

1.4 Is Seed Vigour a GOD's Gift? 9

1.5 Complications Related to Seed Vigour 10

1.6 Seedling Vigour in Field Crops 11

    1.6.1 Cereals 12

        1.6.1.1 Rice (*Oryza sativa*) 13

        1.6.1.2 Wheat (*Triticum aestivum*) 14

        1.6.1.3 Maize (*Zea mays*) 15

        1.6.1.4 Barley (*Hordeum vulgare*) 16

        1.6.1.5 Sorghum (*Sorghum bicolor L.*) 16

    1.6.2 Oil Seeds 17

        1.6.2.1 Groundnut (*Arachis hypogea*) 17

        1.6.2.2 Brassicas 17

        1.6.2.3 Soybean (*Glycine max*) 18

    1.6.3 Cotton (*Gossypium herbaceum*) 18

    1.6.4 Grain legumes 19

        1.6.4.1 Peas (*Pisum sativum*) 19

    1.6.5 Forage Crops 20

    1.6.6 Vegetable Crops 21

    1.6.7 Other Miscellaneous Crops 21

1.7 Policy Issues 22

1.8 Summary 24

1.9 Exercise 25

1.10 References 27

CHAPTER 2

Origin and Meaning of Seed Vigour

(A strong edifice cannot be built with poor bricks)

2.1 Introduction 35

2.2 Origin of Seed Vigour 35

    2.2.1 Genetic Origin 36

    2.2.2 Physiological and Biochemical Origin 37

2.3 Meaning and Definition of Seed Vigour 38

    2.3.1 Differences in Seed Vigour 39

    2.3.2 Vigour Parameters 41

    2.3.3 Adverse Effects of Using Low Vigour Seeds 42

    2.3.4 Relationship of Viability and Germination with Vigour 45

    2.3.5 Seed Viability 49

    2.3.6 Germination 50

        2.3.6.1 Phase –I: Imbibition 52

        2.3.6.2 Phase –II: Active metabolism and Hydrolysis 56

        2.3.6.3 Phase- III: Cell division, Elongation and Growth 57

    2.3.7 Germination and Standard Germination Testing 61

    2.3.8 Basic requirements of a Germination Test 62

    2.3.9 Objectives of Standard Germination Testing 63

    2.3.10 Relevance of Germination Testing 65

    2.3.11 Causes of Variation in Germination Test Results 65

    2.3.12 Limitations of the Germination Test 66

2.4 Summary 68

2.5 Exercise 69

2.6 References 72

CHAPTER 3

Need, History, Definition and Concepts in Seed Vigour and Testing

(It is better to light one candle than to curse the darkness)

3.1	Introduction	77
3.2	What is a Seed Vigour Test?	77
	3.2.1 Essential Characteristics of a Seed Vigour Test	79
3.3	The Need of Seed Vigour Tests	79
3.4	History of Seed Vigour Testing	83
3.5	Definitions of Seed Vigour and Vigour Testing	87
3.6	The Concept of Seed Vigour Testing	90
	3.6.1 Biological Basis for the Concept of Seed Vigour	91
	3.6.2 Modern Basis for the concept of Seed Vigour	92
	3.6.3 The Per-se Concept of Vigour	92
	3.6.4 The Multiple Concept of Seed Vigour	94
	3.6.5 The Molecular basis of Seed Vigour	95
	3.6.6 The Genetic basis of Seed Vigour	97
3.7	Summary	97
3.8	Exercise	98
3.9	References	99

CHAPTER 4

Principles and Elements in Seed Vigour Testing

(Self trust is the first secret of success)

4.1	Introduction	105
4.2	Aims of Vigour Testing	105
4.3	Goals of Seed Vigour Testing	106
	4.3.1 The Ultimate Goal	106
	4.3.2 The Common Goal	106
4.4	Basis of Vigour Testing	106
4.5	Principal Objectives of Seed Vigour Testing	107
	4.5.1 Requirements of Vigour Test	107
	4.5.2 Key Elements of Seed Vigour	107
4.6	Criteria for Vigour Testing	108

4.7 Predominant Views in Seed Vigour Testing 108

4.8 Importance of Seed Vigour and Vigour Testing 109

    4.8.1 Importance of Seed Vigour 109

    4.8.2 Importance of Seed Vigour Testing in Agriculture 110

4.9 Factors Limiting Seed Performance 111

4.10 Why Vigour Testing? 115

4.11 Principles of Seed Vigour Testing 117

4.12 Nature of Vigour Tests 117

4.13 Characteristics of a Seed Vigour Test 117

4.14 Significance of Seed Vigour 119

    4.14.1 Significance of Seed Vigour Tests 119

4.15 Summary 120

4.16 Exercise 121

4.17 References 123

CHAPTER 5

Differences and Manifestation of Seed Vigour

(*Knowing others is wisdom, knowing yourself is enlightenment*)

5.1 Introduction 127

5.2 Differences in Seed Vigour 128

    5.2.1 Morphological Maturity Indices 128

5.3 Factors Responsible for Differences in Seed Vigour 130

    5.3.1 Genetic Factors 130

        5.3.1.1 Within a Genotype 130

        5.3.1.2 At Individual Seed Level 134

        5.3.1.3 At Seed Population Level 135

    5.3.2 Environmental Factors 137

        5.3.2.1 On Mother Plant 137

    5.3.3 Differences in Seed Vigour due to Storage Related Factors 141

        5.3.3.1 Physiological Damage 142

        5.3.3.2 Pathological Damage 142

    5.3.4 Differences in Seed Vigour due to Ageing of Seeds 143

    5.3.5 Differences in Seed Vigour due to Imbibition Damage and
        Vigour Loss 144

    5.3.6 Differences in Seed Vigour due to Interaction of Ageing and
        Imbibition Damage 145

5.3.7 Other Factors 146

    5.3.7.1 Effect of Tillage 146

    5.3.7.2 In the Field after Sowing 146

    5.2.7.3 Effect of Chemicals and Fertilizers 146

    5.3.7.4 Effect of Soil Moisture and Temperature 147

5.4 Manifestation of Differences in Seed Vigour 147

    5.4.1 Germination 147

    5.4.2 Emergence.. 148

    5.4.3 Establishment 149

    5.4.4 Growth 149

    5.4.5 Morphological Abnormalities................................. 150

    5.4.6 Longevity of Seeds 150

5.5 Summary 150

5.6 Exercise 152

5.6 References 155

CHAPTER 6

Relationship of Seed Vigour with Yield

(The hands that help are holier than the lips that pray)

6.1 Introduction 163

6.2 Factors Responsible for the Effect of Poor Seed Vigour on Crop Yield.................. 164

    6.2.1 Initial Crop Growth 164

    6.2.2 Seed Size.. 164

    6.2.3 Environment.. 164

        6.2.3.1 Environmental Stress during Vegetative Growth........ 164

        6.2.3.2 Environment Stress during Seed Filling..................... 165

        6.2.3.3 Environment Stress during Seed Development........... 165

6.3 Direct and Indirect Effects of Seed Vigour on Crop Yield 165

    6.3.1 Direct effects of Seed Vigour on Crop Yield 166

    6.3.2 Indirect effects of Seed Vigour on Crop Yield...... 166

6.4 Effect of Seed Size on Seed Vigour and Crop Yield.. 166

    6.4.1 Effect of Large Size of Seeds on Seedling Vigour........................... 167

    6.4.2 Effect of Small Seeds on Seedling Vigour 168

6.5 Relationship between Seed Vigour and Crop Yield 168

6.6 Relationship between Seed Vigour and Plant Growth 169

    6.6.1 Effect of Seed Vigour on Early Plant Growth........ 169

        6.6.1.1 Initial Seed Growth... 169

6.6.1.2 Efficient Resource Capture 170

6.6.1.3 High Productivity 170

6.6.1.4 Canopy Development 170

6.6.2 Effect of Seed Vigour on Late Plant Growth 170

6.7 Effect of Seed Vigour on Crop Yield 171

6.7.1 Seeds with Low Vigour 172

6.7.2 Seeds with High Vigour 172

6.8 Summary 173

6.9 Exercise 174

6.10 References 175

CHAPTER 7

Seed Deterioration and Mechanism of Seed Vigour

(The impossible is often untried)

7.1 Introduction 181

7.2 Seed Deterioration 182

7.2.1 Seed Viability 183

7.2.1.1 Classification of Seeds 184

7.3 Storage Requirements of Seeds 187

7.3.1 Relationship between Longevity-Temperature and
Moisture Content of Seeds (The Viability Equation) 189

7.4 Manifestation of Seed Vigour 190

7.4.1 Physiological Manifestations of Seed Vigour 190

7.4.2 Biochemical Manifestations of Seed Vigour 190

7.5 Physical and Chemical Reactions involved in Seed Deterioration ... 191

7.5.1 Physical Reactions 191

7.5.2 Chemical Reactions 191

7.5.2.1 Hydrolytic Reactions 191

7.5.2.2 Oxidative Reactions 191

7.5.2.3 Peroxidation Reactions 192

7.5.2.4 Millard Reaction 192

7.5.2.5 Amadori Reaction 192

7.6 Changes in the Ultra-structural Properties of Organelles 193

7.6.1 Nucleus 193

7.6.2 Mitochondria 193

7.6.3 Golgicomplex 193

7.6.4 Plastids 193

7.6.5 Ribosomes and Lysosomes 193

7.6.6 Endoplasmic Reticulum 194

7.6.7 Cell Membranes 194

7.7 Basis of Seed Deterioration 194

7.7.1 Changes at Genetic Level 194

7.7.2 Basis of Seed Deterioration - Changes at Physiological Level 197

7.7.2.1 Changes in Enzyme Activity 197

7.7.2.2 Changes in Respiration Rate 198

7.7.2.3 Changes in the Quality of Chemical Constituents 198

7.7.2.4 Impairment of Membrane Integrity 198

7.8 Basis of Seed Deterioration - Changes at Biochemical and Molecular Level 199

7.8.1 Release of Volatile Compounds 199

7.8.2 Free Radicals 199

7.8.3 Auto –oxidation 200

7.9 Basis of Seed Deterioration and Aageing- Changes at Molecular Level 200

7.9.1 Changes at Protein Level 200

7.9.1.1 Protein Synthesis 200

7.9.1.2 Protein Inactivation 201

7.9.1.3 Protein Modification 201

7.9.1.4 Changes at DNA Level 201

7.10 Theories proposed on Seed Deterioration 202

7.10.1 Depletion of Food Reserves 202

7.10.2 Reducing Sugars 203

7.10.3 Loss of Glossy State 203

7.10.4 Accumulation and Repair of Cellular Lesions Model 204

7.10.5 Loss of Antioxidant Molecules 204

7.10.6 Acquisition of Desiccation Tolerance Model 205

7.10.6.1 Desiccation Damage 205

7.10.6.2 Desiccation Intolerance or Sensitivity 205

7.10.6.3 Desiccation Tolerance 205

7.10.6.4 Acquisition of Desiccation Tolerance 206

7.10.6.5 Desiccation Tolerance- Protection 206

7.10.6.6 Desiccation Tolerance- Repair Mechanism 207

7.10.6.7 Desiccation Tolerance – Response of Metabolism 208

7.10.6.8 Lipid Peroxidation Model 208

7.10.6.9 Biological Effects of Lipid Peroxidation 209

7.10.7 Monitoring Lipid Loss 210

7.10.8 Detection of Free Radicals 211

7.10.9 Detection of Hydro-peroxides 211

7.10.10 Detection of Secondary Products 211

7.11 Strategies to Limit Lipid Peroxidation in Seeds 211

7.12 Ageing / Repair Hypothesis 213

7.13 Vijay- Dadlani Model of Seed Ageing (2003) 214

7.14 Summary 214

7.15 Exercise 215

7.16 References 218

CHAPTER 8

Requirements and Procedures for Standardization of Seed Vigour Testing

(A fault once denied is twice committed)

8.1 Introduction 227

8.2 Aspects of Performance Associated with Seed Vigour 228

    8.2.1 At the Time of Sowing 229

    8.2.2 At and During Storage 229

8.3 Standardization of Vigour Test Procedures 229

    8.3.1 How far a "Vigour Test" is Worthwhile? 232

    8.3.2 Need of Uniform Seed Testing 232

    8.3.3 Need of Reproducibility of Vigour Test Results 232

8.4 Reasons for Lack of Standardization among Laboratories identified by AOSA 233

8.5 Development and Validation of Vigour Tests 234

    8.5.1 Stages in Test Development 234

    8.5.2 Research 234

    8.5.3 Development 234

    8.5.4 Validation 234

        8.5.4.1 Comparative Testing 235

        8.5.4.2 Referee Testing 235

8.6 Recent Classification of Vigour Tests 235

    8.6.1 Validated (Recommended) Vigour Tests 236

    8.6.2 Suggested Vigour Tests 237

    8.6.3 Proficiency Testing 237

        8.6.3.1 The ISTA Vigour Test Committee 238

8.7 Standardization of Vigour Test Protocols 239

    8.7.1 Social and Attitudinal Challenges for Seed Vigour Testing 240

8.7.1.1 Who are going to derive the real benefit from the seed vigour testing information?..240

8.7.1.2 How this information on seed vigour is useful? 240

8.7.1.3 Research leaders...241

8.7.1.4 Seed analysts and research managers......................241

8.7.2 Technical Challenges Encountered during Seed Vigour Testing.......242

8.7.2.1 Vigour Test Requirements..242

8.7.2.3 Variables in Vigour Testing...243

8.8 Summary 245

8.9 Exercise..246

8.10 References 248

CHAPTER 9

Precision of Seed Vigour Testing and Reporting of Seed Vigour Test Results

(Efficiency is the capacity to bring proficiency into expression)

9.1 Introduction.. 251

9.2 Precision is important in Seed Vigour Testing. Why? 252

9.2.1 The Need 252

9.2.2 Why Precission? 252

9.2.3 How to Achieve Precision while Conducting Vigour Tests? 253

9.3 ISTA Vigour Tests 253

9.3.1 Electrical Conductivity Test... 253

9.3.2 Cold Test... 254

9.3.3 Accelerated Ageing Test... 254

9.3.4 Tetrazolium Test 254

9.3.5 Tests based on Seedling Growth... 254

9.4 Reporting and Interpretation of Seed Vigour Test Results 255

9.4.1 Presentation of Vigour Test Results 255

9.4.2 Use of Standards while Presenting Seed Vigour Test Results 256

9.5 Control Samples in Seed Vigour Testing 257

9.5.1 Selection of Control Seed Lots 257

9.5.2 Use of Control Samples... 257

9.5.3 Maintenance and Preservation of Seed of Control Samples.............. 257

9.5.4 Maintenance of Moisture in Control Samples.................................... 258

9.6 Interpretation of Vigour Test Results.. 258

9.7 Summary 259

9.8 Exercise 260

9.9 References 262

CHAPTER 10

Seed Vigour Testing

(If quality is definite, the benefits are infinite)

10.1 Introduction 265

10.2 Applications of Seed Vigour Testing 265

    10.2.1 Plant Breeders 266

        10.2.1.1 Objective 267

        10.2.1.2 Practice in Plant Breeding 267

        10.2.1.3 Examples 267

    10.2.2 Consumers 268

    10.2.3 Quality Seed Production 268

    10.2.4 Seed Quality Control 269

    10.2.5 Seed Storage 271

    10.2.6 Seed Marketing and Promotion 271

10.3 Plug Production in Nurseries for Transplant Production Systems 272

10.4 Uses of Seed Vigour Information 274

10.5 Usefulness of Seed Vigour Tests 274

10.6 Limitations of Seed Vigour Tests 275

10.7 Precautions to be taken while Conducting Seed Vigour Tests 276

10.8 Strategies to Improve Seed Vigour 277

10.9 Summary 279

10.10 Exercise 280

10.11 References 282

ANNEXURE 1

Acronyms

Acronyms 285

ANNEXURE 2

(Strength does not come from physical capacity, it comes from indeomitable will)

Glossary

Glossary 287

Index 297

Seed Vigour *(Quality knows no border)*

1.1 Introduction

The universally agreeable fact about seed is that seed, being a reproductive unit of the plant kingdom, has been bestowed with the twin commitments viz., (i) a biological role and (ii) an ecological role. The former deals with the nourishment of the living cells of embryonic axis, especially during the initial stages of seedling emergence with respect to growth and development[1], whereas, the later deals with the fulfillment and establishment of a new and robust seedling, so as to give rise to the succeeding generation. In other words, the success of seed germination and the establishment of a normal seedling features determines the propagation of any plant species that has both economic as well as ecological significance. It is estimated that approximately 80% of economically significant crops are planted with either direct or indirect use of seeds. In all agricultural systems, the seed used for commercial cultivation is considered as a means of production, because seed is not only a starting point and first determinant of the future plant development, but also the master key to success with its cultivation consequently.

Since, 'Agriculture' has been aptly defined as "interference with nature for the benefit of humankind," probably, during the initial

LEARNING OBJECTIVES

- The twin commitments of seed as a connecting link between two generations and establishment of a robust new seedling.
- Positive and negative selections.
- The need to establish a good seedling character.
- Capacity of a seed to give raise to normal seedlings and healthy plants.
- Which quality aspect of seed is still illusive?
- Which parameter transforms the quality of seed in a specified period of time?
- Which quality parameter of seed influences seed purchasing decision of farmers?
- Why farmers should have easy access to seed with high germination ability coupled with high vigour?
- How far it is correct to say that seed vigour is a GOD's gift?
- How to identify seeds with high vigour?
- Some important accumulations during seed maturation phase.
- Accumulation of degenerative changes during post maturation phase and causes of death of seed at different levels?
- How can you say that seed vigour is a complex phenomenon?
- List of characteristically confronted complications related to seed vigour.
- Vigour status in individual seed crops.
- Why study of seed vigour becomes highly essential?

---

[1]There by serves human nutrition and food security.

stages of development of agriculture itself, our forefathers might have gradually started and gained control over the seeds in the first step followed subsequently by soil and water solely with an intention of feeding themselves, their children and the animals they have domesticated thus far. For achieving this, obviously, they might have depended on a vast variety of pre-existing plant forms, evolved several thousands of years ago, and available that too within their own vicinity. But, even today, presumably ten to twelve thousand years after initiation of agriculture, still we could explore only a fraction of plant species existing around us, because the present dependency is confined exclusively to a mere handful of species necessary only for meeting our daily calorie requirements from plants growing around us either from native domesticated species or by introductions through the means of both positive (picking seeds from a number of desirable plants) and negative (elimination of undesirable plants from the variety) selections.

'Soil', the true nature's gift, and 'seed quality' are considered to be the two most important and basic farming inputs. A wise utilization of these two critical inputs together is expected to provide higher yield and good quality produce very easily. But, in reality matured dry seeds of most of the species require a period of dry storage, also known as after ripening, so as to relieve them from dormancy Iglesias *et al.,* (2011). Dormancy is a physiological condition in which seeds will not germinate even under optimal conditions that are otherwise favorable for their germination, when they become non dormant, Baskin and Baskin (2004).

Seed quality is gradually gaining increasing importance in the light of global climate change. Quality seeds and planting materials are the key agricultural inputs, which not only determine the productivity of crops, but also establish the efficacy of other agricultural inputs such as fertilizers, pesticides and irrigation and accounts for 20-25 per cent of productivity. This clearly shows that the timely availability of quality seed at affordable price to farmers is necessary for achieving higher agricultural productivity and production. The wide diversity prevalent in agro-climatic conditions of the country is suitable for cultivation of a large number of crops and varieties, which in turn necessitates the production of quality seed and planting materials for the diversified range of crops for achieving the targeted production. The organized sector, comprising of both public and private sector, accounts for only 15%-20% of the total seed distributed in India.

 ## 1.2 Seed Quality

"Seed quality" is a multifaceted concept with several components. The journey of a true seed, really starts on the mother plant itself and passes through various stages of pre-harvest, harvest and post-harvest operations such as processing, storage and distribution before being transferred to the field for sowing with a view to establish the succeeding crop. The modern post-harvest processing involves a series of mechanical operations such as drying or cooling, cleaning, grading, chemical treatment and bagging or packaging. At pre-harvest and harvest stages, seeds will be at their maximum dry weight with peak mass or physiological maturity and is considered to be initiation of most crucial period for the deterioration. However, consequent to reaching the harvest maturity, at which the seeds are nearly close to their storage moisture content, suffers from different deterioration related problems such as pre-matured drying, (as a consequence of water stress) or slow drying leading to chlorophyll loss as in windrows and weathering (due to high temperature, rain and humidity).

During the course of cultivation of plants for agricultural purpose, several production factors like geographical area, (environmental conditions and soil characteristics) economical frame work (agronomic management including tillage, irrigation, fertilization and crop protection) and the farmers themselves owing to their skills to take appropriate and timely decisions etc., play a crucial role. Therefore, realization of satisfactory results depends not only on the production of higher yield of valuable products and the economic benefits thus derived, but also depends on the local and global market trends prevailing at that time.

Seed biology in general and germination in specific have drawn the attention of farmers and scientists alike since antiquity and consequently led to the establishment of agriculture. Theophrastus (372-287 BC) the first seed physiologist, studied seed quality in a scientific way considering it as a real problem related to agriculture, Evenari (1984). Seedlings emerging even from the robust and healthy seeds appear to be less likely to be over-come by diseases, weeds and insects when compared to weaker ones and thereby play a vital role not only in the continuation of species in general but also helped a lot in enduring commercial agriculture and allied sectors based production systems in particular. On the other hand rapid and uniform emergence of seedlings of a desired cultivar is the lone key that endures high plant performance by way of uniformity in development, yield and quality of the harvested product. It means, one has to take into consideration the high level of vulnerability to which the seed gets exposed to is determined by the process of germination, which in turn is the most crucial phase in the life cycle of any plant.

In India, a large majority of seeds sown are produced by farmers themselves and is usually low in germination. But in reality, the contribution of unorganized sector, comprising of mainly farm saved seeds is in the order of 80%-85% of total seed planted in the country, produced traditionally by farmers themselves in the previous season. These seeds are usually low in their germination because of the poor prevailing storage conditions. Therefore, good seedling establishment character becomes not only desirable but also essential for maintenance of optimum seed population, if the modern crop production has to be both resources efficient and cost effective.

No doubt, the evaluation of germination and the identification of seed lots with high and outstanding performance is an important initiative during the course of successful crop production. It is also anticipated that, the information generated in appropriate seed testing laboratory must not only detect the difference in physiological potential of seed lots accurately, but also demonstrate the potential of seeds both during storage and after sowing in such a way as to indicate the adequacy of procedures used in evaluating the seeds / seed lots. In addition to these, there are several other equally important components of seed quality that needs much emphasis. These quality parameters can be divided into three broad groups. viz., (i) Cultivar (uniformity, seed weight and purity), (ii) Hygiene (seed health, contamination from other noxious weeds, insects, termites, storage fungi, virus and nematodes) and (iii) Potential performance, which include germination, seed and seedling vigour, moisture, and uniformity in field emergence, Hampton, (1994).

Seeds developed under favorable environmental conditions and managed appropriately from physiological / harvest maturity to commercialization constitutes a solid basis for enhanced field stand establishment and higher crop productivity. The ultimately theoretical capacities of seeds to express their vital function both under favorable and unfavorable conditions are governed by their physiological potential, which includes viability and vigour.

But in reality, the results of seed testing laboratories are reproducible easily in the field only when the environment is favorable both for the emergence and initial seedling growth. Therefore, this particular situation obviously necessitates the use of other approaches for estimating the performance of seed even under less favorable field conditions, as deviations in ideal field conditions are the most common particularly in a country like India with diversified agro-ecological situations. However, one should always keep in mind that the concept of suboptimal conditions varies with species, cultivar, and stage of development in particular and availability of water, temperature, soil characteristics and plant protection management practices in general.

Prudent mechanism for seed certification, testing, labeling and enforcement is necessary to maintain the seed quality. Around 80-90 per cent of all seed quality tests conducted or performed annually by the seed testing laboratories across the globe are purely confined to the estimation of either purity and /or germination only. The methods for evaluating the germination have been designed to have high levels of reproducibility, or repeatability and reliability. However, worse than optimum conditions will only be prevailing under the real field situation. Under these circumstances, seed lots not differing in germination may differ in their emergence and storage potential, Kolasinka *et al.,* (2000). Seed vigour can be evaluated by rapid and homogeneous germination, as tolerance to stress after sowing and length of germination duration. Black and Bewley, (2000). Powell *et al.,* (2001) regarded seed vigour as an additional aspect to the physiological quality of germinable seeds. Bowtright *et al.,* (2002) reported that the expression of vigour in field conditions and its translation in to higher yield depends purely on the prevailing environment.

Fig. 1.1 Use of quality seed determines the initial plant stand establishment

In order to have a vibrant seed industry, varietal development, plant variety protection, seed production, quality assurance, creation of infrastructure for seeds, transgenic and imported materials on one hand and export of seed and promotion of domestic seed industry on the other hand are the prime necessities. Introduction of The Seeds Bill 2004 in the Indian Parliament not only overcomes the limitations of The Seeds Act 1966, but also provides for the necessary regulation of seed quality and planting material of all agri-horticultural crops to Indian farmers. Besides this, this Bill also offers to curb the sale of spurious and poor quality seed; protects the rights of farmers; enhances private participation in seed production, distribution and seed testing; liberalizes import of seeds and planting materials in tune with WTO standards.

In addition to this, on one hand, stress should be laid on enhancing the seed multiplication ratio from Breeder seed to Foundation seed and from Foundation seed to Certified seed by seed producing agencies engaged both in public and private sectors. On the other hand, stress should also be laid to ensure adequate and timely supply of quality seed at the door step of farmers through appropriate tie-ups with National and State Seed Corporations. Adoption of Good Agricultural Practices, enhanced seed replacement (SRR) rate popularizing new technologies such as ridge and furrow sowing, opening dead furrows and deep ploughing, zero tillage, seed treatment and integrated nutrient and pest management are some of the recommendations for enhancement of the productivity in different crops.

1.2.1 Components of Seed Quality

There are four chief components of seed quality, viz., genetical, physical, physiological and soundness. These aspects get adversely affected during seed production, processing, storage and transport. In order, to ensure the production of quality seed, the seed technologists have already directed their efforts in suggesting appropriate corrective measures to the problems originating during seed production such as roguing to maintain genitical quality; use of precision equipment to maintain physical purity and soundness. Seed vigour has been identified as one of the important physiological characteristics of seed quality, reflecting potential seed germination, seedling growth, seed longevity and tolerance to adversity (Sun *et al.,* 2007). On the other hand, seeds with strong vigour significantly improves the speed and uniformity of seed germination and final percentage germination and ultimately led to perfect field emergence, good crop performance and even higher yield under different conditions. Foolad *et al.,* (2007).

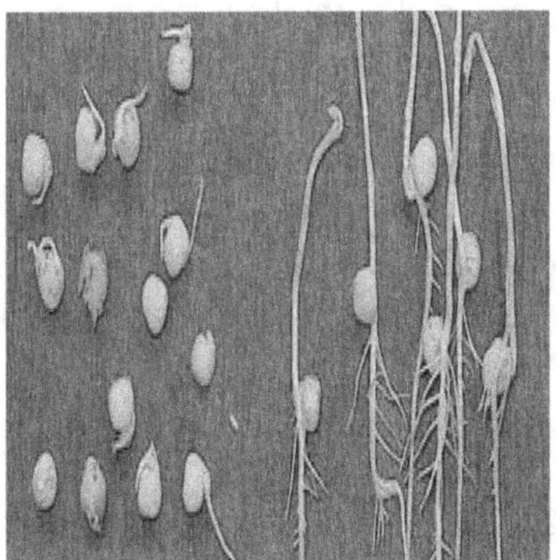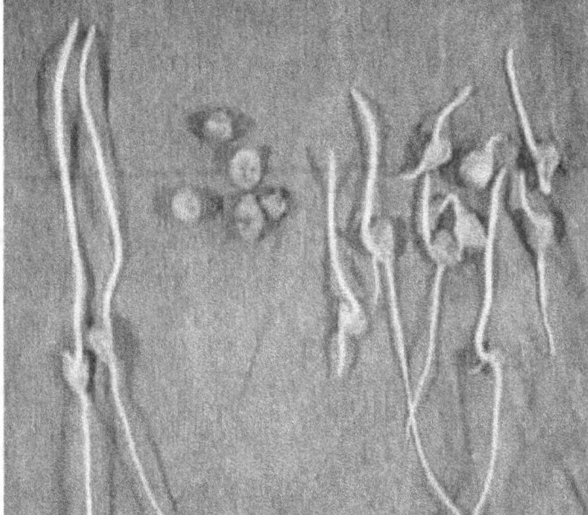

Fig. 1.2 Seeds showing poor potential germination

In addition to the above mentioned inherent and environmental parameters, the incidence of seed borne pathogens on one hand, coupled with seed ageing at physiological, biochemical and molecular levels on the other hand, permits gradual accumulation of degenerative changes in the seed, and thereby ultimately resulting in the deterioration or death of the seed, which is usually expressed in the form of failure or complete loss in ability of germination, (Fig.1.2). Deterioration in seeds can occur in the field after they attain physiological maturity, particularly when the harvesting is delayed in wet weather and is commonly referred to as weathering, a condition which is as bad for seed quality as bad storage conditions. But before this stage is attained, the individual seeds lose their vigour at different rates. This is what we generally observe as increased sensitivity to adverse storage conditions, resulting in changes at both physiological, biochemical and molecular levels as detailed below.

1.2.2 Major Reasons for Seed Deterioration in Storage Level

 (i)   Too humid or too hot environment unsuitable for seed storage,
 (ii) Storing of seeds at high moisture content,
 (iii) Placing low quality seed in storage,
 (iv) Insecure warehousing facilities such as poor ventilation, prone to heating,
 (v)   Exposure to moisture penetration,
 (vi) Non adoption of first in first out policy leading some seed to prolonged storage,
 (vii) Kind of crop seed stored that are naturally prone to rapid deterioration such as groundnut and soybean,
 (viii) Relative humidity of > 60% facilitates rapid seed deterioration in storage houses.

1.2.3 At Physiological Level

The changes in seed quality at physiological level are represented by

 (i)   Slower rate of seedling growth and development and less uniform germination,
 (ii) Reduced tolerance to biotic and a biotic stresses and early seedling growth
 (iii) Lack of hypocotyl elongation and stunting of radical
 (iv) Ultimately, an unexpected increase in the number of abnormal seedlings.

1.2.4 At Biochemical Level

The earliest possible changes that occur chiefly at sub-cellular level lead to

 (a) Reduced membrane integrity, resulting in enhanced solute leakage from imbibed seed which effects several metabolic activities including respiration rate in respect of gaseous exchange, enhanced RQ, reduced P/O ratio, enzyme activity in respiratory pathways.

 (b) Deterioration of membranes,

 (c) Decreased ability to retain inorganic electrolytes organic solutes, viz., aminoacids, organic acids and water soluble sugars

 (d) Reduced enzyme activity of dehydrogenases, carboxylases, amylases, lipases, esterases was recorded during seed germinaton.Increased in the activity of degradatve enzymes like protelylases, phopspholipses with seed ageing was reported by Basavrajjapa et al., 1991. And inhibits the activities of Peroxidase scavenging enzymes like peroxidase, catalase, SOD and free radicals play a protective role.

 (e) Decreased respiration rates.

1.2.5 At Molecular Level

The two pronounced changes at molecular level are represented by

(i) Impairment of synthesis of macromolecules particularly protein metabolism, initiation and extent of de novo synthesis results in loss of vigour.

(ii) Which in turn leads to an increase in the incidence of chromosomal aberrations, alteration in polypeptide profiles due to protein cross linking, glycation of functional proteins. The translational efficiency, considered to be the ability to mobilize seed storage reserves in a controlled manner and detoxification efficiency are said to be the essential mechanisms for seed vigour. Volatile aldehydes together with acyl radical are the principle agents of aberrations at nucleic acid level enhance bring alteration in the structure by cross-linking with micro molecules. In addition to this free radicals too attack nucleotide bases and result in increased propetency of genetic mutations. Further ageing of seeds result in cleavage of DNA in embryonic tissues which need to be repaired during pre-germination hydration phase-II. Similarly impairment of key enzymes like DNA ligase, DNA polymerase and accumulation of cellular lesions beyond the threshold levels result in failure of seed germination and seedling development.

Thus seed vigour is a very highly complex phenomenon, operating both at biochemical and molecular levels, duly involving all the energy and biosynthetic metabolisms, besides co-ordination of cellular activities, for transportation, breakdown and utilization of reserve foods at germination level and therefore, it not only involves in the speed and totality of germination, but also determines the pushing power of the seedlings to the above ground level, that too in the presence of a wide array of abiotic and biotic stress situations under which germination usually occurs under field conditions.

1.3 Seed Vigour

1.3.1 What it is?

Seed produced for commercial utilization from different sources under optimal conditions usually results in similarly high levels of germination under controlled laboratory conditions. However, the same seeds under more stressful conditions in the field may be experiencing and exhibiting vastly contrasting abilities to establish plants due to differences in their vigour (Fig. 1.3). Although this is a commonly observed phenomenon, seed vigour has proven difficult even to define precisely. A widely accepted definition of vigour is "the sum total of those properties of the seed that determine the potential level of activity and performance of the seed during germination and seedling emergence" (Perry, 1978, 1980). This has evolved and extended to the current International Seed Testing Association (ISTA) definition: "Seed vigour is the sum of those properties that determine the activity and performance of seed lots of acceptable germination in a wide range of environments" (ISTA, 2015). Thus, seed vigour is not a single measurable property, but a concept associated with aspects of seed performance that include: rate and uniformity of seed germination and seedling growth; emergence ability of seeds under unfavourable environmental conditions; and performance after storage, particularly the retention of the ability to germinate. Vigour can therefore be considered as the potential performance of viable seeds in agricultural practice and this is determined by the complex interaction between genetic and environmental components (Whittington, 1973; Hodgkin and

Hegarty, 1978). However, the reasons for variation in this performance are complex and remain little understood.

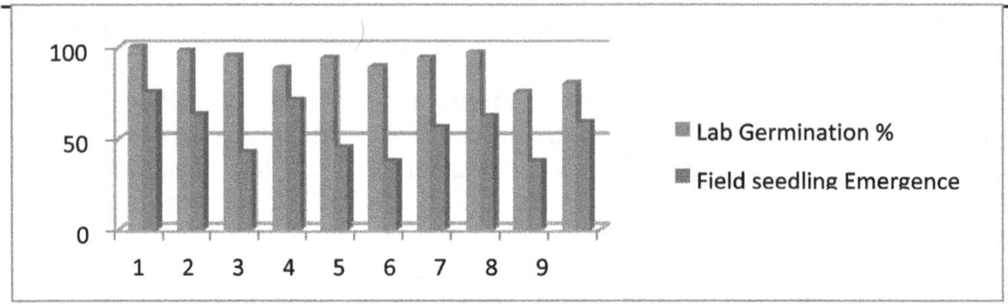

Fig. 1.3 Seed germination under laboratory and field conditions after storage

1.3.2 A Critical Factor

The chief goal of agriculture for realizing improved crop production is to have a rapid and uniform germination and seedling establishment once the seed is sown in the field. The capacity of a seed to give rise to normal seedlings and establish a healthy plant is usually referred to as Seed Vigour, which in simple terms of normal seedling morphology is the rate or speed at which seeds germinate and grow during their early stages of seedling establishment. Therefore, the vigour of seeds that usually refers to and lies not only in the inherent strength but also in the ability to germinate and establish normal seedlings successfully. Therefore, Finch - Savage (1995) aptly reported that successful crop stand establishment depends largely on high seed vigour. Beyond doubt, seed vigour has been proved at several occasions and widely recognized as a critically important seed quality trait. Despite the persistent efforts and perfections brought out by Seed Technologists on the one hand, Seed Certification and Law Enforcement Authorities on the other hand together with the strong will and determination of seed growers and seed producers for the past few decades, the aspect of seed quality, with particular reference to seed vigour still remains mostly misleading and highly illusive.

During the preliminary stages of investigation, seed vigour was considered only as an indicator of seedling growth rate under field conditions, Frank (1950). If it is so, then why the seeds harvested afresh even from the same field on the same day and time differ in their yield potentiality despite having the same genetic background? This only points out and clearly demonstrates its dependency to a larger extent on the adoption of different methods and history of seed production in addition to subsequent post harvest handling including storage and transport to till the seed is used for rising the ensuing generation. Therefore, vigour of seeds, the great indiscernible parameter, thus transforms the quality of seed only over a specified period of time and thereby exerts greater influence while making seed purchasing decisions by the poor and uneducated farmer.

When it comes to commercial cultivation, selection of a right variety is extremely important, because different varieties behave differently in accordance with the soil, climate, irrigation and timely supply of other inputs in an area. It means, a variety may produce excellent quality seed, which can fetch a premium price in one area, and the same variety may register total failure and rejection in another area. For a moment, let us assume that a few seed growers in general, an uneducated and elite farmers alike in specific, by utilizing all resources available at their disposal, if only produces a bad crop, it only reflects precisely and exclusively on the quality of seeds used by them as planting material. Since, in simple economic terms, seed is considered as the cheapest of all inputs, barring in those crops where the seed multiplication ratio is very low such as Groundnut (1:8) or where the production of hybrid seed is cumbersome due to the incorporation of high-end technologies like Bt. Cotton etc., in realizing higher productivity. Therefore, this clearly validates the requirements that the end users or farmers should have easy access to seed with good germination ability besides high seed vigour.

 ## 1.4 Is Seed Vigour a GOD's Gift?

Even today, neither the biochemistry nor the biophysics of protoplasm present in the seed is well understood. In addition to this, only a little is known about either the effects of environment on seed development and maturation, mechanism of attack of various kinds of fungi in innumerable ways, several enzymes and their interaction with the environment, effects of injury caused to embryo, seed coat and cotyledons on short and long term basis makes seed vigour a more complex phenomenon.

At micro level, the accumulation of sugars, antioxidants, oligosaccharides and late embryogenesis abundant proteins etc., in seeds during their development on the mother plant is expected to serve two basic purposes, viz. (i) survival from desiccation and (ii) permits the cells of the seeds to persist in dry state for longer period of time. The layers of seed coat or testa, pericarp and endosperm or perisperm etc. not only cover the embryo but also protect the embryo from physical damage in addition to aiding in the dispersal mechanism. The testa becomes a dead tissue in most cases by the time seed attain their maturity. It is relatively permeable in most seeds; cracks readily upon hydration and facilitates embryo swelling. In case of leguminous seeds, once the seed is dried to low moisture contents, the seed coat becomes hard and acts as an impermeable barrier, even restricting the entry of water and there by prevents germination.

It is widely accepted that, the intrinsic genotypic potential expression of seed depends to a large extent on the interaction between its internal as well as external environments. The internal environment, also known as seed production environment, is the one in which the seed develops gradually on the mother plant and attains its maximum physiological maturity and there after declines slowly till planted in the subsequent season. It means, the vigour of seed, is not a **God's gift,** as thought of earlier. But it is the one, which is gradually acquired by the seed itself over a period of time in its own production environment. Further, on one hand, in the production environment, the seed health needs to be managed judiciously not only during seed production cycles but also in the post harvest-handling environment including the storage and transportation. Because, this particular stage, at times, is considered not only very much crucial but also renders the seed survival highly vulnerable. On the other hand, the external environment too plays a vital role, by modifying the vigour, both at micro and macro levels. Further, the soil conditions both during planting, and even during pre-emergence are often may not be optimum and congenial. A seed population is believed to launch an optimum plant stand under both optimal as well as suboptimal soil environments so as to meet the needs of cost-effective exploitation.

The Seedling Vigour Classification Test (SVCT) is merely an extension of germination test where in the normal seedlings obtained are further classified into strong and weak categories.

Usually seedlings are evaluated for vigour at four points Vig., root system, hypocotyl, cotyledons and epicotyl. If all these four areas as well developed and free from defects, an indication of satisfactory performance in a wide range of climatic conditions as classified as strong. Whereas normal seedlings with mere deficiencies such as missing part of root / breaks / necrotic lesions / twisting / curling etc. are classified as weak, while those that do not show any satisfactory growth are classified as abnormal (Fig:1.3).

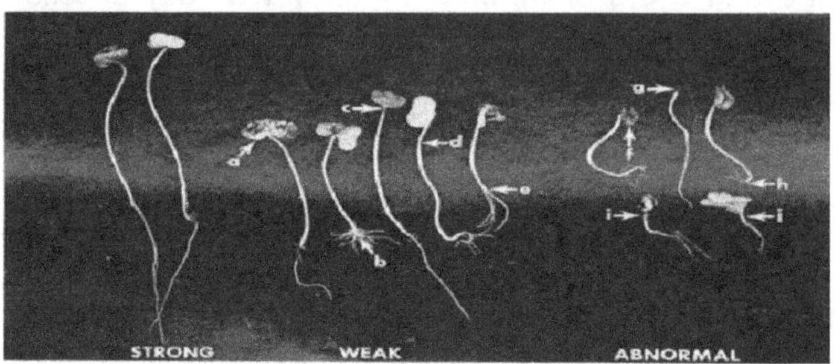

Fig. 1.3 Seedling Vigour Classification Test (SVCT)

1.5 Complications Related to Seed Vigour

Some of the important and characteristically confronted complications that are related to seed vigour are listed below.

1. Germination is seriously affected in a number of situations, viz
    (a) Both in the field after sowing and in storage places before sowing,
    (b) Both at the godowns of retail shops and in farmer's store rooms,
    (c) At all places wherever the seed gets exposed to chemicals, fungicides, pesticides, fertilizers, micronutrients and weedicides *etc.,*
    (d) Seeds stored in cold storage facilities wherever potatoes were stored previously,
    (e) Storing the seed along with chilli both under cold and normal storage facilities,
    (f) Accidental spray or use of excessive dose of herbicide glyphosate, may not kill the seed immediately, but only exerts influence in the consequent germination. The seeds may be alive and do germinate but give only abnormal seedlings, weak and unhealthy plants in the field under predominantly suboptimal conditions.

2. Mechanical injury is a common phenomenon in dry areas. Further, seeds harvested, threshed and processed with high-speed machinery that obviously makes the seed more susceptible because of the enhanced respiration rate due to the exposure of embryo of the seed.

3. Carry over seed of all species, if not stored under optimum conditions prescribed for that species is likely to downgrade with respect to germination quality.

4. The seedling diseases of wheat and barley normally affect germination test functioning more during wet years than when compared to dry years

5. Early season dormancy makes the assessment of germination more difficult.

6. Seed quality of field-bean is highly variable due to mechanical damage and the situation will be worse especially in dry years.

7. Prevalence of excessive hot weather during harvesting time; heat generated during artificial drying; heat generated by the highly moist seed as a consequence of higher rate of respiration; higher rate of CO_2 evolved by seeds; and enhanced activity of microbes during storage, is a common problem in all kinds of seeds.

8. A number of factors including heat damage, mechanical damage, chemical damage and occasionally diseases deteriorate the quality of seed quickly.

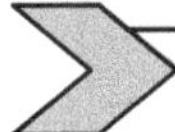 ## 1.6 Seedling Vigour in Field Crops

Seed size is an important physical indicator, one of the important components and widely accepted measure of seed quality that affects vegetative growth and is frequently related to yield, market grade factors and harvest efficiency, Ambika *et al.,* (2014). Genetic variation is the chief cause for variation in size of seed between varieties. Depending on the crop, seed can be classified on the basis of size as very large, large, medium, small and very small. This variation could be attributed to the flow of nutrients in to the seed at the time of maturity from the mother plant. A wide array of different effects of seed size has been reported on germination, emergence and related agronomic aspects in several crop species, Kaydan and Yagmur (2008). Seeds of a seed lot are bound to vary by size, weight and density due to production environment and cultivation practices. Large seeds in general have high seedling survival growth and establishment, Jerlin and Vadivelu, (2004).A brief review of effect of seed size on seedling vigour and seed yield is presented Table 1.1.

Table 1.1 Effect of seed size on seedling vigour and seed yield

Crop	Effect	Reported by
Rice	Germination rate and seedling vigour index values increased with increase in seed size suggesting selection of larger seeds for stand establishment	**Roy *et al.,* 1996**
Wheat	Seed size is positively correlated with seed vigour, larger seeds tends to produce more vigorous seedlings	**Ries and Everson, 1973, Cookson et al., 2001**
	18% increase in yield with larger sized seeds due to more food storage for embryo growth resulted in vigorous growth of seedlings before weeds emerge and compete.	**Stougaard and Xue, 2005**
	Crop growth rate was higher at initial stages in seedlings arising from larger seeds.	**Lima *et al .,* 2005**
	Crop grown from larger kernels consistently yielded higher than crops grown from small kernels of same cultivar	**Stobbe *et al* 2008**
	Increase in biological yield by increasing seed size was related to higher seedling weight and weight of 100 plants	**Zareian *et al.,* 2013**
Maize	Large seed size of elite had higher seed quality and higher seed yield compared to medium and small sized seeds.	**Adejare 2010**

Table 1.1 *Contd…*

Crop	Effect	Reported by
Sunflower	Higher germination percentage, seedling length, seeding vigour index, dry weight, field emergence in large seeds compared to small seeds.	**Nagaraju 2001**
Soybean	Smaller seeds resulted in lower yield than medium and large unscreened seeds	**Morrison and Xue, 2007**
	Tropical soybean seedlots with small seed size had maximum germination and emergence while those with large seed size produced the highest seed per plant, pods per plant and seed yield per plant.	**Adebisi et al., 2013**
Mannea surega	Higher and quicker germination in bigger sized seeds could be due to the presence of higher carbohydrates and other nutrients than in small and medium sized seeds.	**Gunaga et al., 2011**
Chickpea & Lentil	Plants from large seeds yielded 6% and 10 % respectively from medium and mixed seeds.	**Bicer 2009**
Mung beans	Large seeds produced larger sprouts mass and head diameter.	**Chiamai et al., 2010**
Soybean	Small seeds had better germination, uniformity and getting reserves more faster than larger ones to seedlings.	**Rastegar and Kandi 2011**
Safflower	Small seeds germinated faster and grew higher under saline conditions and gave better stands.	**Farhoudi and Motamedi 2004**
Pea	Ciltivars with low 100 seed weight had higher germination percentage than larger sized seeds.	**Peksen _et al., 2004_**
Multipurpose tree species	Small to medium sized seeds have produced better germination and seedling vigour than bigger seeds	**Dar et al., 2002**

1.6.1 Cereals

Germination and seedling emergence are the two most critical stages in any plant's life cycle. Insufficient seedling emergence and inappropriate stand establishment are the main constraints in the production of rain fed crops. At times, farmers do not have sufficient resources to meet the requirement of seedbed preparation for sowing and they are at more risk as compared to progressive farmers. On the other hand good establishment increases competitiveness against weeds, increases tolerance to drought period, increases yield and avoids the time consuming need for re-sowing that is costly too. Stand establishment is of primary importance for optimizing field production of any crop plant. At suboptimal environment conditions, poor seed germination and subsequently poor field establishment is a common phenomenon. TeKrony and Egli (1991) had identified that seed vigour is important contributor to the high yield in a number of field crops. Seed vigour is less relevant than seedling vigour in transplanted crops. It has been reported that one of the major obstacles to high yield and production of crop plants is the lack of synchronized crop establishment due to poor weather and soil conditions, Mwale et al., (2003).

TeKrony and Egli (1991) has identified that seed vigour is an important contributor to high yield in a number of field crops. However, in case of transplanted crops, seed vigour is possibly less relevant than seedling vigour.

1.6.1.1 Rice (*Oryza sativa*)

Rice is the prominent cereal foodstuff and staple food for more than half of the world's population, especially in tropical Latin America and East, South and Southeast Asia (Seck *et al.*2012). Transplanting is the major rice establishment method and 77 % of rice is transplanted globally (Rao *et al.* (2007). The practical difficulties encountered on day to day basis in the transplanting system, together with shifts in rainfall pattern, delayed arrival or early withdrawal of monsoon, coupled with delayed scheduling and release of water from the canal system, necessity of transplanting over aged seedlings, crop failures at nursery stage itself due to early withdrawal of monsoon and shortage of labour during peak operations led the farmers to bring in a shift in the crop establishing methods for low land rice from transplanting to direct seeding system, despite the advantages of easy seedling establishment, restrained weed growth, enhanced nutrient availability, (Singh *et al.,* 2001). The transplanting puddle rice system requires large amount of water, labour, and energy and shortage of these inputs renders rice production more expensive, less profitable, and unsustainable. (Farook *et al,* 2011). Dry direct-seeded rice has emerged as an alternative option for rice production and refers to the process of establishing the crop from seeds sown in non-puddled and unsaturated soil rather than transplanting seedlings in a puddled field (Liu *et al.*2015). Direct seeded rice cultivation system is widely used in Asian countries. The direct seeded rice suffers from such disadvantages as non-uniform germination / emergence, uneven population with high weed growth, (Chauhan and Abuho, 2013).

Seedlings with rapid uniform emergence and strong seedling vigour always have more uniform emergence than those with low vigour, Egli and Rucker, (2013). Delayed planting on one hand and broadcasting the germinated seeds on the surface of flooded and water saturated soil after puddling the field on the other hand will lead to both hypoxia (low O_2) and anoxia (absence of O_2) conditions, wherein seeds grow poorly, Jones (1933). Under such conditions seedling vigour plays an important role in crop establishment because seedling establishment vary with each cultivar in and anaerobic soil conditions was well established by Yamauchi and Winn (1996), Ismail *et al.,* (2009) and Muthalagan (2009). In case of rice, the increase in straw and grain yield to the tune of 20% was initially ascribed to either an increase in seed nutrient content or increase in seedling vigour. The seedlings from nurseries with higher nutrient application rates have produced more vigorous seedlings and *vice-verse* despite loss of 40-60 per cent of root system during transplanting. For example, Lemont a Japonica rice cultivar is insensitive to temperature for seedling vigour, while Tequing, an indica cultivar exhibited high seedling vigour under normal temperature but quite sensitive to low temperature, Zhang *et al.,* (2005). Rice gene pool is rich in genetic variation for early uniform emergence as well as early vigor (Redona and Mackill, 1996; Zhao et al., 2006; Namuco et al., 2009) and this variation could be exploited in identifying and introgressing favorable alleles for selected traits through marker-assisted breeding. Uniform emergence and early vigor are complex traits which are influenced by different factors such as seed vigor (ability of the seed to grow), heterotrophic (seed reserve) and autotrophic growth (photosynthesis), and environmental conditions, among others. Seed priming is used to enhance the seedling quality and vigour and weed suppressing ability and yield in rice Jurami et al., 2012 and enhancing resistance to environmental stress, Goswami, et al. 2013. Therefore, for bringing improvement in rice crop early vigour was recognized as the most promising and effective strategy to relieve the constraint. Early rice vigour is a combination of the ability of the seed to germinate uniformly and emerge after planting and the young plants ability to grow after emergence, Chen et al, 2015. Rice plants originated from high vigour seeds were superior as for leaf area, tillering and grain yield. On the other hand, less vigorous seed lots present lower emergence percentage and speed, compromising the establishment of the desired plant density, mainly when the edaphoclimatic conditions are not favorable and the used sowing density is close to the minimum recommended limit, Marcos- Filho, (2015). Response of rice yield to nursery treatment is largely due to increased seedling vigour and can be affected in a range of nutritional and non-nutritional treatments of seedlings that increase in seedling dry matter, nutrient content and nutrient concentration.

Direct seeded rice cultivation system is widely used in Asian countries owing to delayed arrival or early withdrawal of monsoon. Hence, strong seedling vigour has become a desirable trait in direct seeded rice system not only for enhancing crop establishment but also to increase the ability to compete against the weeds. The practical difficulties encountered on day-to-day basis in the transplanting system, together with shift in rainfall pattern coupled with delayed scheduling of water release from the canal system need to transplant over aged seedlings resulting in crop failure at nursery stage itself and shortage of labour during peak operations led the farmer to bring in a shift in the crop establishment methods from low land rice from transplanting to direct seeding system. Despite the advantages of early seedling establishment restrained weed growth, enhanced nutrient availability. Singh *et al.,* 2001. The transplanting puddled rice system requires large amount of water, labour, energy and shortage of these inputs renders rice production more expensive, less profitable and unsustainable. Fraooq et al., 2011. Direct seede rice has the disadvantage of non-uniform germination / emergence, uneven population coupled with high weed growth. Chouchan and Abaugho, 2013. Seedlings with rapid uniform emergence with strong seedling vigour always has more uniform emergence than those with low vigour. Eglie and Rucker., 2012.

Poor crop establishment is one of the major constraints in the direct seeded rice systems especially under advance growing environments like drought (Du and Tuong, 2002, Kumar *et al* 2009).

1.6.1.2 Wheat (*Triticum aestivum*)

Wheat (*Tritium aestivum* L.) is the second most important crop in the world, offering around 20 per cent of the energy to human food (Ahmadi *et al.*, 2004; Shewry, 2009). Seed quality is very important for realizing optimum growth and yield. Production is under the influence of several factors such as genetic characteristics, viability, germination per cent, vigor, moisture content, storage conditions, survival ability and seed health. However, germination per cent and vigour are most important, (Akbari et al., 2004). Whereas, factors like genetic structure, environment and parental nutrition, maturity stage in harvest time, mechanical damages, seed storages, age and aging and pathogens, affect seed germination and vigor. Amount of dry matter (seed storages) or the seed weight is one of the important criteria of seed vigour. Germination and seedling emergence demands a lot of energy derived through the oxidation of seed storages. Thousand grains weight (tes weight) is one of important scales in seed quality. It depends on embryo size and seed storages for germination and emergence. High thousand seeds weight will increase germination per cent, seedling emergence, tiller ring, density, spike and yield (Noor- Mohammadi et al., 2000; Cordazzo, 2002). Thus seed weight or thousand grain weight has a large effect on seed germination, seed vigor, seedling establishment and yield production. Nedeva and Nicolova (1999) reported that after flowering and during grain filling period of wheat, decreased the moisture percent and increased the dry matter per cent (dry grain weight) and % germination. High quality seeds of wheat are associated with the performance of the field cultivation practices because they promote the establishment of stands, the growth and development of plants and the scope of high productivities, Lima *et al.,* (2006), Franca- Neto *et al.,* 2010). Seed lots coming from the same cultivar, with similar germination capacity, may present distinct behaviors in emergence of seedlings on the field, since they have different vigor levels, Carvalho and Nakagawa, (2012). The genetic makeup of variety, environment, parental nutrition received at maturity stage, mechanical damages at harvesting and post harvest processing stages, seed storages, age of pathogens and ageing of seeds are bound to affect seed germination and vigour. Similarly the amount of dry matter (seed storages) or the seed weight is one of the most important criteria of seed vigour in wheat as germination and seedling emergence obviously demand a lot of energy derived through the oxidation of seed storages.

Hence, in wheat crop, thousand seed weight or test weight assumes one of the important scales in seed quality and depends on embryo size and seed storages required for germination and emergence. Therefore, high thousand weight of wheat seeds will increase germination percentage, seedling emergence, tillering density, spike and yield. Noor Mohammadi *et al.,* (2000); Cordazzo, (2002). Uniform stand establishment and early vigor are the principal determinants of wheat crop performance (Chivasa *et al.,* 1998). The factors limiting the uniform stand establishment are use of poor quality seed (Radford, 1983), conventional sowing methods (Radford, 1983), poor seedbed preparation (Joshi, 1987), low moisture at sowing time (Harris, 1996), late sowing and sub-optimum temperature at sowing (Farooq *et al.*2008). Late planting affects the growth, quality and reduces wheat yield at the rate of 36 kg ha^{-1} day^{-1} (Hussain *et al.,* 1998). In late planted wheat, low soil temperature, prevailing below 10°C during germination substantially affects the germination and seedling emergence. Germination is a critical process, as temperature below 12°C result in poor and uneven emergence (Timmermans *et al.,* 2007). Therefore, the rate of emergence and final emergence percentage are important factors to determine the wheat crop potential. Poor crop establishment results in few tillers and finally decreased grain yield (Farooq *et al.,* 2008). Short period of high temperature stress i.e., ≥35°C at reproductive stage in wheat decreases grain weight (Wardlaw & Wrigley, 1994) and reduces grain quality (Randall & Moss,1990; Savin *et al.,* 1996).

1.6.1.3 Maize (*Zea mays*)

In seed corn, physiological maturity occurs at high speed moisture content which precludes harvest by mechanical means. The extent of deterioration in maize varies significantly with different stages of pre and post harvest handling. It requires at least one month in the field to bring down the moisture content to 13%. Under these circumstances, the seed corn industry has resolved this issue by harvesting the entire ear of corn near physiological maturity and conditioning or sorting of seed in a controlled environment, thus allowing corn seed producers to minimize the risks of physiological seed deterioration due to field losses, an early frost and weathering. Dry threshing also minimizes the extent of loss. On the other hand, the post harvest processing though cause damage, once dried the seeds store well and extent of damage is minimized. However, if stored for more than one year, though germination will be maintained at more than 90%, the seeds will be low in their vigour and accordingly emergence in the cool soils will also be reduced.

The impact of plant- to-plant variability on maize grain yield at canopy level by Tollenaar *et al.,* (2006) concluded that, non-uniform spatial distribution of plants along the planting row as well as uneven seedling emergence results in variability of biomass, reduction in yield. Hofs *et al.,* (2004) reported that maize plants originating from seeds with high physiological potential show higher efficiency in biomass production. Egli and Rucker

(2012) reported that by combining conditions favoring slow emergence with use of low vigour seeds could result in unevenness on the plant stand sufficient to reduce crop yield. Maize plants exert interference on growth and reproduction of weeds, a direct method of weed control, and further reported that seed vigour had direct effects on plant initial growth, reflected in the competitive capacity of plants with weeds that usually have lower growth rate than maize, Dias *et al.,* (2010). Callaway,(1990) reported varietal differences in weed suppression ability in maize, soybean, potato, cotton crops. Mondo *et al.,* (2012) reported that use of high vigour seeds of maize can be an important strategy for the integrated weed management systems that help farmers in minimizing the amounts of herbicides used in their fields. High initial growth of maize plants originated from high vigour seeds could be identified as a direct result of increased grain yield as a consequence of enhanced uptake of natural resources. On the other hand, the higher growth helps maize plants competing with weeds as well as helping growers make maize production more environmentally suitable by reducing herbicide usage, Begna *et al.,* (2001). Mondo *et al.,* (2013) reported that the increase on percentage of high vigour seeds in a given seed lot improves the initial growth of corn crop until the stage of eight leaves mainly on plant height, stem diameter and leaf area index.

1.6.1.4 Barley (*Hordeum vulgare*)

Seed vigour is not a problem in barley crop, because barley seed is physiologically sound, properly matured and handled very carefully. Hence, produces very vigorous, uniform germinating seed with good early seedling growth. Heat treatment for 2-3 weeks after harvesting at 40^0C increases rapid, vigorous and uniform germination. Malting industry needs homogenous and rapid germination. According to Riis *et al.* (1991), grain with low vigour needed by 42% longer malting than grain with high vigour even if the final germination percentage was nearly the same. The vigour was also related to bread quality of wheat (Chloupek *et al.*, 2008). Grain samples with 80%-90% vigour produced the greatest bread volume. A lower vigour was correlated with a high occurrence of fungi (as indicated by ergosterol assays) and to lower field emergence rates of the samples. (Chloupek *et al.*, 2003). Moderate benefit from selection can be expected from seed vigour (Ullmannová *et al.*, 2013). But the expression of vigour in field conditions and the translation to higher yields depend on the environment (Botwright *et al.*, 2002). The relative contributions of the variety to the total variation have been reported to be higher for vigour than for germination (Chloupek *et al.*, 1997). Bonder *et al.* (2013) reported that, the germination and vigour were related to the parameters that are important for malting in barley. The germination capacity of all lines was higher than their vigour and germination energy, by 2.9% higher than vigour and by 4.6% higher than germination energy on average. This finding has confirmed opinions of many authors who had reported that seed performance under optimal conditions is often higher in comparison to the seed performance in vigour experiments under stress and field conditions. Moreover, it has been confirmed that samples of the same germination capacity may have different vigour and storage potential.

1.6.1.5 Sorghum (*Sorghum bicolor L.*)

In sorghum, Camargo (1971) reported that, the speed of germination has more accurately differentiated the degree of deterioration of the seed lots and was closely correlated to yield. Adverse effect of planting low vigor sorghum seeds was found to be related to plant height, panicle exertion, anthesis, tillering capacity, length of the panicle, and yield. Less vigours seeds produced plants which were not able to "catch up" to the vigorous ones. These slow growing plants were significantly inferior in panicle exertion to those produced from more vigorous seeds. Haya Gelmond *et al.,* (1978) reported that though the seed lots showed 88 to 92 per cent germination in case of sorghum in the laboratory, but poorly correlated with vigour. Whereas, seed vigour in sorghum correlated well with both the rates of germination and root emergence when tested in the laboratory as well as emergence in the field.

1.6.2 Oil Seeds

1.6.2.1 Groundnut (*Arachis hypogea*)

Groundnut is cultivated predominantly in the tropics and subtropics, under rain-fed systems, where the availability of water is a major constraint on yield, Nautiyal *et al.,* (1994). Being an oilseed crop, groundnut seed contains about 35% to 50% oil content and deteriorates rapidly during storage, resulting in loss of viability and vigour, Nautiyal *et al., (*1997). In addition, groundnut seed vigour is influenced by environmental conditions during curing/ drying, Nautiyal and Zala, (1991; Nautiyal and Zala, (2004) and storage, Nautiyal and Ravindra, (1996); Nautiyal *et al.,* (2004). Persistence of fresh seed dormancy in groundnut also has a significant influence on crop establishment and seed vigour, though it depends on cultural practices followed; Nautiyal *et al., (*2001). There is little information on the influence of seed size on field emergence, crop stand and productivity, Singh *et al. ,* (1998); Devi Dayal *et al. ,* (1999).

Rain-dependent groundnut cultivation systems require high seed vigour to produce seedlings which are more likely to biotic and abiotic stresses under adverse environmental conditions. For example, during seedlings growth may vary. This may adversely affect seedling establishment as well as its distribution over time. Moreover, poor germination capacity of groundnut seed is likely the cause of poor crop stand establishment. Moreover, field emergence and crop establishment seems to be the result of interactions among ambient weather conditions, soil, seed and seedling characteristics, Finch-Savage *et al.,* 2001). Since, soil and climatic factors are not easily manipulated, genetic manipulation of seedling vigour is a promising tool to improve crop stand. Due to lack of information on groundnut our understanding of the factors that govern seed performance under field situations remains unclear.

1.6.2.2 Brassicas

In Brassicas, seedling vigour is a problem. Brassica crops have indeterminate growth and flower over an extended period of time. Harvested seed is therefore comprised of seed of varying degrees of physiological maturity and quality, David (1999). A seed that has reached its maximum viability and vigor is defined as being physiologically mature. Mass maturity is the point of maximum seed dry weight; physiological maturity occurs at or after mass maturity in brassicas and many other species (Still and Bradford, 1998). The environment in which a seed matures while on the mother plant has a significant effect on seed germination characteristics and environmental after effects can be manifested in the progeny (Stoehr *et al.,* 1998; Wulff, 1995). Surprisingly little is known about the physiological, biochemical, genetic and molecular mechanisms contributing to seed quality traits. This is due to the strong environmental influence on seed development and the fact that these traits are likely quantitative. By quantifying specific components of seed quality it is possible to investigate its underlying biological basis. Brassicas are an ideal model system in which to study seed quality because seed at several different stages of maturity can be harvested at any given time. Seed developmental patterns are similar among brassicas making it possible to utilize rapid cycling brassicas as a model system (B. napus cultivars; Still and Bradford, 1998). Sowing of high vigor seeds would result in high performance of crops in the field via improving seedling vigor, seedling establishment and winter survival of seedlings (Ghassemi Golezani *et al.*, 2008a). Rapid emergence of seedlings from high vigor seed lots could lead to the production of large and vigorous seedlings (Ghassemi-Golezani *et al.*, 2008a) with high leaf chlorophyll content (Ghassemi-Golezani *et al.*, 2008b).

1.6.2.3 Soybean (*Glycine max*)

Development and maturation of soybean seed has been linked to several physiological and biochemical processes and physiological maturity also occurs at high seed moisture content (50%) and required 2-3 weeks drying time in the field for attaining 13% moisture content. Greater extent of damage is possible when the seed with below 20% moisture content is exposed to rain and the same can be measured with the help of tetrazolium test. Soybeans are more susceptible inherently to mechanical damage than maize seeds particularly at high moisture contents. The extent of damage to soybean seeds will be more when combine-harvested at a moisture content ranging between 12% to 18%. Therefore, it is suggested to use gentler combine for harvesting or else avoid harvesting at more than 12% moisture content. Being an oil seed crop, soybeans are more prone to lipid per oxidation, particularly when stored at high temperature and relative humidity, Steward and Bewley, (1980). Harvesting and drying soybean seed in intact pods not only minimizes the negative effects of harvesting at high moisture but also helped in maintaining seed quality aspects of viability and vigour, Samarah *et al.,* (2009), despite the limitations of range of drying temperatures and varietal differences. The approach of corn seed industry can be applied to soybean seed production also.

The increased importance of soybean seed quality could be attributed to the continued emphasis on early planting. As such most of the plant breeding programmes on soybean spend a great deal of amount and time to breed seed that emerges rapidly and uniformly from cool and wet soils. In order to receive maximum return from investment, soybean seed producers constantly look for ways to increase or maintain seed quality while minimizing field losses.

1.6.3 Cotton (*Gossypium herbaceum*)

The vigour of seeds, in case of cotton, starts to decline after attaining physiological maturity and is followed by the rapid loss of viability. A uniform stand of healthy, vigorous seedlings is essential if growers are to achieve the yields and quality needed for profitable crop production. It is important for growers to plant high quality seed of varieties adapted to their farm situations, management styles, and intended market uses. However, obtaining adequate stands is not always easy. Cotton seedlings often encounter multiple adverse stress conditions at the onset of the growing season. While high seedling vigor may not mitigate all of these factors, it can definitely help. The failure of seeds to germinate or the failure of seedlings to survive the initial few weeks of growth can be caused by a number of factors, many of which can be managed by cotton producers. The problem of 'Hardheadedness' can be eliminated easily by way of the adoption of selection procedures of plant breeding. Cotton seeds are extremely sensitive to cool, wet soils during the early phases of germination and seedling growth. If the stress is severe, germination can be delayed or may not occur. Young, tender seedlings also may be damaged or killed if exposed to prolonged periods of cool, wet conditions. Growers should not be tempted to plant cotton when cool, wet weather is expected. Planting under these conditions can lead to poor stands and may result in the need to replant. Chilling injury is most likely from the time the seed imbibes water and for the next 2 days as the radical emerges. Growers may want to cease planting 2 days prior to a predicted wet and cool spell. Chilling injury can result in seedling death, poor seedling development, J- rooting, and tap root abortion. The abortion of the tap root can decrease the ability of the plant to perform well in dry periods later in the season.

1.6.4 Grain legumes

The seeds of grain legume cultivars with partially or completely unpigmented testa have been reported to be susceptible to imbibition damage when compared with their counterparts having pigmented testa in Soybean (Tully *et al.* 1981), Phaseolus Vulgaris (Powell *et al.*, (1986 a, 1986 b), Chickpea (Legesse (1991) and long bean, Abdullah et al (1991). This rapidity and grater levels of imbibition damage resulting in poor vigour and is expressed in the form of poor field emergence has been demonstrated successfully in Pea isogenic lines by Powell *et al.*, (1989) and studies on pea seed development by Legesse and Powell (1996). In close adherence of pigmented testa to the cotyledons appears to limit the rate of water movement has been reported by Chick pea seed, by Legesse and Powell (1996) and Phaselous Vulgaris (Powell *et al*, 1986 b)

Investigations have clearly demonstrated that seed colour exert influence of water uptake (Powell *et al* 1986) gas diffusion (West and Harris, 1963, seed dormancy (Baskin *et al* 2000) seed quality and seedling emergence (Mavi 2011) in some crop plants. Demur 1996 has reported injuries to water uptake in species and legume genes due to differences in seed colour texture.

In grain legumes a close relationship between seed coat permeability and imbibition damage was often reported. Cultivars with high seed coat permeability and high incidence of testa injury imbibe more rapidly and may show extensive imbibition damage. Legesse and Powell 1996 in Chick pea.

1.6.4.1 Peas (*Pisum sativum*)

The loss of vigour in case of wrinkled pea seeds will be much earlier and faster than when compared to round and flat seeds. This could be attributed to seed coat cracking, which allows rapid inrush of water into the cells of cotyledons and cause them to rupture. Cavitation or Hallow Heart is a physiological disorder, observed in peas. Where in the cells in the center of adaxial surface of cotyledons are dead. Upon imbibition, the center of seed contains a cavity. The dead cells present there will produce a growth toxin, which inhibits germination and thereby reduces the survival rate of seedlings. It can be attributed to the presence of high temperature during seed maturation.

Keeping in view all these problems listed above, it becomes very much essential to deal with the problem of seed vigour very carefully and thereby ensuring the highest possible vigour to a seed lot offered for sale. In order to achieve this objective, the focus then should be clearly laid on both pre and post harvest management aspects of the seed crop. It means, the seed crop should be provided with conditions congenial not only for crop growth, but also for harvesting the crop as early as possible that too either immediately after attaining physiological maturity and /or at least at mass maturity. This in turn depends to a large extent on the crop species and various methods of handling the seed. Therefore, these operations have to be carried out in such a way that cause minimum mechanical damage during processing and conditioning and also minimizes physiological damage during storage. This becomes necessary either to reduce or slow down the process of ageing. Otherwise, this leads to the death of seed and ultimately culminates in the failure of the valuable seed to germinate. Therefore, in order to realize the basic objective of obtaining maximum yields, the seed growers' prerequisite will be to lay hands on the seed with a capacity to establish optimum plant stand rapidly and that too under diversified environmental conditions. Since, germination tests refer only to the production of normal seedlings under favorable conditions; understandably it becomes a poor indicator under unfavorable conditions.

1.6.5 Forage Crops

Despite significant advances, plant stand establishment continues to be one of the most common production problems affecting forage species worldwide. This in turn demands a need for research on stand establishment of forage crops against abiotic and biotic stress. Without a good plant stand, the effectiveness of other agronomic inputs is drastically reduced, and usually such inputs can never compensate for the negative impact of a poor stand. No doubt, the forage seed industry produces and markets seed of high quality. However, new methods for assessing seed vigor are the need of the hour. Seed treatment and seed coating are used in the forage seed industry, with strong benefit in some environments. There is an increase in no-tillage seeding of forage crops, but improvements in the no-tillage planting equipment are needed to make them better suited to small seeds. Other recent developments in seeding techniques include broadcasting seed with dry granular and fluid fertilizers, which improves the efficiency of the seeding operation. In case of perennial forage crops, failure to achieve a dense stand at initial planting can have long-lasting consequences and is particularly true for seeded annual and perennial forages that exist as discrete plants (e.g., alfalfa [*Medicago sativa* L.] and love grass [*Eragrostis* spp.]), Miller & Stritzke, (1995). Obtaining dense stands from initial plantings of forages which vegetatively propagate (e.g., through rhizomes or stolons) is less critical; however, establishment of healthy plants that are able to compete and withstand abiotic and biotic stress is still very important with those species.

Research on vigour testing of forage seed lots has received little attention compared with that conducted on other species. Several researchers have evaluated the use of accelerated aging and the conductivity test for forage seeds, Clark, (1982); Hall & Wiesner, (1990); Hampton & Bell, (1989); Helmer *et al.,* (1962); Hampton & TeKrony, (1995); Wang and Hampton, (1993). Standardized procedures for these tests have been developed for several forage species, and hopefully their use will increase and assist the industry in assessing seed lot planting value of seeds for different markets. Hall & Wiesner (1990) concluded that a combination of tests which include both physiological and biochemical measures of seed quality should be used in adequately assessing seed vigour. This promises to be a fruitful area of research with potential to provide real and practical benefits to forage producers.

Seed dormancy in forage grasses and hard seed content in forage legumes are two important factors affecting the stand establishment. In nature, these mechanisms ensure a broad, dormant soil seed bank capable of germinating when favourable conditions are encountered for growth and establishment, (Mc Donald *et al.,* (1996). They also ensure that not all seeds germinate solely on a single environmental cue, but that continued emergence and establishment of the species occurs under a variety of habitats and environments.

In forage crops, though seed dormancy is advantageous in nature, production agriculture demands that seeds planted germinate uniformly and quickly to produce a productive stand. Therefore, various mechanisms have been devised for breaking seed dormancy. Mc Donald *et al.* (1996). Similarly, various methods have been suggested for decreasing hard-seededness in forage legumes (Bass *et al.,* 1988). However, mechanical scarification is probably the most commonly used method by the processing industry. Mechanical scarification injures seeds, thus resulting in higher abnormal seedling counts and deterioration of vigour and viability. A recent study demonstrates that scarification may not always be necessary to ensure adequate field performance of forage legume seed lots with high hard seed content (Albrecht *et al.,* 1996). Alfalfa seed lots with hard seed levels of up to 50% had no significant impact on total seedling emergence, establishment year forage yield, or forage yield the year after establishment. Rate of emergence was reduced slightly by high levels of hard seed. Scarification increased rate of emergence of most seed lots, but had no effect on forage yield. These results indicate that it may notbe necessary to mechanically scarify alfalfa seed lots with moderate levels of hard seed content in order to ensure good performance in the field. There is a need for further study on factors governing hard seed expression under natural field conditions, and methods developed which adequately predict field germination and emergence of forage legume seeds.

1.6.6 Vegetable Crops

The establishment of a predetermined level of plant population is critical particularly in vegetable crops. This is so because, direct drilling of vegetable crops cannot compensate readily as with a tillering habit. Poor germination is a common phenomenon at sub- optimal temperatures, which is a great concern of growers that grow seedlings in late winter and early spring seasons. Optimum seed germination and seedling emergence occur at relatively high temperatures (20-30°C) for several species (e.g., tomato, eggplant, bean, watermelon, cucumber, melon). Rapid and uniform field emergence are two essential prerequisites to increase yield, quality, and ultimately profits in crops. Uniformity and percentage of seedling emergence of direct-seeded crops have a major impact on final yield and quality. Slow emergence results in smaller plants and seedlings, which are more vulnerable to soil-borne diseases? Extended emergence periods predispose the planting bed to deterioration and increased soil compaction, particularly under adverse environmental conditions

1.6.7 Other Miscellaneous Crops

Gorge and Ray (2004) showed that with increase in hundred grains weight of *Parthenium argentatum* L., increased the germination per cent, for example with increasing in hundred grains weight from 0.08 g to 0.1 g, the germination per cent increase from 34 per cent to 80 per cent.

Khan (2003) documented that with increasing in seed weight of *Artocarpos heterophyllus* L., from 4-6 g to 12-14 g, the germination per cent increased from about 15 per cent to about 85 per cent (i.e., there is positive and significant correlation between seed weight and seed germination per cent). Malcolm *et. al.,* (2003) noted that with increase in seed weight and seed size of Peach rootstock increased.

On the other hand, a standard laboratory germination test, frequently used to evaluate seed quality, specifies information on the percentage of normal seedlings obtainable under optimal favorable conditions only stipulated by International Seed Testing Association (ISTA, 2008). Therefore, under unfavorable conditions, understandably it becomes a poor indicator. Hence, it is expected that a seed vigour test should obviously correlate more closely with field emergence. Though several researchers have reported significant correlation coefficients between field emergence and standard germination tests, they have also reported inconsistencies and difficulties with respect to prediction of field emergence, Porhadian and Khajepour, (2010). Several authors have also reported laboratory germination tests to correlate closely with field emergence as in Accelerated Ageing Test for peas, Hampton and TeKrony, (1995), Controlled Deterioration Test and Cool Test for Sugarbeet, Hampton and TeKrony, (1995); Cold Test for Corn (Noli *et al.,* 2008); Conductivity Test for Safflower (Khawari *et al.,* 2009). Though some of these tests are accepted internationally, but still, there is no single vigour test available or recommended for all plant species.

1.7 Policy Issues

In a developing country like India, which accounts for about 2.4 % of the world's geographical area and 4% of its water resources, is expected to serve around 17% of the world's human population and 15% of the livestock population. In India, agriculture is an important sector accounting for 14% of the nation's GDP and 11% of exports. More than half of the country's population still relies predominantly on agriculture as their principal source of income. In addition to this, agriculture is a source of raw material for several industries.

Prior to 19th century, farmers usually produced their own seed, and hence commerce in seed trade is confined to replacement of degraded and mixed stock. The developments in the Indian seed Industry, particularly during the past four decades are very significant, Advent of plant modern plant breeding methods with enhanced sophistication has resulted in the development of seed as a product of commercial value that can be attributed to several quality components that can cater to diversified market niches and made the seed an internationally traded commodity.

Consequent to the liberalization of policies by Indian government, revised seed regulations have rendered open access to the germplasm available any where on the globe for public research programmes. This has automatically created a demand for commercial seed and in turn attracted private investment into the Indian seed Industry. As a result, a number of players specializing in a number of fields such as plant breeding, production of source seed, seed multiplication, conditioning and distribution and quality control emerged in both public and private sectors. A dynamic seed system demands effective cooperation and an efficient flow of information between these diverse players. The private seed sector of India, has concentrated on hybrids and other high value seeds of vegetable crops. But several important crops of India, such as paddy, wheat, pulses, small millets are non-hybrid based high yielding varieties. Hence the farmers can save the seed from the farm produce for sowing in next year.

On All India basis, despite the availability of quality seed of superior genotypes from organized seed industry, the age-old tradition of use of 'Farm Saved Seed' is predominant and continues even today. The developments in the seed industry, particularly during the past four decades are very significant. The landmark policy initiatives in seed sector are presented in Table 1.2. The National Seed Policy, (2002) ensures the importance of providing quality seed to farmers on time to meet country food security. Consequently, in 2005 a National Seed Research and Training Center was established near Varanasi, U.P. to have a separate seed quality control laboratory. This also serves as a Central Seed Testing Laboratory (CSTL) in addition to functioning as a referral lab in case of disputes arising in the court of law pertaining to seed quality. Under the Ministry of Agriculture, at National level, National Seeds Corporation, a "Mini Ratna", the 4th largest in the world, undertakes the responsibility of production, processing and marketing of seeds of cereals, pulses, oilseeds, fodder, fibre and vegetable crops and is instrumental in establishment of State Seed Development Corporations for production and marketing and State Seed Certification Agencies for independent maintenance of quality control in addition to its own International Seed Testing Association (ISTA) accredited seed testing laboratory. Seed testing laboratories are involved in checking the compatibility of different varieties of seeds with local soil and other farming conditions and carryout a wide range of tests on seeds including germination and vigour, disease, purity, genetic traits, health and quality assurance. Similarly, state seed farms have been established in all states not only for multiplication of seed of high yielding varieties meant for distribution to farmers , but also to demonstrate and provide training to farmers. The new seed policy on one hand promotes seed village schemes for production and making available quality seed of desired crop varieties at local level and on the other hand established seed banks in non- traditional areas to make compensate the short fall during natural calamities and exigencies in seed production and supply.

Table 1.2 Policy Initiatives in Seed Sector taken by Government of India

S.No.	Year	Initiative	
1	1966	Enactment of Seeds Act	
2	1968	Seed Review Team	
3	1972	National Commission on Agriculture's seed Group	
4	1975-85	National Seed Programme- World Bank Funded	
5	1977-78	Phase- I	Strengthening of seed infrastructure, turning point in shaping an organized seed Industry,
6	1978-79	Phase –II	
7	1980-81	Phase- III	
8	1983	Seed Control Order	
9	1986	Technology Mission on Oilseeds and Pulses (TMOP) now known as ISOPOM (Integrated Scheme of Oilseeds, Pulses, Oil palm and Maize).	
10	1986	Production and Distribution Subsidy, Distribution of Mini-kits	
11	1987	Seed Transport Subsidy Scheme	
12	1988	New Policy on seed Development: Another significant milestone in the development of Indian Seed Industry. Gave Indian farmers the access of best of seed and planting material available anywhere in the world. Stimulated appreciable private investments in seed sector with emphasis on high value hybrids of cereals, vegetables and hi-tech products like Bt-cotton.opened gates for private sector in seed business	
13	2000	Seed Bank Scheme	
14	2001	PPVFRA Act Protection of Farmers Rights	
14	2002	National Seed Policy-opening up of FDI in seeds, participation in OECD, PPFRVA Act 2004.	
15	2002	Biodiversity Act	
	2002	Establishment of ACABC (Agriclinics and Agribusiness centres)	
15	2004	The Seed's Bill-issues guaranted seed quality	
16	2005	Formulation of National Seed Plan, creation of PPVFRA Authority on 11-11-2005	
17	2007	National Food Security Mission	
18	2007	Rashtriya Krishi Vikas Yojana	

The Indian Seed Improvement programme is a highly vibrant, energetic and well recognized in the international arena not only due to the diverse and ideal agro-climatic conditions distributed throughout the country catering to the needs of temperate, tropical and subtropical crop varietal requirements at competitive prices, but also blessed with an excellent backup by a strong contingent of crop improvement programmes in both public and private sectors spread across the length and breadth of the country. Over the recent past few decades, the country has initiated, developed and perfected the techniques of seed quality with specifications comparable to international standards through a mechanism of voluntary seed certificate and compulsory labeling monitored by provincial level Seed Law Enforcement Agencies.

Today, on global basis, in a good majority of production systems, where, suboptimal conditions are a rule, rather than an exception, the seed processing and conditioning industry together with the seed transport and storage industry must work in close coordination obviously so as to meet the ever increasing requirements of the seed growers on one hand and the commercial seed farmers on the other hand and ultimately this is aimed at fulfilling the requirements of the consumers. In our contemporary society where most of the farmers are buying seed are willing to pay a premium price for better quality.

Therefore, the study and estimation of physical, physiological, biochemical and molecular phenomenon of seed vigour becomes highly essential.

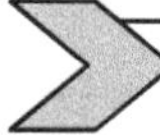

1.8 Summary

Seed, as a connecting link between two generations, has to perform both biological and ecological roles for continuation of species. The seedlings emerging even from robust seeds are most unlikely to be overcome by diseases, weeds, and insects when compared to weaker ones. Therefore, a good seedling establishment character is not only desirable but also essential for a seed population to perform well and flourish.

Seed vigour is the capacity of seeds to give rise to normal seedlings and establish a healthy plant or in other words it is the rate and speed at which the seedlings establish a plant during their early stages of growth. Seeds harvested fresh from same field, same day from the same plant differ in their yielding potential despite having same genetic background and demonstrates the need of adoption of different methods and history of seed production and subsequent handling. The vigour of seed is not a God's gift as thought of earlier, but is acquired gradually in its own production environment. Since 1960 onwards, the potential attributes of seed vigour either as a fundamental physiological characteristic or its association with field stand establishment and ultimately crop productivity gained a worldwide recognition, Julio Marcos Filho, (2015).

Accumulation of sugars, antioxidants etc., into the seed during maturation phase serves twin purposes of survival from desiccation and persists cells in dry state for a longer period of time in addition to protection and dispersal respectively by cotyledons and testa. Genetical, physical, physiological and soundness aspects get adversely affected during pre and post harvest handling stages and cause degenerative changes and deterioration in the absence of appropriate corrective measures and lead to death of seed usually represented by loss of germination. Farmers by utilizing all resources available at their disposal if only produce a bad crop it is purely due to the use of poor quality planting material. There are several characteristically confronted complications associated with seed vigour. Today, on global level where production systems are suboptimal as a rule than exception, study and estimation of seed vigour becomes highly essential. The assessment of seed vigour has several key implications for seed industry such as (i) monitoring the physiological potential of seed lots during various phases of seed production, (ii) a strategic support during the selection of high quality seed lots capable of meeting the requirements of end-users. Selection of suitable production areas will help in minimizing low moisture content of seeds, a crucial pre-harvest factors responsible for deterioration. Similarly, bringing the moisture content to below 12%-13% prior to storage, as well as maintaining the seed in low moisture and temperature both during storage and transportation will also minimize deterioration. On the other hand, continuous seed quality monitoring during storage and limiting the storage time no doubt maintains high germination, but fall in vigour in respect of crops like maize, melons and cotton was also reported in literature.

In tropical and subtropical countries like India, seed storage is a serious problem where high temperature and high relative humidity greatly exert influence on the seed ageing phenomenon and resulting in rapid deterioration and loss of viability of seeds. The problem of retention of seed vigour in India is much more acute because of extremely high relative humidity prevailing during the major part of the year, which is highly conducive for the growth of the microorganisms. Indian seed programme now occupies a decisive place and is poised for continuous growth in the future with a diverse product range to offer in a farmer centric, market driven approach. The gradual expansion in respect of product diversity, volume and value of seed handled, level of seed distributed to the new areas etc., have led to the development of a competent and experienced seed producers, dealers and distributors network for scientifically sound seed improvement and commercially viable technologies.

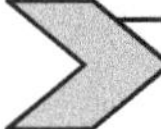 ## 1.9 Exercise

I.   Explain the following in one or two sentences

   1.   What are the twin commitments of a seed?

   2.   Resource efficient and cost effective crop production is the order of the day. Why?

   3.   Seed vigour is not a God's gift. Substantiate.

   4.   List out the causes for the death of seed.

   5.   Why is it necessary to reduce or slowdown the process of ageing?

   6.   Suboptimal conditions are a rule rather than exception at all global level production systems. Why?

   7.   Seed vigour is less relevant than seedling vigour in transplanted crops. Why?

   8.   Briefly explain the components of seed quality.

   9.   List out the key policy issues initiated by Government of India in Seed Sector.

II.   Define the following.

   1.   Weathering.

   2.   Seed Vigour.

   3.   Ageing.

   4.   Hallowheart

III. Answer the following

1. Seed Vigour is a highly complex phenomenon. Comment.
2. List out the characteristically confronted complications related to seed vigour.
3. Why study of seed vigour is highly essential.
4. List out the situations where germination of the seed is affected seriously.
5. Farmers should have easy access to seed with high germination besides high vigour. Why?

IV. Fill in the blanks with suitable words/ phrases.

1. Under unfavorable conditions, -------- test become a poor indicator of seed performance.
2. High temperatures during seed maturation in peas results in -------- a physiological disorder.
3. If not stored under optimum conditions, all lots of carry over seed of all species needs to be downgraded with respect to -------- quality.
4. High temperatures during harvesting time of a highly moist seed results in higher rate of -------- leading to the release of --------.
5. Early season -------- makes the assessment of germination more difficult.
6. Higher dose of herbicide, Glyphosate may not kill the seed instantaneously, but exerts influence on the subsequent --------.
7. A seed vigour test should obviously correlate more closely with --------.
8. --------, an indiscernible parameter transforms the quality of seed over a specified period of time.
9. A crop with low seed multiplication ratio is --------
10. In order to meet the cost effective exploitation, a seed population is expected to give -------- under optimal and suboptimal soil environments.
11. Second significant milestone in the development of Indian seed industry is the pronouncement of -------- in 1988.
12. Seed Bank scheme was initiated in the year --------.
13. National Seed Policy came in to force from the year --------.
14. The Seed's Bill was enacted by the Indian Parliament in the year --------.
15. The period of dry storage required by matured dry seeds of several species to relieve them from dormancy is known as --------.
16. A large majority of seeds sown in India produced in unorganized sector is referred to as --------.
17. Seed quality is a -------- concept with several components.
18. Successful crop stand establishment largely depends on high --------.
19. The -------- environment renders seed survival highly vulnerable.
20. High -------- prevailing during most part of the year is conducive for the growth of microorganisms.

Answers

IV. Fill in the blanks with suitable words/ phrase.

1. Germination	2. Hallow heart
3. Germination	4. Respiration, higher rate of CO_2
5. dormancy	6. Germination
7. Field emergence	8. Seed Vigour
9. Groundnut	10. Optimal plant stand
11. New policy on seed Development	12. 2000
13. 2002	14. 2004
15. after ripening	16. Farm saved seed
17. Multifaceted	18. Seed vigour
19. Production	20. Relative humidity

1.10 References

Abdullah, W.D., Powell, A.A. and Matthews, S., (1991) . Association of differences in seed vigour in long bean (*Vigna sesquipedalis*) with testa colour and imbibition damage. Jour. of Agricultural Science, Cambridge, 116 : 259 –264.

Adebisi, M A., T O Kehinde, AW Salau, LA Okesola, JBO Porbeni, AO Esuruoso and KO Oyekale, (2013). Influence of different seed size fractions on seed germination, seedling emergence and seed yield characters in tropical soybean. Int. J. Agric. Res. 8: 26-33

Adejare, LO, (2010) Effect of seed size and shape on field performance of tropical maize varieties. Project Report. Department of plant breeding and Seed Technology. Federal University of Agriculture, Abeokuta, Nigeria.

Advances in Applied Biology.10 : 217 – 285.

Agric. Sci., 7: 35–47.

Akbari GhA, Ghasemi Pirbalouti M, Najaf-Abadi Farahani M, Shahverdi M. (2004). Effect of harvesting time on soybean seed germination and vigor. J of Agr. 6:9-18.

Albrecht, K.A.; Moutray, J., and Undersander D.J., (1996).Hard seed effects on alfalfa establishment and yield. Agronomy Abstracts, p.116

Almadi A, Yazdi-Samadi B, Zargar-Nataj J. (2004). The effects of low temperature on seed germination and seedling physiological traits in three winter wheat cultivars. Agr. Sci. Nat Res 11:117-126.

Ambika, S., V. Manonmani, and G. Somasundaram. (2014) Reviewon effect f seed size on seedling vigour and seed yield. Res. J. Seed Sci. 7(2): 31-38. Academic journals.Inc.

Baskin, J.M., and Baskin, C.C.,(2004). A classification system for seed Dormancy. Seed Science Research. 14: 1-16.

Bass, L.N., Gunn, C.R., Hesterman, O.B., Roos, E.E. (1988). . Seed physiology, seedling performance, and seed sprouting. In: HANSON et al. (Ed.) Alfalfa and Alfalfa Improvement. Madison: ASA-CSSA-SSSA, Cap.31, p.961-983. (Agronomy Monograph, 29)

Begna, S.H., Hamilton, R.L., Dwyer, L.M., Stewart, D.W., Clouter, D.,(2001). Morphology and yield response to weed pressure by corn hybrids differing in canopy architecture. European J. of Agron. 14(4): 293-302.

Bicer, BT., (2009). The effect of seed size on yield and yield components of chickpea and lentil. African J Biotechnol. 8: 1482-1487.

Bodner, G. – Ullmannová, K. – Středa, T.(2013). Prospects of selection for barley seed vigour as a precondition for stand emergence under dry condition. Kvasny Prum. 59N(9), p. 238–241.

Botwright, T. L., Condon, A. G., Rebetzke, G. J., Richards, R. A., (2002). Field evaluation of early vigour for genetic improvement of grain yield in wheat. Aust. J. Agric. Res., 53(10): 1137– 1145.

Callaway, M.B., (1990). Crop varietal tolerance to weeds. A compilation.. Dept. of Plant Breeding and Biometry Publication series.n.1. Cornell university. Ithaca.

Canmargo, C.P.(1971). Effect of seed vigour up on field performance and yield of grain sorghum. Mississippi State University Publication.

Carvalho, NM and Nakagava, J. , (2012). Sementes: ciência, tecnologia e produção. 5ed. Carvalho, NM and Nakagava, J. , (2012). Sementes: ciência, tecnologia e produção. 5ed.

Chandan Kumar Patil and Aloke Bhattacharjee, (2015). Retention of seed vigour and enhancement of plant potential of safflower species by chemical manipulation. Research Journal of Agriculture and Environmental Management. Vol. 4(6): pp 274- 277.

Chauhan B. S., Abugho S. B. (2013). Effects of water regime, nitrogen fertilization, and rice plant density on growth and reproduction of lowland weed Echinochloa crus- galli. Crop Protoc. 54, 142–147. 10.1016/j.cropro.2013.08.005.

Chen, C. Jiang,Q, Ziska, LH.,Zhu,J., Liu,G. Zhang, J and Zhu, C. (2015). Seed vigour of contrasting rice cultivars in response to elevated Carbon di oxide. Field Crops Research. 178, 63-68.

Chiamai, P N., P. Laosuwa and A. Waranyuwat, (2010). The effects of mungbean seed size on germinability, bean sprout production and agronomic characteristics. Brazil. J. Agriculture. (In press).

Chivasa, W., D. Harris, C. Chiduza, P. Nyamudeza and A.B. Mashingaidze, (1998).Agronomic practices, major crops and farmers' perceptions of the importance of good stand establishment in Musikavanhu Communal Area, Zimbabwe. J. Appl. Sci. South Africa, 4: 9–25

Chloupek, O., Both, Z., Dostál, V., Hrstková, P., Středa, T., Betsche, T., Hrušková, M., Horáková, V., (2008). Better bread from vigorous grain? Czech J. Food Sci., 26(6): 402–412.

Chloupek, O., Ehrenbergerová, J., Ševčík, R., Pařízek, P., (1997). Genetic and non-genetic factors affecting germination and vitality in spring barley seed. Plant Breeding, 116(2): 186–188.

Chloupek, O., Hrstková, P., Jurečka, D., (2003). Tolerance of barley seed germination to cold- and drought-stress expressed as seed vigour. Plant Breeding, 122(3): 199–203.

Clark, S.M. An evaluation of seed quality in crested dog's tail in relation to storability. Seed Science and Technology, v.10, p.517-526, 1982.

Cookson, WR., JS Rowarth and JR Sedcole.(2001) Seed vigour in perennial ryegrass.Effect and cause. Seed Sci. Tech. 29: 255-270.

Cordazzo CV. (2002). Effect of seed mass on germination and growth three dominant species in Southern Brazilian coastal dunes. Bra J Bio 62:427-435.

Crop Science. 31 : 816 - 822.

Dar, F A., M. Gera and N. Gera., (2002) Effect of seed grading on germination pattern of some multipurpose tree species of Jammu Region. Indian For. 128: 509-512.

David W Still (1999). The development of seed quality in Brassicas. Hortechnology. 9(3): 335- 340.

Devi Dayal, Ghosh, P.K. and Ravindra, V. (1999). The influence of seed maturity variations on crop establishment, growth and yield of groundnut (Arachis hypogaea L.). Tropical Agriculture (Trinidad), 76, 151-156.

Dias, M.A.N., Mondo, V.H.V., Cicero, S.M., (2010). Maize seed vigour and weed competition. Brazilian Seed Journal. 32(2): 93-101.

Du, L.V., Tuong. TP. (2002). Enhancing the performance of dry seeded rice effect of seed priming.seedlingrate and time of seedling in Pandey, S. moutimer, wade, L, Tuong, TP, Lopes, K., Hardy , B. (EDS).Direct seeding. Research strategies and oopurtunities. IRRI Phillipines. Pp: 241-246.

Egli, D.B., Rucker, M., (2012). Seed vigour and the uniformity of emergence of corn seedlings.Crop Sci. 52 6): 2774-2782.

Farhoudi, R and M.Motamedi, (2010) Effectof salt stress and seed size on germinatin and early seedling growth of safflower. Seed Sci. Tech. 38: 73-78.

Farooq M., Siddique K. H. M., Rehman H., Aziz T., Lee D. J., Wahid A. (2011). Rice direct seeding: experiences, challenges and opportunities. Soil Till. Res. 111, 87–98. 10.1016/j.still.2010.10.008

Farooq, M., S.M.A. Basra, H. Rehman and B.A. Saleem,(2008). Seed priming enhancement the performance of late sown wheat by improving chilling tolerance. J. Agron. Crop Sci., 194: 55–60.

Finch-Savage, W. (1995). Influence of seed quality on crop establishment, growth and yield. Seed Quality: Basic Mechanisms and Agricultural Implications: 361–384.

Finch-Savage, W.E., Phelps, K., Stickle, J.R.A., Whalley , W.R. and Rowse, H.R. (2001). Seed reserve-dependent growth responses to temperature and water potential in carrot (Daucus carota L.). Journal of Experimental Botany, 52, 2187-2197.

Foolad, M.R., Subbaiah, P., and Zhang, L., (2007). Common QTL affect the rate of tomato seed germination under different stress and non- stress conditions. International Plant Genomics.

Franca-Neto, J.B. Krzyzanowski, F C. and Henning AA. , (2010). A importância do uso de sementes de soja de alta qualidade. Informativo Abrates, v.20, n.1-2, p.37- 38.

Frank, W. J., (1950). Introductory remarks concerning modified working of the International Rules for seed testing on the basis of experience gained after the world war. Proc. ISTA. 16: 405-433.

Ghassemi-Golezani, K., S. Khomari, M. Valizadeh and H. Alyari (2008a). Effect of seed vigor and the duration of cold acclimation on freezing tolerance of winter oilseed rape. Seed Sci. Tech. 36:767-775.

Ghassemi-Golezani, K., S. Khomari, M. Valizadeh and H. Alyari (2008b). Changes in chlorophyll content and fluorescence of leaves of winter rapeseed affected by seedling vigor and cold acclimation duration. J. Food Agric. Environ. 6:196- 199.

Goswami, A., R. Banerjee and S. Raha, 2013. Drought resistance in rice seedlings conferred by seed priming. Protoplasma, 250 (5): 1115-1119. Gupta, P.C., 1993. Seed Vigour Testing. In:

Gunaga,RP., Doddbasava and K. Vasudeva., (2011). Influence of seed sise on germination and seedling growth in Mammaea suriga, karnatakaJ. Agric. Sci. 24: 415-416.

Hall, R.D. and Wiesner, L.E.,(1990). . Relationship between seed vigor tests and field performance of `Regar' meadow bromegrass. Crop Science, v.30, p.967-970.

Hampton, J. G., and TeKrony, D.M. (1995). (ed.) Hand Book of vigour Test Methods.3rd Edition. ISTA Vigour Test Committee.

Hampton, J.G., (1991). Herbage seed lot vigour - do problems start with seed production?

Journal of Applied Seed Production, v.9, p.87-93.

Hampton, J.G., (2002). What is seed quality? Seed Science and Tech. 30: 1-10.

Hampton, J.G., and Bell, D.D. (1989). Seed quality and storage performance of prairie grass (Bromus willdenowii Kunth cv. Grasslands Matua). New Zealand Journal of Agricultural Research, v.17, p.139-143.

Hampton, J.G., and TeKrony, D.M., (1995). (Ed.) Handbook of vigour test methods. 3.ed. Zurich: International Seed Testing Association,.

Harrington, JF., (1972). Seed Storage and Longevity. Academic Press, London. UK.

Harris, D., (1996). The effects of manure, genotype, seed priming, depth and date of sowing on the emergence and early growth of Sorghum bicolor L Moench insemi-arid Botswana.Soil Till.Res., 40: 73–88

Haya Gelmond, Luria, I and Woodstock L. W and Pearl, M. (1978). Effect of accelerated ageing of sorghum seed on seedling vigour. Experimental Botany. 29 : (2) pp 489 - 495.

Hodgkin T, Hegarty TW. 1978. Genetically determined variation in seed germination and field emergence of Brassica oleracea. Annals of Appled Biology 88, 407-413.

Helmer, J.D., Delouche, J.D., and Lienhard, M., 1962. Some indices of vigour and deterioration in seeds of Crimson clover. Lincoln: Association of Official Seed Analysts, 1962. v.52, p.154-161

Hofs, A., Schuch, LOB., Peske, S.t., Barros, ACSA., (2004). Emergence and initial growth of the rice seedlings according to seed vigour.Brazilian Seed journal. 26(1): 92-97.

Hussain, A., M. Maqsood, A. Ahmad, S. Aftab and Z. Ahmad, (1998). Effect of irrigation during various development stages on yield, component of yield and harvest index ofdifferent wheat cultivars. Pakistan J. Agric. Sci., 34: 104–107.

Iglesias- Fernandez,R.,Rodriguez- Gacio,M.C.,Matilla, A.J., (2011). Progress in research on dry after ripening. Seed Sci. Research. 21: 69-80.

Ismail, A.M., Evangelina, S., Ella gGeorgina, Vargara, V., and David, J. MacKill (2009). Mechanisms associated with tolerance to flooding during germination and early seedling growth in rice. (Oryza sativa).Annuals of Botany.103 : 197 – 209.

ISTA, 2015 International Rules for Seed Testing, Basserdorf, Switzerland: International Seed Testing Association (ISTA)

Jaboticabal: FUNEP 590p. Jaboticabal: FUNEP 590p.

Jerlin, R and KK Vadivelu(2004) Effect of fertilizer application in nursery for elite seedling production of Pongamia pinnata L. J Trop. Agric. Res. Extension.7: 69-71.

Jones, J.W. (1933). Effects of reduced oxygen pressure on rice germination. J. Amer. Soc. Agron. 25 : 69-81.

Joshi, N.L., (1987). Seedling emergence and yield of pearl millet on naturally crusted soils in relation to sowing and cultural methods. Soil Till. Res., 10: 103–112

Julio Marcos Filho (2015).Seed Vigour Testing- An overview of the past, present and future perspectives. Scientia Agricola. Vo. 72(4): 363-374.

Juraimi, A.S., M.P. Anwar, A. Selamat, A. Puteh and A. Man. 2012. The influence of seed priming on weed suppression in aerobic rice. Pak. J. Weed Sci. Res., 18: 257-264.

Kaydan, D and M. Yagmur (2008) Germination and seedling growth and relative water content of shoot in different seed sizes of triticale under osmotic stress of water and NaCl. Afr. J. Biotechnol. 7: 2862-2868.

Khawari, F., Ghaderi Far F, and Soltani , E., (2009).Laboratory tests for predicting seedling emergence of safflower (Carthamus tinctorius,L.) cultivars. Seed Technol. 31: 189-193.

Kumar V, Ladha JK. 2011. Direct seeding of rice: Recent developments and future research needs.Advances in Agronomy. 111 :297–413

Kumar V, Ladha JK. And Gathale, M.K. (2009). Direct drill seeded rice. A Need of the day. In annual meeting of Agronomy Society of America. Pitsburg. Nov: 1-9, 2009.

Legesse, N., and Powell, A.A., (1992). Comparison of water uptake and imbibition damage in eleven cowpea cultivars. Seed Science and Technology. 20 : 173 – 180.

Legesse, N., and Powell, A.A., (1996). The association between the development of seed coat pigmentation during maturation of grain legumes and reduced rates of imbibition. Seed Sci. and Tech. 24 : 23 – 32.

Lima, ER., AS Santiago, AP Araujo and MG Teixeria (2005). Effects of the size of sown seed on growth and yield of common bean cultivars of different seed sizes. Brazil. J. Plant Physiol. 7: 273-281.

Lima, T.C.; Medina, P.F.; Fanan, S. (2006)Avaliação do vigor de sementes de trigo pelo teste de envelhecimento acelerado. Revista Brasileira de Sementes , v.28, n.1, p.106-113,

Liu H, Hussain S, Zheng M, Peng S, Huang J, Cui K, Nie L. 2015. Dry direct-seeded rice as an alternative to transplanted-flooded rice in Central China. Agronomy for Sustainable Development 35:285–294.

Marcos – Filho, J.. Fisiologia de sementes de plantas cultivadas. Londrina: ABRATES, 2015.

McDonald, M.B., Copeland, Lo.O., Kanap, A.D., Grabe. D.F., (1996).. Seed development, germination and quality. In: MOSER, L.E. et al. (Ed.) Cool-season forage grasses. Madison: ASA-CSSA-SSSA, 1996. cap.2, p.15-70. (Agronomy Monography, 34)

Miller, D.A., and Stritzke, J.F., (1995). Forage establishment and weed management. In: BARNES, R.F. et al. (Ed.) Forages, an introduction to grassland agriculture. Ames: Iowa State University Press,. v.1, cap.7, p.89-104.

Mondo Vitor Henrique Vaz., Silvo Moure Cicero., Druval Dourado- Neto., Tulio Lourenco Pupim., Marcos Altomani Neves Dias,(2013). Seed vigour and initial growth of corn crop. J.Seed Sci. 35(1):

Mondo, VHV., Cicero, S.M., Dourado- Neto, D., Pupim, T.L., Dias, M.A.N.,(2012). Maize seed vigour and plant performance. Brazilian Seed Journal. 34(1): 143-155.

Morrison, MJ and AG Xue, (2007). The influence of seed size on soybean yield in short season regions. Can. J. Plant Sci. 87: 89-91.

Muthalagan, N., (2009). Studies on genetic variability and association analysis for seedling vigour traits in BC2 F5 generation of rice. M.Sc., Thesis. Annamalai University, Annamalai Nagar, T.N. India.

Mwale, S. S., Hamusimbi, C. and Mwansa, K. (2003). Germination, emergence and growth of sunflower (Helianthus annuus L.) in response to osmotic seed priming. Seed Sci Tech. 31: 199-206.

Nagaraju,S (2010) Influence of seed size and treatments on seed yield and seed quality of sunflowercv. Morden. M.Sc. Thesis. UAS.Dharwad, Karnataka, India.

Namuco O.S., Cairns, J.E., Johnson, D.E. (2009). Investigating early vigour in upland rice (Oryza sativa L.): Part I. Seedling growth and grain yield in competition with weeds. Filed Crop Res.113, 197-206.

Nautiyal, P.C. and Ravindra, V. (1996). Drying and storage method to prolong seed viability and seedling vigour of rabi-summer-produced groundnut. Journal of Agronomy and Crop Science, 177, 123-128.

Nautiyal, P.C. and Zala, P.V. (1991). Effect of drying methods on seed viability and seedling vigour in Spanish groundnut. Seed Science and Technology, 19, 451-459.

Nautiyal, P.C. and Zala, P.V. (2004). Influence of drying method and temperature on germinability and vigour of groundnut (Arachis hypogaea) seed harvested in summer season. Indian Journal of Agriculture Science, 74, 588-593

Nautiyal, P.C., Bandyopadhyay, A. and Misra, R.C. (2004). Drying and storage methods to prolong seed viability of summer groundnut (Arachis hypogaea) in Orissa. Indian Journal of Agriculture Science, 74, 316-320.

Nautiyal, P.C., Bandyopadhyay, A. and Zala, P.V. (2001). In situ sprouting and regulation of fresh seed dormancy in Spanish type groundnut (Arachis hypogaea). Field Crop Research, 70, 233- 241.

Nautiyal, P.C., Joshi Y.C. and Zala, P.V. (1994). Screening of Spanish groundnut cultivars for germination under simulated drought stress. International Arachis Newsletter, ICRISAT, India Centre, Patencheru, Andhra Pradesh.

Nautiyal, P.C., Ravindra, V. and Misra, J.B. (1997). Response of dormant and non-dormant seeds of groundnut (Arachis hypogaea) genotypes to accelerated ageing. Indian Journal of Agriculture Science, 67, 67-70.

Nedeva D, Nicolova A. (1999). Fresh and dry weight changes and germination capacity of natural or premature desiccated developing Wheat seeds. Bulletin. Jour. Phi. 25:3-15.

Noli, E., Casarini, G., Urso, G., Conti, S. (2008). Suitability of three vigour test procedures to predict field performance of early sown maize seed. Seed Sci. Technology. 36: 168-176.

Noor-mohammadi Gh, Siadat A, Kashani A. (2000). Agronomy (cereal). Ahwaz University Press. 446p.

Peksen, E., A. Peksen, H. Bozoglu and A. Gulumser, (2004). Some seed traits and their relationships to seed germination and field emergence in peas. J. Agron. 3: 243-246.

Perry DA. 1978. Report of the vigour test committee 1974-1977. Seed Science and Technology 6, 159-181.

Perry DA. 1980. The concept of seed vigour and its relevance to seed production techniques. In Hebblethwaite PD ed. Seed Production, London: Butterworths 585

Porhadian, H., and Khajepore, M.R., (2010). Relationship between germination test and field emergence of wheat. Asian Journal of Applied. Science.

Powell, A. A., (1989). The importance of genetically determined seed coat characteristics to seed quality in grain legumes. Annals of Botany.63: 169 -175.

Powell, A. A., Matthews, S., Oliveria, M., de, A., (1984). Seed quality in grain legumes.

Powell, A. A., Oliveria, M., de, A., Matthews, S., (1986a). Seed vigour in cultivars of French bean (Phaseolus vulgaris) in relation to the colour of testa. Jour. Of Agricultural Sciences, Cambridge, 109: 419 – 425.

Powell, A. A., Oliveria, M., de, A., Matthews, S., (1986b). The role of imbibitions damage in determining the vigour of white and coloured seed lots of dwarf French Bean (Phaseolus vulgaris). Journal of Experimental Botany.57: 716 – 722.

Radford, B.J., (1983). Sowing techniques: effects on crop establishment. J. Australian Inst.

Randall, P.J. and H.J. Moss, (1990). Some effects of temperature regime during grain filling on wheat quality. Australian J. Agric. Res., 41: 603–617

Rao AN, Johnson DE, Sivaprasad B, Ladha JK, Mortimer AM. 2007. Weed management in direct-seeded rice. Advances in Agronomy, 93:155–255

Rastegar,Z., and MAS Khandi, (2011). The effectof salinity and seed size on seed reserve utilization and seedling growth of soybean. Int. J. Agron. Plant. Prod, 2: 1-4.

Redona, E.D., Mackill,D.J. (1996). Genetic variation for seedling vigour traits in rice. Crop Sci. 36, 258-290.

Ries, SK., and EH Everson, (1973).Protein content and seed size relationship with seedling vigour of wheat cultivars. Agron. J. 65: 884-886.

Riis, P., Bang-Olsen, K., (1991). Breeding high vigour barley – a short cut to fast maltings. Proceedings of the European Brewing Convention Congress, Lisbon, IRL Press: Oxford, 1991, 101–108.

Ros, C., Bell, R W and White , P .F. (2003) Seedling vigour and the early growth of transplanted rice. Plant and Soil. Vol. 252 (2): Pp 325 – 337.

Roy, SKS, A. Hamid, MG Miah, and A. Hashem(1996). Seed size variation and its effects on the germination and seedling vigour in Rice.J. Agron. Crop Sci. 176: 79-82.

Samarah, N.H., Mullen, R.E., Golgi, S., and Gaul, A.,(2009). Effect of drying treatment and temperature on soybean seed quality during maturation. Seed Sci. and technol. 37: 469-473.

Savin, R., P.J. Stone and M.E. Nicolas, (1996). Response of grain growth and malting quality of barley to short periods of high temperature in field studies using portable chambers. Australian J. Agric. Res., 47: 465–477

Savin, R., P.J. Stone and M.E. Nicolas, (1996). Response of grain growth and malting quality of barley to short periods of high temperature in field studies using portable chambers. Australian J. Agric. Res., 47: 465–477

Sheaffer, C.C. hall, M.H., Martin, N.P. Rabas, D.L., Ford, JH., Warnes, D.D.,(1988). . Effects of seed coating on forage legume establishment in Minnesota. University of Minnesota. Agricultural Experimental Station Bulletin, n.584

Shewry PR. (2009). Wheat. J of Exp Bot. 60:1537-1553.

Singh S., Sharma S. N., Prasad R. (2001). The effect of seeding and tillage methods on productivity of rice–wheat cropping system. Soil. Till. Res. 61, 125–131.

Singh, A.L., Nautiyal, P.C. and Zala, P.V. (1998). Growth and yield of groundnut varieties as influenced by seed size. Tropical Science, 38, 48-56.

Stewart, R R C. and Bewley, J.D., ((1980). Lipid peroxidation associated with accelerated ageing of soybean axes. Plant Physiol. 65: 245-248.

Still, D.W. and K.J. Bradford. (1998). Using hydro time and ABA-time models to quantify seed quality of brassicas during development. J. Amer. Soc. Hort. Sci. 123:692–699.

Stobbe, E ., J . Moes, Y Gan, H. Ngoma and L Bourgeca, (2008). Seeds, seedvigour and seedling research report, Department of Plant Science, NDSU, Agriculture and University Extension, North Dakota, USA.

Stoehr, M.U., S.J. L'Hirondelle, W.D. Binder, and J.E. Webber. 1998. Parental environment aftereffects on germination, growth, and adaptive traits in selected white spruce families. Can. J. For. Res. 28:418–426.

Stougaard RN and QW Xue(2005). Quality versus quantity: Spring wheat seed size and seedling rate of effets on Avena fatua intereference, economic returns and economic threholds. Weed Res.. 45: 351-360 .

Sun, Q., Wang, J.H., Sun, B.Q., (2007). Advances in seed vigour, Physiological and Genetic Mechanisms. Agril Sci. China. 6 (9): 1060 - 1066.

TeKrony, D. M., and Egli, D. B., (1991). Relationship of seed vigour to crop yield, A Review.

Timmermans, B.G.H., J. Vos, J.V. Nieuwburg, T.J. Stomph, P.E.L. Putten and L.P.G.Molendijk, (2007). Field performance of Solanum sisymbriifolium, a trap crop for potato cyst nematodes. I. Dry matter accumulation in relation to sowing time, location, season and plant density. Annl. Appl. Biol., 150: 89–97

Tollenaar, M., Deen, W., Eichart,L., Liu, W., (2006). Effect of crowding stress on dry matter accumulation and harvest index. Agronomy J. 98(4): 930-937.

Tully, R.E., Musgrave, M.E., and Leopold, A.C., (1981). The seed coat as a control of imbibition chilling injury. Crop Sci. 21 : 312 – 317.

Ullmannová, K., Středa, T., Chloupek, O.,(2013). Use of barley seed vigour to discriminate drought and cold tolerance in crop years with high seed vigour and low trait variation. Plant Breeding, 132(3): 295–298

USDA- National Agricultural Statistical Service. (2010). Quick stats.1.0. Average U.S.soybean seed prices.USDA- NASS, Washington, DC

Wang, YR., Hampton, JG., (1993). Seed vigour in lucerne and white clover. In: International Grass land Congress., 17., 1993. Proceedings.. p.1868-1869.

Wardlaw, I.F. and C.W. Wrigley, (1994). Heat tolerance in temperate cereals. An overview. Australian J. Plant Physiol.,21: 695–703

Whittington WJ. 1973. Genetic regulation of germination. In: Heydecker W, ed. Seed Ecology. London: Butterworths 5-30.

Wulff, R.D.(1995). Environmental maternal effects on seed quality and germination, p. 491–505. In: J. Kigel and G. Galili (eds.). Seed development and germination. Marcel Dekker, New York

Yamauchi, M and Winn, T.,(1996). Rice seed vigour and seedling establishment in anaerobic soil. Crop. Sci. 36 : 680- 686.

Zareian,A , A. hamedi, H. Sadeghi and MR Jazaeri,(2013).Effect of seed size on some germination germination characteristics, seedling emergence percentage and yield of three wheat cultivars in laboratory and field . Middle East. J. Sci. Res. 13: 1126-1131.

Zhang, Z.H., Yu, S.B., Yu, T., (2005). Mapping QTLs for seedling vigour using RILs of rice. Field Crop Research. 91 (2-3): 161-170.

Zhao, D.L., Atlin, G.N., Bastiaans L., Spiertz, J.H. (2006). Comparing rice germplasm groups for growth, grain yield and weeed suppressive ability under aerobic soil conditions. Weed Res. 46.444-452.

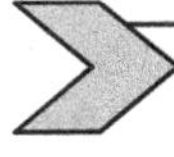

CHAPTER 2

Origin and Meaning of Seed Vigour

(A strong edifice cannot be built with poor bricks)

2.1 Introduction

A good majority of plants are propagated from seed either for food, fibre, or ornamental than any other method of propagation. Therefore, propagation through seed remains the corner stone for producing plant species not only of agronomic importance but also of vegetable, forestry as well as plants of aesthetic significance. It is well known that the final destination of seed is the field and hence establishment of suitable plant stand will be the top priority for producers. As such, in order to ensure seedling emergence in the field, it becomes sufficient to assess the seed vigour previously aiming at identifying seed lots that present higher chance of adequately establishing in the field and provide the expected returns, Marcos Filho (2005).

- Origin of seed vigour at genetic, physiological and biochemical levels.
- Meaning of seed vigour, Differences in seed vigour, Different vigour parameters.
- Adverse effects of using low vigour seeds.
- Relationship between viability and germination with vigour.
- Standard germination testing, Basic requirements of a germination test.
- Objectives of standard germination testing.
- Prerequisites and relevance of germination testing.
- Causes of variation in germination test results- Limitations of the germination test.
- Seed vigour - Differences between germinability and seed vigour.
- Components of seed vigour and Different phases of germination.

Among the diverse parameters used to assess the quality of a seed lot, the physiological potential is assumed to be highly essential. The assessment of seed physiological potential is usually performed through germination and vigour test. The vigour tests are regarded as important for revealing relatively narrow variation in different deterioration stages of a seed lot, Baalbaki *et al.,* (2009), whereas the germination test indicate only sharp differences on seed deterioration degree. Hence, the use of vigour tests is of large usefulness on monitoring seed performance during storage, since the drop of seed vigour precedes the viability loss.

2.2 Origin of Seed Vigour

In the recorded history, probably for the first time, Kolreuter, (1761-66) reported the existence of increased vigour in the hybrids of *Nicotiana, Dianthus, Datura* and other species. This phenomenon was subsequently observed and reported in different species by several prominent other researchers like Sagaret (1826); Lecoq (1845); Naudin (1865); Darwin (1876); Beal (1880); Johnson, (1891), Shull and East (1905-12) and Xiao *et al.,* (1995) etc., are only a few to name. In reality, plants exhibit vigour throughout their life cycle, but with varying degrees during different growth stages. However, only four different stages in the life cycle of a plant viz., (i) storage and survival, (ii) field emergence, (iii) establishment of mature plants and (iv) production and yield of the crops where seed or seedling vigour exerts direct effect, Heydecker, (1972).

2.2.1   Genetic Origin

In respects of seeds, the origin of vigour is genetical, as observed in heterosis. Because, 'Heterosis' represents a measurable superiority of the hybrid progeny over their respective inbred parents, be it, either over the superior parent or over the mean value of both the parents. In some plant species, either, it could be due to the presence of efficient mitochondria or efficient enzyme system for carbon assimilation or even it could also be due to the mechanical damage. It means vigour is the difference between two genetic lines or in other words the difference in vigour between two genotypes of a crop. Four decades ago, Nass and Crane (1970) have identified certain genes responsible for producing more vigorous seed. Similarly, Potts *et al.,* (1978) also reported that hardseededness though genetically controlled, is an undesirable trait that offers protection to the seeds from ageing. Further, factors like chemical composition of seed and susceptibility of seed to damage *etc.,* were also found to be under the genetic control.

It is not an exaggeration to state that in several cases of genes for nutritional quality, for example, high sugar maize lines, almost ended up in poor quality (shriveled, shrunken, not so appealing) of seed. The "Dominance Hypothesis " states that increased vigour is due either to the accumulation of favourable dominant genes or masking of deleterious recessive genes in a hybrid and was found to be consistent with the recent genomic evidence of differences in genetic content between maize inbred lines, Fu and Dooner (2002) and has been demonstrated as an underlying cause of heterotic response for grain yield in QTL mapping study by Graham *et al,* (1997).

On the other hand the 'Over Dominance' theory argues that the heterozygous combination of alleles at a single locus is superior to either of the homozygous combinations. The elimination of hardseededness through the procedures of breeding and selection in pima cotton decreased the impermeability of water that prevents germination. Similarly the open pollinated varieties and double cross hybrids of maize also exhibited good seed vigour when compared to single cross hybrids produced on relatively weak inbred parents.

Application of quantitative genetic studies in the field of genetic architecture of seed germination under various stresses led to the identification of QTLs controlling this trait, Dias *et al .,* (2011); Galpaz and Reymond,(2010). Recent studies at molecular level also indicated the potential for genetical improvement of seed vigour and is evidenced by the identification of QTLs pertaining to germination rate in *Helianthus annuus,* Al- Charani *et al.,* (2005); in *Brassica oleracia,* Finch –Savage *et al.,* (2010), and in *Medicgo truncatula,* Vandecasteele *et al.,* (2011). Further, the fine mapping and cloning of these QTLs with DOG1, dormancy controlling QTL of Arabidopsis, revealed molecular mechanisms governing the response of seed germination to environmental stresses, Bentsink *et al.,* (2006), Chiang *et al.,* (2011) and Kendal *et al.,* (2011). The latest technologies in post genomics field helped Carrera *et al.,* (2007) in generating a list of genes with differential expression in germinating seeds and have been referred to as germination signature, and a seed specific gene ontology named TAGGIT to help in describing this signature. Further, Bassel *et al.,* (2011) through their genome wide network model reported that the decision of seed to remain either dormant or enter in to germination depends on the net activity of germination promoting signals overcoming their inhibiting counterparts.

A great deal of knowledge on seed dormancy and germination has been made available through the well developed model species *Arabidopsis*, (Finch-Savage and Leubner- Metzger, 2006; Holdsworth *et al.*, 2008a; Finkelstein *et al*, 2008; North *et al.*, 2010; Weitbrecht *et al.*, 2011; Rajjou *et al.*, 2012; Rodriguez *et al.*, 2015). The efforts thus far made are confined to the consequences of environmental variation rather than focusing on understanding the mechanisms to determine the initial vigour of seed in an agricultural context.

This small seeded annual has adapted to a wide range of natural environments, but has not been subjected to the selection pressures of crop domestication and therefore its' seeds have not been challenged to perform in an agricultural context. We suggest that adaptation to the natural environment with emphasis on post-shedding dormancy to time germination to seasonal changes is very different from the agricultural necessity to germinate and progress to seedling emergence with the minimum delay following sowing (Section 5). This provides a focus for the difference since both situations require seeds to be dormant before shedding (harvest) to avoid germination on the plant leading to pre-harvest sprouting (Paulsen and Auld, 2004), but their subsequent behaviour is sharply contrasting.

After summarising why seed vigour is important, below we review current understanding of seed behaviour, ageing and, its application in seed technology (pre-and post-harvest). Assuming this knowledge is fully utilised to optimize production practice, minimise negative environmental effects, and limit seed deterioration during storage, the factors that determine seed performance during crop establishment and the potential vigour of the seed are discussed. In doing this we consider what is required of a robust vigorous seed, how to achieve it, and what the underlying mechanisms may be. It is not possible here to be comprehensive in covering the literature; our aim is to provide useful illustrations of the wide range of subject areas covered. Seeds, even those of crop species, are morphologically and physiologically diverse and the review cannot be comprehensive in addressing all the specific issues. Where necessary, we therefore focus on small seeded vegetable crops where seed performance in determining timing, uniformity as well as extent of seedling emergence is particularly crucial to crop production (discussed in section 2). Nevertheless, most issues discussed are equally relevant to larger-seeded and grain crops.

2.2.2 Physiological and Biochemical Origin

The origin of vigour, in addition to the above, is physiological and biochemical too, as these reactions can create and cause the differences in vigour between two seed lots belonging to the same genetic line. For the first time, Fredrich Nobbe, introduced the term *'triebkraft'* way back in 1876, to represent the "driving force" or "shooting strength' of the emerging seedlings. He used this term, to convey the idea that, in addition to germination, speed and uniformity are the two other important factors that determine the quality of seeds; therefore, vigour is physiological and biochemical too. Since, the physiological and biochemical vigour has its basis always lying in the genetic vigour, all these reactions are obviously under the genetic control. The process of germination and subsequent establishment of seedling, the two independent but inter related components of seed vigour and can only be summed up as an act of balance between the thrust generated by the embryo (resulting from the osmotic uptake of water) and the resistance offered by the tissues surrounding the embryo. Most cucurbit seed lots contain high percentage of osmotically distended (OD) seeds. When the harvest is delayed seeds have advanced from physiological maturity to accelerated ageing and can be considered as dead seeds.

Seed vigour is not only greatly influenced by seed production and storage environment but also has an important genetic basis, Betty *et al.,* (2000). Several technological solutions viz., (i) use of improved seed beds, (ii) selection of seed (iii) appropriate seed treatments including conditioning used for improving plant stand establishment etc., are only aimed at tackling the negative effects of seed production and storage environments and hence it remains variable. The recent developments in genetic resources management are providing an opportunity to identify the key seed vigour genes through a trait led approach for use in breeding programmes. Therefore, the development of seed vigour testing protocols is most likely to solve this perennial problem. Because the identification of genes responsible for vigour will not only provide the practical benefits, but also stipulate the required opportunity to develop a scientific understanding for explaining the basis of seed vigour differences, which is lacking at present.

2.3   Meaning and Definition of Seed Vigour

There are diverse thoughts and opinions in the scientific field about seed vigour, Most of the vigour experts often disagree on how to spell the word Vigour or Vigour.

The dictionary meaning for "Vigour" indicates physical or mental strength, energy and forcefulness. Though the word vigour can be described synoptically with viability, robustness, sturdiness, integrity, adaptability, resilience, none of them are acceptable scientifically. In the published literature several synonymous words are used for seed vigour. These are "Pushing power, "Nobbie (1876) ; "Field or Planting value" and Physiological Pre-determination by Kidd and Wett (1918); "Vitality" by Munn (1935); "Driving force" by Heydecker (1965); Potential low planting value" by Matthews and Bradnock (1967) and Field Emergence Potential by Bradnock and Matthews (1970). So, the meaning of vigour of a seed or seed lot, in simple terms, is a state of good health and natural robustness of the seeds, which, upon planting gives rise to rapid germination, uniform stand establishment coupled with normal growth under a wide range of environmental conditions. A seed is said to be vigorous, only when it shows uniformity in germination and normal plant growth and development even under sub optimal conditions.

Of late, the interest in seed vigour has assumed considerably more prominent in at least on two circumstances. First of all, it is expected to meet the requirements of the chain of super markets coupled with the demands from the transporters and consumers alike, to have a product of uniform quality. This, in turn has created a great demand for uniformity in the product, which obviously necessitated uniform rate of growth and maturity at the same time duly facilitating mechanized harvesting. Despite controlling micro-environmental conditions by way of providing (a) uniform soil conditions, (b) depth of planting, (c) best management efforts on soil fertility and moisture during planting time etc., a good number of unproductive plants still existed under the field conditions. This particular situation is nothing but the real expression of abnormally low vigour seeds present in the material offered for sale. As such, by enhancing the vigour of these seeds by way of pre and post harvest management practices, realization of still higher yield with a little bit more additional cost per unit area is not impossible. According to Sun *et al.,* (2007), seed vigour has been known as a comprehensive character affected by many factors, such as genetic background, environmental factors during seed maturation and development stages, and conditions prevalent during storage, which makes the genetic analysis of seed vigour even more complex and difficult. Secondly, due to varied soil, seed bank and environmental conditions seed vigour cannot be an absolute value representing field emergence. However, it can provide information about field emergence, storage potential and consistently ranking these features, which are not offered in germination test.

Recently, the term vigour of viability is gaining prominence to describe the physiological characteristics of seeds that control its ability to germinate rapidly in the soil and to tolerate various mostly negative environmental factors. So in respect of seeds, viability is the state of being alive, while vigour is related to denote the degree of their aliveness and hence vigour can be considered as the closest measure of potential performance. Hence, seed vigour is the capacity of seed to emerge from the soil and survival under potentially stressful field conditions and includes revival of growth under favourable conditions.

Seed vigour can be defined in terms of normal seedling morphology as the rate at which the seeds germinate and grow in the early stages. It means, possessing strong vigour by seeds is likely to offer several advantages to the grower because the seeds are less likely to be overtaken by weaker ones, weeds, diseases and insects etc. According to the definition adopted by AOSA (1979), 'Seed vigour comprises of those seed properties which determine the potential for rapid, uniform emergence and development of normal seedlings under a wide range of field conditions'.

2.3.1   Differences in Seed Vigour

In respect of maintenance of quality of seeds in general and germination in particular, Standards have already been prescribed below which the seed cannot be offered for sale commercially. Seed germination test together with purity estimation provides the planting value of seed lots, which needs to be presented on seed labels as mandated by law. This automatically ensures that, the failure of seed to emerge is not due to the inability of the seed alone. Despite this, differences in emergence of germinable seeds do occur when sown in poor field conditions and this was termed as seed vigour by Perry (1970). From this, it is very much clear that germination and vigour are not the same, i.e., vigour is only an indication of the ability of seed / seed lots to germinate and establish seedlings even under suboptimal conditions. Therefore vigour gives only an assurance as to the reliability of emergence especially for costly seeds of new varieties even in less than ideal sowing conditions. It means, different seed lots of a genotype or variety, harvested from the same area on the same date and time and registering similar high laboratory germination percentage are bound to differ significantly with respect to the levels of vigour and ultimately the plant population and yielding ability.

Vigour pertaining to seed or a seed lot is simply, the ability of germination, emergence and seedling growth rate controlled by a co-adaptive genotype in a given environment. In other words, seed vigour is merely a relative superiority in performance of a seed lot of a particular genotype over the seeds of the same or other genotypes in a given environment under well-defined experimental conditions. The meaning of seed germination from a 'Crop Botanists' or 'Plant Physiologist's' view is simply related to the emergence of radical, contrary to this, a 'Seed Technologist' views it as whether or not the development of all essential structures that lead to the production of the seedling, which in turn will produce a normal plant. So, the difference between these two definitions can be called as vigour. In other words, the failure of the germination test to precisely predict the differences in the field emergence potential, particularly under poor field conditions, reveals the existence of another physiological aspect of seed quality and this has come to be referred to as seed vigour. Therefore, seed lots having poor emergence percentage in the field despite high laboratory germination are referred to as low vigour seeds, where as those giving good emergence are termed as high vigour seeds (Fig. 2.1).

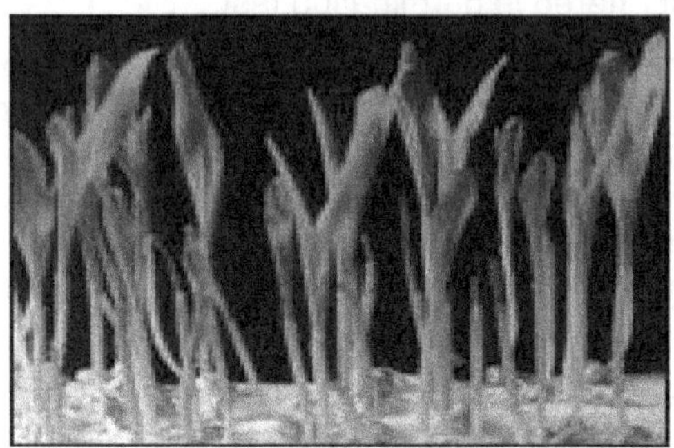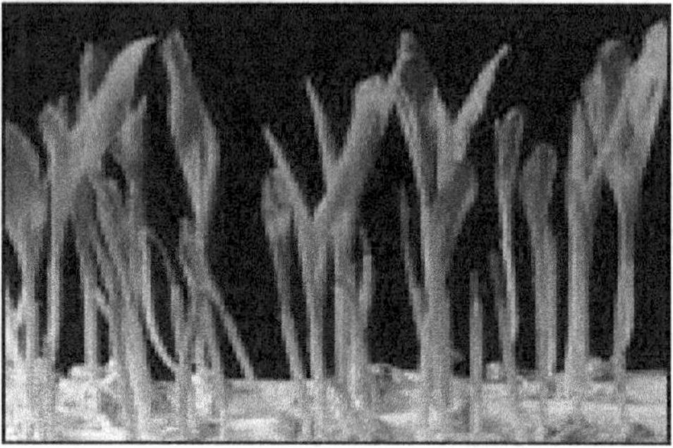

Fig. 2.1 Uniformity in seedlings emergence

Usually, the seeds in general and the seed lots in particular are expected to lose their ability to carry out their normal physiological functions assigned to them in a stipulated period of time. This period of time varies significantly between varieties of a species and among different species in a genera, and the process is called as physiological ageing, which at times, depending on the species and variety, starts while the seed is still intact on the mother plant itself and is likely to continue through different stages of post production like harvesting, processing, conditioning and storage or even till planted in the soil for growing the succeeding generation. The progressive reduction in the performance capabilities of seeds could be attributed either due to (i) the changes occurring in the integrity of cell membrane(s), or (ii) the level of activity of various enzymes and ultimately the synthesis of appropriate proteins. Further, the changes at biochemical level may also occur either very rapidly in few days, or more slowly and may take even years, depending again up on several factors like (a) the genetic makeup of the species, (b) production environment and (c) the interaction of these two, which, of course, are not yet fully understood. Generally, the end point of this deterioration ultimately leads either to the death of seed or ends up in complete loss of germination. Usually, the seeds lose their vigour much before they lose their ability to germinate. For this reason, the seed lots that have registered high similar germination values differ significantly with respect to the level and extent of deterioration. Hence, the seed lots differ significantly in their seed vigour and therefore, differ in their ability to perform ultimately.

Similarly, differences in vigour of seed lots are also reflected in seed longevity. According to Delouche and Baskin (1973), a highly vigorous seed lot is the one, which will have a good storage potential and hence retains good germination capacity. While clarifying the meaning of vigour in terms of seed, seedling and performance of plant, Copeland (1976) suggested that, speed of germination, ability to emerge through crusted soil, uniformity in germination and development even under non-uniform conditions (like cold, wet, pathogen infested soil etc.,) and establishment of normal seedlings, normal crop yield, good shelf life and higher proportion of food reserves etc., need to be taken into consideration. In order to ensure the highest possible vigour of a seed lot, one has to put efforts with great focus on (i) creating a positive growth environment such that it facilitates the development of vigorous seeds, (ii) harvesting the seed crop at the most appropriate stage, be it the physiological maturity or even after that at harvest maturity depending on the species, and (iii) planning and execution of safe handling and storage conditions so as to maintain the seed in such a way that minimizes the damage or even slowdowns the process of deterioration.

The terms, vigour and viability are frequently misused at different levels. For some seed Technologists, viability is the capacity of a seed to germinate and produce normal seedling in one sense, and in another sense, it is the degree to which seed is active and possess enzymes necessary for catalyzing metabolic reactions needed for germination and seedling growth. In this particular context, since a seed comprises of both living and non-living tissues, the seed may or may not be capable of germination, and hence, may or may not be viable.

Vigour differences present in the seed populations clearly exhibit significant disparities with respect to crop stand establishment and at the same time, may be these dissimilarities are not always reflected in the yield. This could be justified simply and attributed to the capacity of a single plant in a reduced plant stand which is likely to compensate and produce only a little less when compared to a denser crop. Delouche and Caldwell, (1960) reported that, differences in emergence under poor field conditions is not the only characteristic of seeds differing in vigour, it is also reflected by the rate of germination and seedling growth and applies to both favourable and unfavorable conditions for germination and emergence. This can best be explained with maize, a widely spaced and non tillering crop sensitive to cold temperature at 8%-12°C, The low or poor emergence due to cold temperature or low vigour result in reduced yield, where as contrasting result can be observed in case of wheat, a closely spaced and tillering type. Powell (2001) proposed three aspects of performance of seeds associated with seed vigour such as (i) emergence ability of seed, even under unfavourable environmental conditions. It is also called as rate of emergence. Differences in seedling emergence time results in variable plant size. (ii) Rate of uniformity of seed germination and seedling growth. Vigour being a characteristic of seed lots with high germination exerts influence on the rate, uniformity and final emergence level in the field as well as glass house condition and (iii) ability to perform/germinate especially after storage.

2.3.2 Vigour Parameters

The encyclopedia of seeds has considered (a) mean rate of germination, (b) spread of germination time, (c) mean seedling size, (d) emergence potential, and (e) storage potential as the key parameters of seed vigour. These parameters with high and low levels of vigour are presented in table 2.1.

Table 2.1 Comparison of different vigour parameters. (Encyclopedia of Seeds)

S.No.	Vigour Parameters	Level of vigour	
		High	Low
1	Mean rate of germination	Fast	Slow
2	Spread of germination time	Narrow / Good	Wide / poor
3	Mean seedling size	Large, uniform	Small, variable
4	Emergence potential	Good in most soil conditions	Poor in less than optimum soil conditions
5	Storage potential	Good	Poor

2.3.3 Adverse Effects of Using Low Vigour Seeds

Decline in germination speed of viable seeds is followed by reduced seedling size despite the maintenance of high primary root protrusion rate and enhanced abnormal seedlings are the chief manifestations of initial seed ageing. The low germination rate can be attributed to early signs of membrane disorganization, where as occurrence of seedling abnormalities particularly when the seeds are in their final stages of deterioration could be attached with the significant death of tissues in different parts of the specially in the meristematic tissues. Low vigour seeds are expected to decrease the percentage of emerged seedlings and seedling emergence speed, initial growth, leaf area, and leaf dry mass accumulation. Schuch *et al.,* (2000). Differences in initial crop growth related to seed vigour have been published for rice, Hofs *et al.,* (2004); Melo *et al.,* (2006); Mielerzski *et al.,* (2008); Dias *et al .,* (2010) in maize,; Schuch *et al.,* (2000) in black oats; Mondo *et al.,* (2012a), Dias *et al.,* (2011), Schuch *et al.,* (2009) and Kolchinski *et al.,* (2006) in soybean. Egli and Rucker (2012) have confirmed that high vigour maize seeds always have more uniform emergence than low vigour seeds. Delayed emergence at any stage will automatically result in lower plant development than other plants with early emergence. As a consequence of initial uneven growth compete for light and other resources such as water, nutrients of soil and weeds.

Seed lots with low seed vigour exhibit poor storage potential and hence shows rapid decline in germination, (Hampton, 1992), especially when stored under suboptimal storage conditions. Assuming that low vigour seeds are planted in a field, the following twelve adverse effects were reported by various researchers from time to time.

1. Sub-optimal plant population,
2. Delayed and reduced emergence of seedlings,
3. More prone to stress,
4. Inferior early performance-plants become dwarf with thin stems,
5. Reduced number of nodes and leaf area
6. Poor tillering,
7. Reduction in panicle length,
8. Incomplete panicle exertion
9. Delayed anthesis,
10. Increase in the number of barren plants
11. Decreased yields, and
12. Delayed crop maturity.

Differences in initial emergence obtained from seeds of same age from lower and higher vigour seed lots is presented in Fig.2.2. Whereas, initial development of soybean seedlings of same age from a lower (A) and higher (B) vigour seed lots are presented in Fig. 2.3.Similarly, Seeds from low and high vigour seed lots sown on same date in the field under direct sown production system are presented in Fig. 2.4. Further, seeds emerging from high and low vigour seed lots sown on same date in transplant production system under controlled conditions are presented in Fig. 2.5. The inferior early performance derived from low vigour seeds will disappear automatically over a period of time and hence there may not be any significant reduction in final yields. Similarly, reduced plant population up to a manageable limit can be compensated in thickly sown crops by producing yields close to normal stands. The vigour differences thus pose a great problem particularly with expensive F_1 hybrid vegetable seeds or when the production system demands a uniform crop stand to derive uniform produce or to facilitate the use of mechanical pre-post harvesting operations aimed at generating economic gains by way of producing a uniform sized material for early markets.

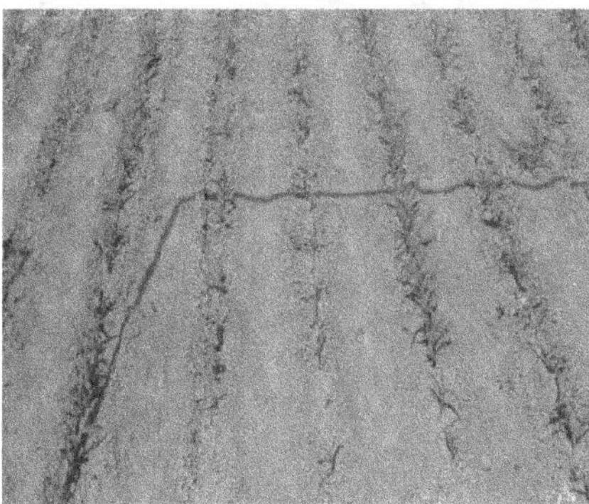

Fig. 2.2 Differences in initial emergence from seedlings of same age from lower and higher vigour seed lots

Fig. 2.3 Initial development of soybean seedlings of same age from a lower (A) and higher (B) vigour seed lots

Fig. 2.4 Seeds from low and high vigour seed lots sown on same date in the field under direct sown production system

Fig. 2.5 Seeds from high and low vigour seed lots sown on same date for transplant production system under controlled conditions

By examining the relationship between seed vigour and yield, Willy and Heath (1969) were able to separate the effects of vigour of individual plant performance from those of poor plant population with the help of the following equations.

$$l/w = a + bp,$$

where,

w = mean yield of individual plants and

p = Number of plants/unit area.,

a and b are constants derived from the linear regression between yield and plant population.

$$Y = p / (a + bp)$$

where, Y = yield /unit area, is a function of w and p.

For example, when the crop is sown at normal sowing rate, with the strong inter plant competition between high and low vigour seeds the negative effects of low vigour seeds on final yield is cancelled. This is so because, the yield per unit area fluctuates between 1/a and 1/b depending on plant population per unit area. As such, it will be nearer to 1/a at lower population densities and closer to 1/b at higher plant densities.

Based on the vigour studies conducted on the seed lots of rice, maize, sorghum and cotton as planting material. The Mississippi State University has concluded that following are the ten most common adverse effects of using low vigour seed lots.

   (i)   Reduction in the seed bearing capacity of plant stand,

   (ii)   Delayed and reduced emergence of the seedlings,

   (iii) Plants became dwarf with thin stems,

   (iv)   Reduced node number,

   (v)   Reduced leaf area and tillering,

   (vi)   Reduced panicle length,

   (vii) Delayed panicle exertion and anthesis,

(viii) Increase in the number of barren / unproductive plants,

(ix) Delayed maturity in the crop and

(x) Decreased yield.

2.3.4 Relationship of Viability and Germination with Vigour

Seed viability, germination and vigour are the three different but unique parameters of quality of a seed lot or a seed population. The information pertaining to the vigour of variety can be inferred partially from the germination data available only when one is familiar with the general relationship between viability and vigour. Therefore, this renders the data of germination without vigour data and vice versa is less meaningful than when considered together. Therefore, in order to have a holistic approach, these parameters need to be understood together only to describe the quality of a seed population more accurately and precisely. Initially or immediately after harvesting, the vigour and viability of seed lots will be 100 per cent and decreases gradually with the passage of time.

The relationship is usually expressed as per cent germination under a set of conditions after a period of ageing. Generally the viability curve represents germination under optimal conditions.

Contrary to this, the vigour curve expresses germination only under stressful conditions. For example, after eight months of storage, germination under optimal conditions is about 80 per cent where as under stress conditions it is only 30 per cent.

The vigour has been differentiated from germination in several aspects viz.,

(i) the desire to measure uniform and rapid seedling emergence,

(ii) with a strong focus on germination potential,

(iii) emergence potential in a wide variety of both optimal and suboptimal field conditions and

(iv) hence more relevant to typical agronomic conditions. Seed vigor is the capacity of seed to emerge from the soil and survive under potentially stressful field conditions and to grow rapidly under favorable conditions. The term vigour of viability is being used recently to describe the physiological characteristics of seeds that control its ability to germinate rapidly in the soil and to tolerate various, mostly negative environmental factors. Hence, Seed Vigour can be used to denote the degree of the seed's aliveness. As such, seed vigour can be considered as the closest measure of potential field performance.

In general, the relationship between vigour and viability is similar except that vigour declines much before viability. The precise relationship between Viability, germination and vigour of a seed during its development and attaining physiological quality will be exactly the opposite after some initial storage depending on the species and is presented in detail in the following Fig. 2.7. i.e., during deterioration physiological quality changes in inverse order of seed development, with seed vigour declining first followed by loss of germination and viability. But the decline in vigour is much faster than the decline in seed viability. The viability curve represents germination under normal/ optimal conditions where as the vigour curve represents germination under stressed conditions. For example, after storing the seeds for about eight months, germination under optimal conditions is about 80 percent, whereas the same under stressed conditions is around 30 per cent only.

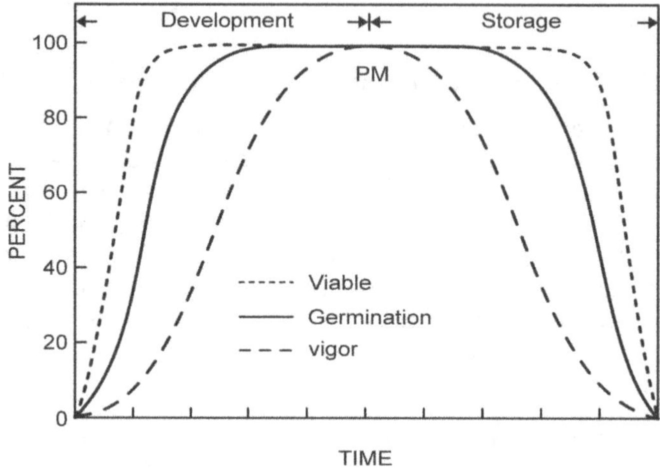

Fig. 2.7 Relationship between viability, germination and vigour over a period of time.

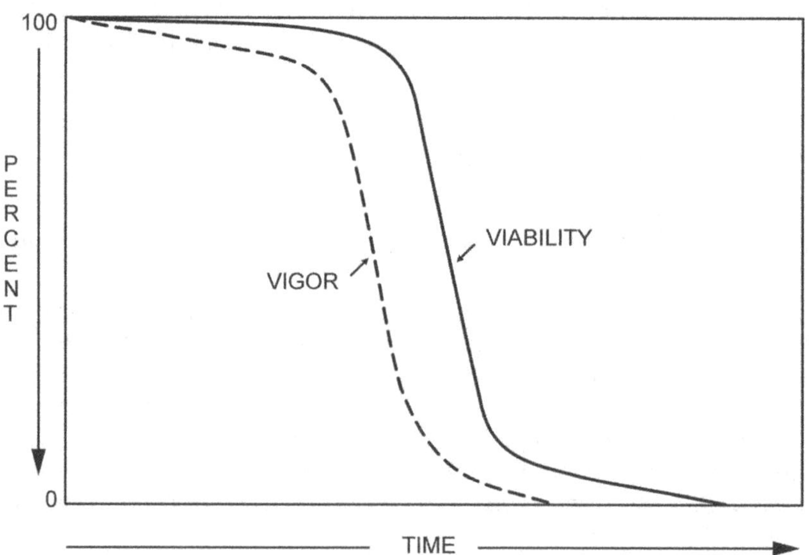

Fig. 2.8 A general relationship proposed by Harri gton, (1972) between decline in vigour and viability.

The information pertaining to the vigour of a variety can be inferred partially from the germination data if one is familiar with the general relationship between these two. Usually the relationship between viability and vigour is expressed as per cent of germination under a given set of conditions after a certain period of ageing or time. The relationship of seed vigour with viability can be explained with the help of a sigmoid survival curve as shown in the Figure. 2.8. which is a generlized one proposed by Harrington (1972). The relationship between viability and vigour is similar except that Sed decline of vigour is much earlier than the viability and this relationship can be explained with three distinct phases as follows.

Phase-I: When the first germination is approximately 80 per cent or above, then the seed is both viable and vigorous.

Phase-II: During the second phase, deterioration progresses very rapidly and

Phase-III: In third phase, the rate of deteriorations slows down to approximately 20 per cent or even below and eventually all seeds die slowly. But in reality, significant differences are likely to occur in different types of seed and among different varieties with respect to seed viability and vigour. At times, depending on the substandard storage conditions and especially when the relative humidity is congenial for the growth of storage molds, the differences between vigour and viability show up more predominantly. Differences between seed vigour and germinability are presented in table 2.2.

Table 2.2 Differences between Seed Vigour and Germinability

Seed Vigour	Germinability
Vigour is a qualitative character, which cannot be defined but only compared and reported on individual seed as a unit.	Germinability is a quantitative character that can be defined and reported on lot as a unit.
It is controlled by several factors like genetic makeup, seed size, nutrition provided by the mother plant, mechanical injury, pre harvest maturation to pre sowing environment, deterioration and ageing, soil moisture and fertility after sowing *etc.,*	The germination test conducted is under the influence of controlled conditions maintained in the test chamber.
Vigour test gives more sensitive index of quality than germination test.	It is less sensitive and often gives unreliable results.
It is expressed in several units like seed lot with maximum seed weight, seedling length, seedling dry weight, desi-siemens/g of seed, CO_2 released etc.	Unit of expression is percentage.
Barring highly problematic soils usually correlates with field emergence.	It may or may not correlate with field emergence.
Standard evaluation techniques are not available; techniques vary with crop and are crop specific like cold test in corn, electrical conductivity test of peas; Accelerated ageing test of soybean *etc.*	Standard evaluation techniques available are unique to crops
Loss of vigour is invisible and is expressed at physical, physiological and biochemical levels.	Loss in germination is visible and is expressed only physiologically.
Usually vigour loss precedes germination loss during deterioration.	During deterioration, germination loss occurs at subsequent stages.
This test decides the vigour status of individual seeds, ensures enhanced storage potential, makes the seeds resistant to adverse soil conditions and biotic stresses, advances uniform flowering and improves seed yield through uniformity in emergence and hence will show higher seed quality.	The test gives information only on the probable number of seeds that can germinate physiologically but does not give any guarantee neither the probable number of seedlings established nor their uniformity.

The Components of Seed Vigour

Longevity, germination and seedling growth are the three major components of seed vigour. The process of germination can be divided into three distinct but overlapping and interacting phases, viz.,

(i)   Reactivation of pre-existing system,

(ii)   synthesis of enzymes necessary for the breakdown of stored food reserves, their mobilization to the organelles and the synthesis of new cellular compounds. Similarly, the initiation of seedling growth, the second component of seed vigour, that usually accompanies germination sharply can also be divided in to three interacting and overlapping phases like (i) cell division, (ii) differentiation and (iii) development of functionally and structurally diverse tissues and organs.

The seed material, after harvesting, threshing and processing will be usually kept in storage for some time, wherein a gradual decline in germination and vigour occurs, before being utilized. Any seed company that holds the stock of seeds in storage has to test the seeds typically twice a year depending on the conditions of the storage. But, even when the seed is stored at 14°C or at subfreezing temperatures also needs testing, albeit in a longer cycle, depending on the method of seed drying adopted and the initial temperature at which the seed was stored. Obviously, germination tests needs to be conducted at regular intervals, wherein, the percentage germination and test date only are recorded precisely on the label of the stock container, but never the degree of seedling vigour. It is always better to have a germination test number recorded on the label so as to consult the test log for any concerns about the vigour. The loss of seedling vigour will be apparent in the germination test, either, as the seeds usually take longer time to germinate than usual or the seedlings are smaller in size and of course they may be malformed at times.

The angiosperm seeds, during the course of their development and as they approach maturity, physiological or harvest as the case may be, accumulate their characteristic soluble sugars like sucrose, oligosaccharides, antioxidants and late embryogenesis abundant proteins etc., (Amuti and Pollard, 1977). Accumulation of these compounds are expected to serve two basic purposes, viz., (i) to contribute to the development of tolerance to desiccation as well as to heat denaturation (Black *et al.*, 1979); and permits the cells of seeds to persist in dry state for varying periods of time i.e., longevity itself (Sun and Copeland, 1993; Bernal Lugo and Leopold 1995). Preferably, it is possible either through the maintenance of structural integrity of membranes or by providing optimum stability to the macromolecules like proteins (Crowe and Crowe, 1986) or through the formation of a glassy state, a consequence of withdrawal of water from a solution of sugars that ultimately serves as a stabilizer (Franks *et al.*, 1991). Further, this could be attributed to the formation of an extremely high viscosity of sugar solution that either prevents or suppresses the deteriorative reactions. But according to Leprince and Walters-Vertucci (1995), the role of glassy state in seed storage stability is not proven. Whereas Sun (1997) examined the dynamics of glass transition and concluded that the plasticization effect of water on seed glasses contributes to the loss of relative storage stability among seeds at different water contents. Deterioration of repair enzymes with ageing (Osborne 1983) like DNA ligase (Elder et al.1987) or aspertyl methyl transferase, involved in the repair of proteins damaged by ageing (Mudgett and Clarke 1993) or free co-operativity among deteriorative reactions like synergism between free radicals and Millard reaction (Kristal and Yu 1992) etc., contribute to seed mortality. Seed ageing results in impairment of enzyme systems involved in the elimination of toxic O_2 intermediates generated either during storage or germination better explains why intact non germinable seeds are often metabolically active, yet the seed dies a day or two after hydration.

2.3.5 Seed Viability

Technically, seed can be considered as a connecting link between two generations of a plant species. Then, obviously, the seeds must remain viable or capable of germination in order to save the species from extinction. For this purpose, majority of the seeds have been uniquely equipped to survive as viable till the right time and right place are available for beginning a new generation. This does not necessarily mean that the seeds will retain their viability indefinitely. Just like all other living forms, seeds too will deteriorate and die eventually. Under a given set of conditions, the time taken by a seed population or seed lot to become non-germinable has been referred to as longevity. For example, even in a freshly harvested seed lot, a good majority of seeds exhibit good quality while the remaining few seeds can be further subdivided into seeds with intermediate quality and with varying degrees of poor to very poor quality. Depending on the species, after exposing the seed lot to a specified storage period, now the proportion of poor seeds will be high, followed by intermediate and good seeds. This clearly shows us that different seeds of a seed lot die early, mid and late duly showing a normal distribution (Fig. 2.6).

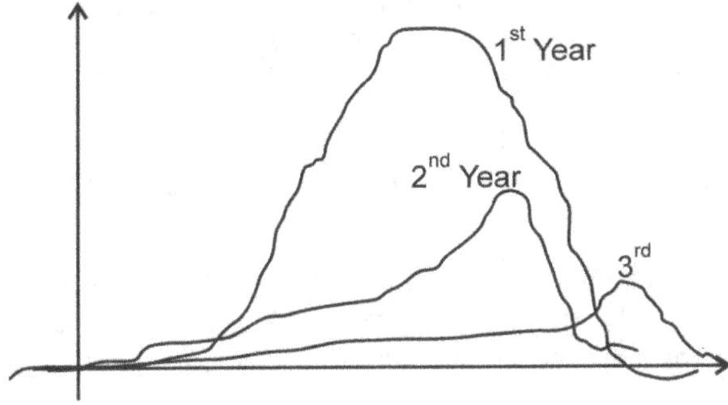

Fig. 2.6 Viability

Ellis and Roberts (1980) proposed a useful model, where in the model allows the prediction of seed storage as a function of environmental factors and a population of seeds stored in constant environmental conditions lose their viability in a declining sigmoid curve. According to this model, stored sample of dry seeds has a particular mean viability period. Since, this model does not accommodate the seeds with a substantial initial period of limited deterioration needs to be applied with caution to seeds which have already begun to lose their viability.

The time required to germination increases prior to loss of viability of seeds because seeds deteriorate during ageing. In order to overcome this problem, Bradford *et al.* (1993) proposed a threshold model based on population to describe the relationship between germination rates and seed deterioration. This model quantifies and predicts seed germination times and percentages after varying periods of ageing under constant conditions. The model assumes that each seed has a maximum potential lifetime. Therefore, the values vary in normal distribution among individual seeds. Hence, the time for germination of a particular seed is inversely proportional to the remaining differences between the accumulated ageing period and the maximum lifetime of that seed.

The loss of viability of seeds in dry storage is assumed to have been controlled by same limiting factors, the seed survival curve should be Sigmoid. However, an examination of time courses of seed survival in dry storage reveals that deterioration in some species is not simply sigmoid due to an initial period of little or no loss of germinability changes in presence of two sectors in survival curves could be attribute to either rate controlling processes during seed ageing or differences in deteriorative mechanisms existing between early and late stages of deterioration. Lugo and Leopold, 1998 concluded that glassy state in seeds serves as a stabilizer and contribute to the period of relative storability. Therefore the rapid progress in deterioration during dynamic period only represents a weakening of glassy state duly facilitating hydrolysis of sugar components through an aarray of oxidative process.

The meaning of viability, from a Seed Technologist's viewpoint simply refers to "the capacity and capability of a seed to germinate and produce a normal seedling." Whereas seed viability, from the seed Physiologist's perspective refers only to whether or not a seed contain metabolically active tissues possessing energy reserves and enzymes that have the capacity to sustain the living cells. So, a viable seed from Crop Physiologist's view may be non-viable from a Seed Technologist's view because of its inability to produce a normal seedling. In respect of seeds, viability can be summed up as the state of being alive.

Comparable longevities can be achieved under non-refrigerated conditions if seeds are sufficiently dried (Zheng et al 1998). The longevity of seeds even at optimum water content decreases as storage temperature increases. Therefore, there is no substitution for storage while trying to prolong the lifespan of seeds

The value of critical water content is not likely to vary greatly among different storage temperatures. So the optimum water contents must be different for different temperatures. Walters, 1998ab.

Bernal Lugo and Leopold (1998) observed the existence of two phases of seed survival, i.e., an initial period of slow mortality in Barley, Tomato, Wheat, Maize etc., where as the seeds of Onion, Lettuce and Blue grass etc. exhibited sigmoid phase of viability at once.

2.3.6 Germination

Most seeds begin to germinate or resume the activity soon after being exposed to or planted in a moist, warm soil or germinating medium. The seeds of every species, at cultivar or ecotype level, have their own unique requirements for germination. The process of germination begins with a swelling of the dry seed as it picks up or imbibes moisture and terminating with the protrusion of the radicle through the seed coats. Moisture is the most important prerequisite for triggering the process of germination. Some seeds require very little moisture whereas other seeds like water lilly and other aquatic plants must be completely submerged in water so as to initiate the process of germination. Once the process of germination begins, an adequate level of moisture content must be maintained, or made available to the seed, so as to prevent the death of seed or seedling due to temporary drying. Contrary to this, too much moisture content present either in the soil or germinating medium also causes saturation and deprives off the seed with oxygen again leading to the death of the seed. The uptake of water is determined not only on the availability within the vicinity but to a large extent is also governed by the size of seed and size of solid soil particles, nature and form of mucilage content present on the surface of the seed and soil particles. Irrespective of soil type and water content of the soil, seed can only absorb moisture from a short distance of say not more than 10 mm. Assuming the water potential of pure water as 0 MPa while the air-dry seeds will have a matrix potential of -3MPa. As a result, the air dry seed will pickup moisture till water potential increases to a level, where the water potential is same between the seed and its immediate surrounding substrate. This rapid inrush of water into the tissues of the seed will be drastic to both vigour and viability of seed, usually referred to as imbibition damage. Assuming adequate availability of water and oxygen, the rate and extent of germination in non-dormant seeds are governed by temperature and light. Each species have their unique requirement of a minimum, maximum and an optimum temperature range at which germination is highest. Some seeds even require darkness to germinate as they are inhibited by light. Further, Bradford (1995) proposed three different models to study the germination and field emergence of the seeds viz., Thermal Time Model (to study the response of temperature on seed germination), Hydro Time Model (to study and characterize effects of reduced water potentials on germination) and Hydrothermal Time Models (to study the pooled effects of temperature and water potentials on seeds).

The germination vigor is driven by the plant embryo, embedded within the seed, to resume its metabolic activity in a coordinated and sequential manner.

The ability of the seed to absorb water and the capacity of the embryo to expand consequent to imbibition are the two basic physical factors that determine the process of initiation of germination. Orthodox seeds permit desiccation during maturation and allow metabolic activity to be interrupted and restarted after imbibition during germination and thereby preserves the embryonic cell viability in dry state that can be extended to over several centuries, Farrant and More, (2011), Walters *et al.,* (2010). This simply implies the existence of specific mechanisms to maintain the state of metabolic quiescence in mature dry seeds duly maintaining their integrity. Similarly, the constituents of seed like sucrose, oligosaccharides, antioxidants and late embryogenesis abundant proteins etc., deposited during the terminal stages of seed development and maturation together with pericarp, testa and endosperm or perisperm respectively helps the cells of seed to persist in dry state for longer periods also provide protection against physical damage besides aiding in dispersal. Similarly, the components of protein synthesizing complex like r- RNA, t-RNA, m-RNA, protein initiation and elongation factors are also get deposited in the seed before maturation. The seed coat or testa becomes permeable to water in most cases and cracks readily upon hydration so as to facilitate swelling of embryo. In case of legumes, the seed coat becomes an impermeable barrier to penetration of water, and thereby effectively prevents germination. So, in addition to the above physical and chemical factors, permeability of the testa and rate of water uptake become the two major factors that directly exert significant effect on the process of germination.

In case of large sized crop seeds like beans and maize, too rapid absorption of water or gushing of water, especially at low temperatures leads to imbibitional damage. This could be attributed to the uneven stress created by differential swelling of different stored food components of the seed. Proteins swell more quickly upon hydration than carbohydrates and lipids. The phospholipids comprising of cellular bilayer membranes with hydrophilic head groups oriented towards the external cell wall whereas the hydrophobic fatty acid chains mostly associated with the middle layer of the membrane bilayer. These membranes undergo structural changes with respect to temperature and water content. Hence, one should always remember that, the membranes would be in gel phase when cool and dry wherein they not only allow passage to cellular constituents but also permit rapid water uptake and thereby brings changes in the membranes to shift into a liquid crystalline state. This phase transition is again temperature dependent, i.e., low temperatures and low water contents promote the gel phase while the opposite results in crystalline phase.

The Water uptake by the seeds occurs over a specified period of time ranging from several hours to many weeks and varies with the prevailing soil conditions and plant species. In the absence of permeability barriers during germination the uptake of water usually follows a three-phase model described by Bewley and Black, (1985) and Bewley (1997) is presented in the Fig. 2.9. In dry seeds, usually the metabolism commences immediately after hydration with simultaneous leakage of solutes on one hand and initiation of respiration and synthesis of proteins from the pre-existing materials on the other hand.

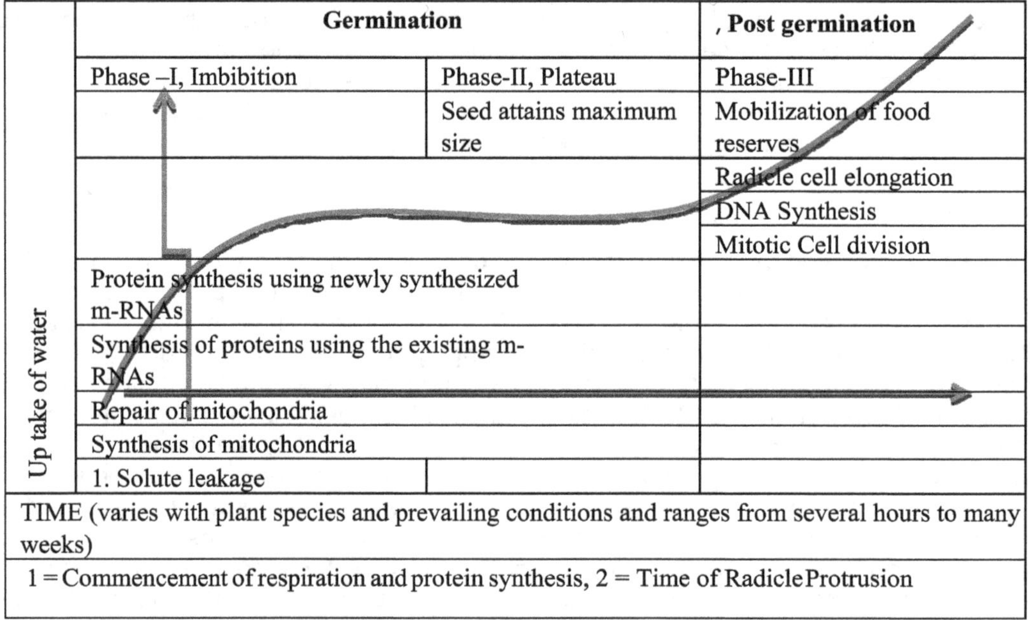

	Germination		, Post germination
Phase –I, Imbibition		Phase-II, Plateau	Phase-III
		Seed attains maximum size	Mobilization of food reserves
			Radicle cell elongation
			DNA Synthesis
			Mitotic Cell division
Protein synthesis using newly synthesized m-RNAs			
Synthesis of proteins using the existing m-RNAs			
Repair of mitochondria			
Synthesis of mitochondria			
1. Solute leakage			
TIME (varies with plant species and prevailing conditions and ranges from several hours to many weeks)			
1 = Commencement of respiration and protein synthesis, 2 = Time of Radicle Protrusion			

(Vertical axis label: Up take of water)

Fig. 2.9 The Tri-phasic diagram of seed germination and growth

2.3.6.1 Phase –I: Imbibition

This first phase occurs in both living and dead seeds. Usually, dead seeds absorb more water than viable seeds, because the membranes in dead seeds are not intact, hence development of turgor pressure to offset the osmotic gradient for water uptake does not arise in the cells. In case of seeds like muskmelon and lettuce where the embryo is enclosed within endosperm, which retains the solutions released from embryonic axis and accordingly swells. As a consequence, this increased extracellular volume makes the water content of the dead seeds usually higher than those of the living seeds (Bradford, 1995).

Generally, the dry seeds will have a very high negative water potential ranging from - 100 to - 200 MPa, and most probably this could be attributed to the colloidal properties of the seed coat. Initially, the surface proteins, cellulose and starch first become hydrated. Since, the uptake water at this stage is physical and more rapid, because of larger differences in water potential gradient between internal and external solution together with the matric absorption (absorption of water by constituents of seed). The selectively permeable membranes attract water osmotically in to the cell and this process continues till the cells are constrained with an opposite and negative force from the relatively rigid cell walls. This results in the development of turgor pressure, which in turn results in the softening / rupturing of the seed coat possibly due to the increase in the size/ volume of the seed. As the seed imbibes water, initially the tissues located at the periphery of the seed i.e., tissues that are immediately below the seed coat, begin to swell while the internal tissues still remain dry. As a consequence, this results in the cracking of dry tissues and thereby results in the extrusion of cellular contents from the partially hydrated cells and ultimately leads to imbibitional damage especially during rapid water uptake.

Phase- 1 does not simply correspond to a mere physical process of imbibition only because within 15 minutes after imbibition changes in gene expression were detected in Arabidopsis seed, Preston *et al.,* (2009). Similarly, rapid changes in the levels of some metabolites were also observed in rice seeds within 1 hour of imbibition, Howell *et al.,* (2009). During this first phase, usually proteins are synthesized from the existing mRNA and DNA. Repair of mitochondria takes place in this phase only.

This initial phase of germination is basically a reversible process wherein the maturation programme is reiterated. Under conditions of water stress, the process of growth and development of a germinating embryo into an autotrophic seedling gets arrested by the action of ABA dependent checkpoint of germination due to the involvement of transcription factor AB15, Lopez *et al.,* (2001, 2002). But, the genetic analysis has already revealed that embryo genic AB13 factor is essential for the expression of AB 15, also required to promote the growth arrest. Since the activity of AB13 factor is essential at the end of the maturation phase for encoding LEA proteins, and its occurrence and accumulation at the beginning of the initial germination phase in the presence of ABA or osmotic stress only reflects the reinduction of developmental processes taking place normally at the end of the maturation phase. Therefore, Lopez *et al.,* (2001) proposed that the formation of embryos arrested in their progress towards germination corresponds to an adaptive mechanism that increases the survival of seeds under conditions of water stress encountered by seed in the soil.

The phase-I of germination process basically comprises of reactivation of the pre- existing and conserved systems of the seed and is initiated with the basal cellular metabolism comprising of carryover type of maintenance together with initiation of respiration, protein synthesis, Glycolysis, oxidation of fatty acids, solute and ion transport, cytoplasmic streaming as detailed below in a sequential manner.

2.3.6.1.1   Respiration

Up on hydration of seed, the three respiratory pathways, Glycolsis, PPP and Citric Acid Cycle begin to operate both during and after germination so as to produce the key intermediates of metabolism like (raffinose series of oligosaccharides along with sucrose and all those enzymes necessary for conversion of these sugars like fructose and glucose that enter glycolysis) together with required energy in the form of ATP and reducing powers in the form of reduced pyridine nucleotides like NADH and NADPH. But, in case of seeds of aquatic environment like rice, bulrush and barnyard millet etc., anaerobic respiration is predominant, wherein seeds complete their germination under water where the concentrations of oxygen are either low or even under extremely controlled conditions of no oxygen. This particular situation may lead to temporary production of lactate or ethanol, which are consumed by enzymes lactate dehydrogenase or alcohol dehydrogenase. Under controlled conditions of oxygen supply, seeds of submerged rice may develop abnormal mitochondria. Contrary to this, seedlings will not get established even after germination and perish ultimately in case of seeds of most of the terrestrial species because of the stay of seeds under water or gases for extended periods. On the other hand, at times, seeds germinating even in the soil also undergo temporary anaerobiosis due to the insufficient uptake of oxygen by the seed coat to fulfill the requirements of respiratory demand. As a consequence of maturation drying and imbibition, the mitochondria usually exhibit damage and can get repaired in two different ways in oil storing and starch storing seeds. The site of initial ATP synthesis in germinating seeds though not clearly understood, but it could be either the oxidative phosphorylation in active mitochondria, which determines the effective synthesis of ATP. Contrary to this, in certain species, in case of inefficient functioning of mitochondria, in the presence of an alternative oxidase, viz., cyanide insensitive oxidative pathway operates during germination and diverts the flow of electrons from NADH to O_2 and thereby prevents the synthesis of ATP.

2.3.6.1.2   Synthesis of Proteins and RNA during Germination

Upon hydration of seed, the metabolic activity in the form of the initial synthesis of protein resumes just within 20 minutes duly involving the translation of pre existing m- RNA messages stored in the dry seed in to polysomes. Proteomic studies have revealed that the success of germination depends on the quality of messenger RNA stored during embryo maturation on mother plant in addition to protestasis and DNA integrity. The surface amino acid pathway due to its close relationship with hormone signalling pathway not only regulates seed germination but also represents a key biochemical determinant of a seed to initiate its development towards germination.

2.3.6.1.3   Pre-translational Processes

The translational processes play an important role in seed germination (35, 43, 60, 97, 117). Selective translation of mRNAs is a characteristic mechanism of gene expression regulation during germination. The messenger RNAs, in association with the protective proteins forms a messenger- ribo-nucleoprotein (m-RNP) complex. These stored messages are released from the protein-synthesizing complex over a period of several hours of time for protein synthesis comprising of ribosomal (mRNA), transfer (tRNA), messenger (mRNA), specific cytoplasmic proteins like initiation and elongation factors etc., are conserved in stable form and stored in the dry seed. A commitment towards germination obviously requires the transcription of genes of imbibed seed to discriminate between those mRNAs that will be utilized in germination and those that will be destroyed. A good majority of these degraded messages are replaced by newly synthesized mRNAs through a process known as mRNA turnover and thereby provide evidence that protein synthesis is required for the completion of germination especially from the stored mRNA templates whose stability conditions seed vigour, Sen and Osborne (1977) and Smith and Bray 1982).

Several proteins synthesized are unique to germination and are called as housekeeping or growth maintenance proteins because these proteins are required for maintaining the metabolic integrity of the cells of germinating seed. Methionine metabolism is a house keeping mechanism. In addition to this, in some species, several proteins thus synthesized are associated with the hydrolysis of cell walls of endosperm surrounding the radicle. In the germinating seeds, usually, both ribosomal RNA synthesis and synthesis of ribosomal proteins occur in the presence of RNA polymerases. The conservation of stored mRNAs in the dicot and monocot dry seeds only reinforces the hypothesis that those transcripts and pathways they code are functionally important to seed germination in all species, An and Lin (2011).

For obtaining physiological germination, most seeds require oxygen in aerobic respiration. Usually, the pattern of oxygen uptake is similar to that of water uptake, i.e., very rapid in phase-1, slowing down in phase-II (Lag Phase), and again rises very rapidly in Phase-III. Similarly, RNA polymerases, the enzymes necessary of the synthesis of t- RNAs have also been identified in seeds during the process of initiation of germination. In contrast to what occurs at the end of germination, Kimura and Nambara (2010), Rajjou *et al .,* (2004)., *de novo* protein synthesis is not required for gene expression during early stages of imbibition because the seed stored transcriptional machinery is sufficient, Kimura and Nambara (2010). Further, low translational activity during the first few hours after imbibition only reflects the use of stored proteins. Therefore, it appears that, the stock of proteins stored in the seed is utilized during the initial stages to restart cellular activity, which is followed by mobilization of the stored mRNA and thereby offers additional evidence in support of the concept that seed germination is prepared well during seed maturation stage itself. These findings were similar to those reported for barley seed germination, Sreenivasulu *et al.,* (2008).The protein "Germin", once considered to be an integral part of germination has been identified now as an enzyme, 'oxalate oxidase', which is synthesized only during post germination.

2.3.6.1.4   Post- translational Processes

Seed proteins are also subjected to a large number of post translational modifications that affect their function has been revealed through proteomic investigations. Thioredoxin, a disulfide protein, acts as early signal during germination and helps in the mobilization of reserves by (i) reducing storage proteins for enhancing solubility and susceptibility to proteolysis, (ii) reducing and inactivating disulfide proteins that inhibit amylases and proteases to facilitate breakdown of stored starch and proteins, and (iii) activating individual enzymes functioning in germination, Buchanan and Balmer (2005). Carbonylation of seed storage proteins plays an important role in scavenging ROS produced during germination and also facilitates mobilization of seed storage proteins during seedling establishment by destabilizing their structure and promoting proteolytic attack, Job *et al.,(2005)*. Phosphorylation and dephosphorylation are the two regulatory mechanisms controlling seed germination through modulation of ABA signaling, Brock et al.,(2010). Though nitric oxide modulates the metabolic activity in seeds, its mode of action in the control of germination was not documented. The sulphur amino acid pathway during metabolism simply represents a key biochemical role and determines the commitment of seed and imitates the development towards germination.

2.3.6.1.5   Synthesis of DNA during Germination

It is well established that seeds contain functional DNA dependent RNA polymerases, Jendrisak and Burgess (1975). Upon hydration of seed, the initial synthesis of DNA (including mitochondrial DNA) occurs in the cells of the radicle. Studies on radioactive thymidine incorporated into DNA revealed two types of DNA synthesis. The unscheduled DNA synthesis occurring immediately after imbibition is thought to be through the repair of DNA damaged during maturation drying and rehydration. The damage could be due to the attack of either free radicles or endonuclease cleavage and could also be due to the uninterrupted loss of bases. The repair process occurs in the presence of pre-existing Thymidine kinase enzyme present in small quantities in dry seed at the time of germination. A part of this repair may also occur due to the synthesis of telomeric DNA in the presence of newly synthesized telomerase enzyme. Telomeres are nothing but the nucleotide sequences present at the each end of a DNA molecule. These telomeres get reduced in size due to successive replications of DNA and hence needs to be extended by the enzyme telomerase periodically. The activities of pre-existing enzymes like DNA polymerases and Ligases also effectively repair the single strand breaks. The delay in the repair of DNA may lead to reduction in the ability to replicate and produce RNA and there by affects the speed and efficiency of germination. DNA polymerases, necessary for the replication and repair are present in the dry seed. DNA replication usually commences just few hours before completion of germination in tune with the requirements of radicle and plumule growth during seedling development.

The source of water or water potential of the seed's surroundings also exerts direct pressure on the potential growth of the embryo. It means the high source water potential tends to promote germination whereas the low water potential initially delays and prevents germination. Seeds are considered to be very sensitive to water availability and perhaps may not germinate at water potentials at which their seedlings will not grow. Germination is prevented in case of majority of seeds at water potentials approximately around −1.0MPa.

2.3.6.2 Phase –II: Active metabolism and Hydrolysis

In this second phase, the water content of seeds increases gradually and becomes relatively stable. The uninterrupted water uptake not only activates the stored enzymes but also stimulates the synthesis of new ones. As a consequence, several new physiological activities associated with germination, including protein synthesis by translation of new mRNAs and synthesis of new mitochondria. The repair mechanisms get activated and initiation of growth begins in this stage. These enzymes thus hydrolyze and transform some of the stored food reserves of the seed as per the instructions received from the pre-existing m-RNA / hormone passed on by the embryo or embryonic axis, into energy and produce lower molecular weight soluble compounds, usually meant for utilization in the production of more cells and tissues. These metabolic processes in turn lower the water potential of embryo and its surrounding tissues. This phase is basically aimed at and dealing with the synthesis and supply of ATP content required for carrying out various synthetic activities during Glycolysis, fatty acid oxidation, and respiration including the synthesis of ADP for the growth of embryonic axis. The length of this phase generally varies widely with species and largely depends on the age of seeds, temperature, presence or absence of dormancy and the water potential of the surrounding medium. This phase appears to be a lag phase between uptake and growth.

During this stage, the rate of water absorption is governed by the internal osmotic potential. The duration of lag phase usually represents the time required for weakening the restraints on embryo enlargement imposed by surrounding tissues to a degree necessary to allow further water uptake by the embryo. Presence of several proteins associated with subsequent stages like metabolic control in seedling establishment is dependent on gibberellic acid, which in turn control the abundance of cell wall hydrolase, α-glucosidase etc., which are likely to mediate embryo cell was loosening during cell elongation and radicle extension. During phase –II exogenously applied hormones like gibberellins and ABA may assist with weakening or strengthening respectively of the tissues surrounding the embryo. Generally, the slow response of gibberellin treatment is probably related to the time lag for enzyme synthesis. A proteomic study of GA deficient *Arabidopsis* mutant *ga1* indicated late implication of gibberellins in germination, a stage coinciding close to radical emergence, Gallardo *et al.,* (2002), and further reported that though gibberellins are required for the completion of germination, they are not involved in many processes taking place during germination like mobilization of seed storage proteins and lipids. Out of 46 protein changes detected during germination, only α 2,4, tubulin is dependent on the action of gibberellin and is confirmed by transcryptomic analysis of gibberellin related genes, *Ogawa et al.,* (2003).

2.3.6.3 Phase- III: Cell division, Elongation and Growth

During this phase obviously the water uptake is usually associated with the growth of the embryo and marks the completion of germination. In general, the onset of this phase coincides with the protrusion of the radicle possibly due to the greater availability of oxygen. Upon hydration, when the accumulated damage gets repaired, and only upon sensing a favourable environment, the physiological processes directed towards growth are initiated. In case of seeds with dormancy, the water content reaches a level and remains there indefinitely. Contrary to this, in germinable seeds, the completion of germination is marked by the expansion and growth of embryo and the tissues enclosing it. The initial expansion required for completion of germination occurs even without cell division in majority of seeds. The cells of dry seed embryo comprises of two sets of double stranded DNA in 2C state. DNA synthesis occurs without cell division and produces DNA in 4C state. In order to continue the organ growth, root and shoot meristems must begin the cell division quickly. Generally, the timing when the cell cycle begins is relative to radicle emergence and varies among different seeds. But one thing is certain that, the seeds shed with immature embryos need more extensive cell division associated with embryogenesis before completion of germination. Contrary to this, cell division is initiated in seeds with mature embryos only after the emergence of radicle.

Usually, phase-III occurs in storage organs of the seed so as to provide sufficient substrate for respiration and protein synthesis in tune with the instructions, either preexisting (in the form of hormone, or long lined mRNA) or coming from embryo or embryonic axis. This stage-III can be identified and defined by rapid growth of radicle and shoot. For this to occur, the water potential of the external solution should not be lower than − 0.2 to − 0.3MPa. Germinating solutions with water potential of − 0.45 to − 0.80MPa noticeably slow down radicle emergence where as the solutions of − 1.0 MPa or even lower severely restricts the expansion of radicle cells necessary for radicle protrusion. Radicle emergence in most situations is an irreversible commitment by the seed to complete germination. Usually, in some species, radicle emergence commences around 60 hours after initiation of imbibition. Similarly, initiation of embryo growth consequent to the weakening/ softening or splitting of the testa or covering tissues is another essential process necessary for completion of germination, i.e., to allow the radicle emergence to occur. It means, the embryo also must be capable of exerting a growth force against the weakened and restraining tissues.

The cell walls of endosperm cap tissues, being rich in galactomannans, act as a barrier to radicle emergence. At maturity, as the testa is a dead tissue, the enzymes secreted by the embryo together with the active participation of living endosperm cells that produce hydrolases under the influence of GA are responsible for degradation and subsequent weakening of galactomannan rich cell walls. The ABA, though supposed to block germination, does not prevent the induction of hydrolases in the endosperm. Therefore, the process of germination can be summed up as an act of balance between the thrust generated by the embryo (resulting from the osmotic uptake of water) and the resistance offered by the tissues surrounding it.

2.3.6.3.1 Cell Division

The completion of germination thus marks the beginning of cell division. During early stages of germination, the enzyme thymidine kinase is present in small quantities and is involved in the repair of DNA only, but increases its activity abundantly during the subsequent DNA synthesis. The re-synthesis of DNA in phase-III usually corresponds with an increase in total DNA content and cell division. In the living cells of some seeds DNA replication in endosperm occurs without cell division, i.e., in a triploid endosperm cell with 3C DNA. The DNA continues to duplicate without simultaneous cell division due to endo-reduplication and consequently some cells often reach 8C status and above. Similarly the replication of DNA in living endosperm cells also occurs without cell division. Therefore, the elongation of cells of the radicle thus merrily becomes a turgor pressure driven process. The enzymes, endo-trans-glycosylase and xyloglucan and the protein Expansin have been implicated for cell wall extension and elongation in vegetative tissues through loosening of hydrogen bonds linking different components of cell wall. The expansins modify the cell walls in unique ways and result in the desired characteristics through regulation of their growth and development. The same expansin is also involved in the degradation and weakening of tissues so as to permit radicle emergence. Simultaneously another expansin gene turns on in the cortical tissue of the radicle as it emerges and grows. Further, the involvement of additional expansins in the mobilization of food reserves cannot be ruled out. The cell walls of endosperm being rich in galactomannans, polysaccharides etc., that contribute to the rigidity of the tissue. The endosperm cap, due to the initiation of endo - β - mannanase activity, allows the penetration of radicle into endosperm and leads to radical emergence, Nonogaki *et al.,* (2000). In addition to this, a number of cell wall hydrolases like Cellulose (β 1,4, glucanase), Polygalacturonase, Xyloglucan and Endotransglucosylase etc., have been identified for their role in germination. A single copy gene, CAMATOSE (CTS) in *Arabidopsis*, encoding for ATP binding cassette (ABC) carrier involved in reserve lipid mobilization, is required for seed germination because of the assumption that it has to play a direct role just before the emergence of the radicle, rather than inhibiting dormancy release during after-ripening, Carrera *et al.,* (2007). Accumulation of a large number of oxylipins (a group of oxygenated products formed from fatty acids) in mutant *cts* seeds during late maturation behaves as strong germination inhibitors, Dave *et al.,* (2011). Therefore, the coordinated action between these hydrolases and expansins is supposed to bring in considerable changes in the cell wall structure and composition especially with respect to stretching under tension. This can be achieved by way of breaking the hydrogen bonds between wall polymers.

Methionine besides being a building block in protein synthesis, is also a universal methyl group donor and a precursor of polyamines and vitamin biotin, Ravanel *et al.,* (2008) plays a vital role in seed germination. Ethylene, regulates seed germination by promoting the weakening and rupture of seed tissues surrounding and enclosing radicle and by counteracting the inhibitory action of ABA on these processes, Linkies *et al.,* (2009). Evidence in support of interaction between ethylene, ABA, Cytokinin, and ROS signaling in controlling germination during early seedling development was provided by Apel and Hirt (2004); Subbaiah and Reddy (2010). Biotin, well known for its role in lipid synthesis is also needed for germination. A seed specific Biotinylated protein accumulates in the later stages of seed maturation and disappears very rapidly during seed germination.

2.3.6.3.2 Growth

The roots from the seeds grew down regardless of the initial orientation of the embryo. Similarly the shoots that emerge a bit later also grew up. This ability of the seedlings to orient their growth is the consequence of a process called as gravitropism, by which plants sense and respond to the direction of gravity. As the seedlings emerge from the soil, they are already beginning to make chlorophyll and turn green instantaneously. The cotyledons and apical hook unfold rapidly as the plant emerges in to the light and the seedlings continue to elongate as the nutational movement or the rotation of the stem becomes more prominent. Since light is the source of energy for plant growth, different plants have evolved several highly sensitive mechanisms not only for regulating the developmental changes but also for efficient utilization of light for photosynthesis, a process by which plants control their development through light which is referred to as photo- morphogenesis. Germination rate is the reciprocal of the time taken starting from time of sowing for the process of germination to complete. The number of seed that are able to complete germination in a population or seed lot is referred to as germination capacity.

Types of Germination

The following types of germination are found in plants.

1. **Hypogeal germination:** In this type of germination both the cotyledons remain beneath the soil surface. The hypocotyls remain short and compact whereas, the epicotyls expand to give rise to first true leaf above the soil surface. (Fig. 2.10).

   Dicot species: Redgram, chickpea, fieldpea etc., Monocot species: All cereals

2. **Epigeal germination:** In this kind of germination due to the elongation of hypocotyls the cotyledon(s) gets raised out of the soil surface and work as true leaves. (Fig. 2.11).

   Dicot species: Oilseed crops, Black gram, Greengram Monocot species: Onion

3. **Epi-hypogeal germination:** In this kind of germination, one cotyledon remains beneath the soil surface as in hypogeal germination and another cotyledon comes out of the soil as in epigeal germination. Eg. *Peperomia peruvianum*, a dicot species.

4. **Pre-harvest sprouting:** In the presence of high moisture content matured non-dormant seeds sprout while still intact on the mother plant.

5. **Vivipary:** It is a condition in which germination occurs in fruits even without compleating the requirements of maturation. Eg. Mangroove species.

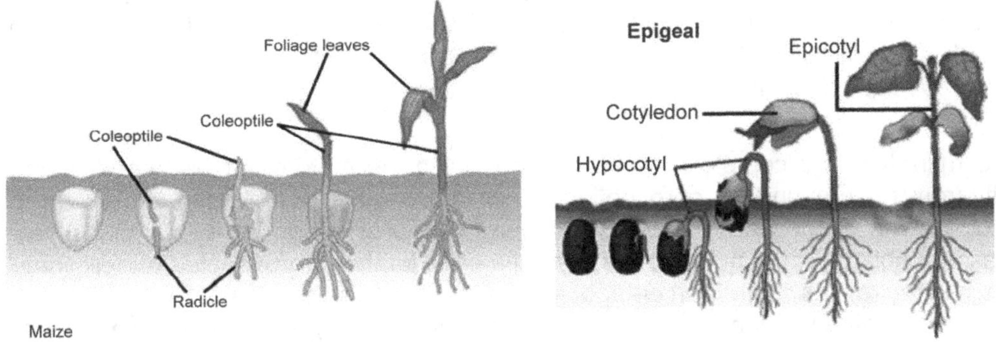

Fig. 2.10 Hypogeal germination in monocot seed.   **Fig. 2.11** Epigeal germination in dicot seed

Orthodox seeds of higher plants can interrupt their development on the mother plant and restart their cellular activity after imbition, through their sophisticated biological system through intense dessication.

A substantial part of seed germination process is determined by the maturation programme lay way of storing specific m-RNAs and proteins in the mature seed to as to initiate cellular actvity after imbition.

Seeds usally maintain a flexible and reverisbel status before radicle emergnece. After initial imbition it the seeds fail to get condition necessary for seedling growht, they will recapitulate their maiuration programme by reversing the physiological processes.

Similarly strigolactomes produced hyroots of platns affected by parasitic needs of orobanche family have been identifed to act as stimulants for germination.

RBA in initial stage and GA in final stae are involved in germination process. Yet other phyto hormones like ethylene brassino steroids, salicylic acid cytolinin auxin, jasmonic acid, oxylipins, and free radilces like reactive oxygen and nitorgen too stages a key note in the process of germination control especially in response to environmental constraints. The role of sulfur amino acid metabolism in linking house keeping metabolic activity and hormonal repulation is very important in seed germination process very recently Kalrikins have been established as a clear of signalling molecules present in smoke arising form burning of veretction that trigger germination several giosperms.

All of us an aware that most flowering plants reproduce by seed production therefore, germination and establishment of normal seedling determines its ecological and economic importance. The high level of vulnerability to injury, disease water and environmental stresses, germination is considered as the most crucial phase in the life cycle of a plant. This process, by definition incorporates several events initiatory with uptake of moisture by mature dry seed and ferminates with the protrusion of radical through seed coats.

The ability of a seed to germinate has been defined differently in physiological and seed technological perspectives. The AOSA, defines germination from a Seed Technologist's point of view as "the emergence and development from the seed embryo of those essential structures, which for the kind of seed in question, are indicative of the ability to produce a normal plant under favourable conditions". Contrary to this, a Plant/Crop Physiologists simply defines germination as a protrusion of radicle through the seed coat, irrespective of whether or not the seeds in a population are alive. Hence, the definition of a Seed Technologist appears to be more appropriate and relevant from a seed grower's point of view because (i) the seed has to produce both normal root and shoot under a given set of conditions prescribed by AOSA for each seed species and (ii) attainment of high germination percentage under the field environments.

Seed germination is a complex process during which the mature imbibed seed undergoes rapid shifts from maturation to a germination driven programme of development and prepares for seedling growth, Nonogaki *et al.*, (2010). Germination of seeds is usually measured and reported as percentage, is the number of seeds of a seed lot that are expected to germinate and grow into healthy plants. The percentage of seeds that are capable of producing normal seedlings under optimum conditions is known as laboratory germination. The certification standards below which the seed cannot be offered for sale automatically ensures emergence. It means, failure to germinate is not due to the inability of seeds. In other words, even in germinable seeds, difference in germination may exist because of soil conditions. Similarly, despite having high laboratory germination poor emergence in the field could be attributed to the low vigour of seeds and in this sense too vigour may not be equivalent to germinability.

2.3.7   Germination and Standard Germination Testing

The seeds of every species, be it the cultivars or cultivated varieties of crop plants or the ecotypes of the wild species, have their own set of requirements and / or characteristic conditions for germination. It means, the extent of germination is not only governed by the species, variety / ecotype alone, but also depends to a greater extent on the seed size and to some extent on the specified depth of sowing, irrespective of whether the seed is sown in the cultivated fields or gardens, meadows or forests etc., so as to ensure uniformity in germination. No doubt, germination is considered as an important aspect of all seeds. But still, the farmers worldwide are worried more only about the germination of the seeds they have purchased and planted in their fields rather than either their genetic potentiality and /or physical quality. First of all, in order to germinate, the seeds meant for planting should be viable. A viable seed, by definition is the one, which is alive and has got both the physical and /or genetic capacity to produce enzymes necessary for uninterrupted running of metabolic reactions during germination and seedling growth. But, sometimes, despite the seed being viable, it may not be in a position to germinate. This could be attributed to the presence of either inherent physical barriers and / or hindrances caused by chemical inhibitors duly blocking the process of germination. There are several factors affecting germination inherent to the seed itself, which can be grouped in to (a) the degree of maturity with respect to the quantity of reserve food materials and enzymes accumulated, size, weight and age of the seed, (b) presence of dormant or rudimentary embryo that hinders germination immediately after attaining maturity, (c) impermeability of the seed coat to either water and oxygen or mechanical resistance of the seed coat to expansion (d) a combination of two or more factors mentioned above or (e) the external conditions exerting influence through secondary dormancy on seeds that are ready to germinate.

In addition to the above, seeds may also be get affected strongly by the mechanisms of dispersal and over wintering on one hand and also on the effects of man's collection, extraction and storage on the other hand. As a rule, immature seeds germinate more slowly than fully ripened seeds because they require more after ripening period. Similarly seeds harvested early are also prone to lose their viability much earlier and faster than those fully ripened seeds and hence are most likely to show retarded germination. Further, age of the mother plant (seeds collected from middle aged trees is better than young trees), size and weight of the seed, age of the seed (older seeds germinate much slower than the fresh seeds), vigour of parent trees, geographical origin (due to imperfect fertilization and delayed ripening) and dormancy are some of the other chief causes.

The rate and extent of germination in a non-dormant seed in the presence and availability of optimum water and oxygen, depends to a large extent on the temperature and light also. It means, the seeds will germinate only in a specified minimum-maximum temperature range and the germination rate will be greatest only at a well-defined optimum temperature. Similarly, some species require darkness to germinate as they are inhibited by light. As a consequence of the actions and interactions between innumerable factors like temperature, water availability, oxygen, light, substrate, level of maturity of seed on the mother plant, physiological age of seed and post harvest handling conditions including storage and transportation etc., renders the process of germination more and more complicated. While conducting the germination tests in the laboratory, usually all these factors are optimized in such a way so as to quantify the maximum number of seeds that are capable of producing healthy and well-developed seedlings.

In general, the process of germination is possible only when all those impeding blocks are removed. Therefore, a viable seed is one, which germinates under favourable conditions subjected to the removal of dormancy, if at all present (Roberts, 1977). Germination of seeds, involves several biophysical (initiated with the imbibition of water) and biochemical (gets terminated with the emergence of plumule in dicots and radicle in monocots) changes. Copeland and McDonald, (1985) defined germination as "the ability of the seed to produce a normal seedling". In simple terminology of germination analysis, with respect to individual seeds, one can observe and record, only two possible states, whether or not the seed germinates. At this particular juncture, one should always remember, that the seeds pass through a series of stages or phases of deterioration both before and during storage but before being declared either incapable of germination or pronounced as dead. On the other hand, the seeds sown in the field, despite having the capability to germinate, though germinate albeit slowly, may not grow normally due to low viability and thereby causes loss both in respect of time and money. In the present day seed industry where time is money, in order to save both time and money, germination test is usually conducted in the laboratory well before the seeds are planted in the field. Since a laboratory germination test does not take into account the effects of non-optimal conditions on the seed, the result is viewed as potential rather than absolute emergence.

According to Milosevic *et al.,* (2009) though numerous methods at molecular level have been developed through the biotechnological means for the examination of quality of seed, no modern technique that could be used in routine analysis of seed germination was established until now. The standard tests designed have numerous shortcomings and often do not show the realistic behavior of seed under field conditions, but are largely still in use. For a greater number of crop species found in the production system, germination tests have been standardized. According to ISTA (2009) the terms of seed germination test comprising of temperature, duration of test and pretreatments for breaking dormancy and criteria for seedling estimation have been identified by International Rules for Seed Testing.

Definition

According to ISTA (1985), germination is the "emergence and development of all 'essential structures' giving rise to normal seedlings under favourable conditions"

Essential structures

According to this, the 'essential structures' obviously include cotyledons, root and shoot axes, terminal buds in dicotyledons and coleoptile in monocotyledons.

2.3.8 Basic requirements of a Germination Test

The five basic requirements of a germination test are (i) Rapidity, (ii) Inexpensive Objective oriented, (iv) Reproducible and (v) Uniformly interpretable. Over a period of time, the seed testing associations (ISTA and AOSA) functioning at International level have developed and standardized the protocols / conditions necessary for conducting germination tests for individual crops and published them in the form of their "Official Rules".

2.3.9 Objectives of Standard Germination Testing

A standard laboratory germination test should be relatively cheap and is usually based on either 200 or 400 seeds, preferably 50 or 100 seeds in four replications. The seeds are allowed to germinate under optimum environmental conditions for an optimum period of time specific to each species. The dormancy breaking measures are usually adopted and distinctions are made between normally and abnormally germinated and dead seeds. Usually non-dormant seeds only are assessed. Generally detailed assessments of damage are not included in the test. The results are reported as per cent germination.

Of late, some of the seed testing laboratories in developed countries are even offering a diagnostic germination test with 200 seeds for wheat and barley at a premium price wherein the quality problems of the seed sample are also recorded in detail and the result is reported with respect to (i) per cent germination, (ii) per cent dead seed, (iii) per cent abnormal seedlings (iv) per cent damage due to dormancy, heat, chemicals, sprouting, diseases and (vi) mechanical damage types. The same test when conducted with 400 seeds though gives more accurate results, but it is relatively more expensive.

The chief objective of a standard germination test states that "germination is the emergence and development from the seed embryo of those essential structures which, for the kind of seed in question, are indicative of the ability to produce a normal plant under favourable conditions." Keeping in view all the above requirements, germination testing is obviously conducted for accomplishing the three basic objectives:

Objective 1

The basic goal of seed germination testing is obviously aimed at predicting the field planting value of a seed lot, in such a way, that both the field emergence potential and seed quality can be determined. This in turn facilitates the farmers to assess the seed rate requirements before actually placing seed purchasing decisions.

Pre-requisites

Emergence ability and high germination rate of more than 90 per cent are the two important and minimum requirements from a seed grower's perspective, be it a pasture crop for grazing purpose or commercial seed crop for either human or animal consumption or even for use in further seed multiplication programmes. Therefore, below a critical threshold level, it is considered as not only deleterious but also affects both the quality and quantity of seed despite adjustments made in respect of seed rate for sowing, Roberts and Ellis, (1989).

Objective II

The second most important objective of conducting a germination test is to

(a) Judge the vigour status,

(b) Compare the value of different seed lots to be offered for sale in the market and

(c) Make decisions and to decide on the following,

    1. Whether the seed is eligible for the statutory certification or not?

    2. Whether or not the seed meets the prescribed standards for germination?

3. Whether to pay premium price to the seed producer or to levy penalty for production of substandard seed by the seed grower?
4. Whether seed lot is suitable for export purpose or can only be utilized for meeting the domestic requirements?
5. Whether blending is required to produce a single lot with acceptable level uniformity and germination? and
6. To decide whether or not the seed treatment is required.

Prerequisite

The germination test result, together with physical purity analysis provides a clear cut information on the planting value of seed lot based on which National and International seed business transactions will take place.

Objective III

The most critical and third objective for conducting germination testing is to decide the storage requirements based on duration, season and the end use. Storage time of the seed lots may be weeks, months, and years or even decades, may be either for avoiding the genetic drift or short, medium or long-term conservation.

Pre requisites

The following are some of the important aspects that need to be taken into consideration while conducting a germination test.

1. The seed sample meant for germination testing should be pure both genetically and physically.
2. The size of the sample drawn from a seed lot should be adequate for conducting the test in four replications and if necessary to repeat the test once again.
3. Each seedling should be examined thoroughly and counted before reporting the result.
4. The result should be expressed only as percentage of normal seedlings.
5. The individual seedlings are evaluated for morphological and physiological characters to differentiate between normal and abnormal plants.
6. The seed sample collected for germination testing should be from the lots maintained at optimum conditions of storage like moisture, relative humidity and temperature etc., because, humidity is known to exert direct influence on both per cent and speed of germination.

Heit (1951) observed better results of germination with humidity conditions lower than that recommended. Hunter and Ericksen (1952) reported that the seeds of beet, maize and soybean have attained 30 per cent, 30.5 per cent and 50.0 per cent moisture respectively before initiating germination. Delouche (1953) observed that variation in water content of test material gave diverse germination test results. The speed of germination was decreased due to loss of soil during testing period. Several researchers have reported on the hostile effects of temperature on germination per cent and speed. Optimum temperature for watermelon is 30°C, Kotovski (1926); Davis (1937). Whereas Harrington (1923) reported that germination accelerates at 20-35 °C.

2.3.10 Relevance of Germination Testing

From the foregoing discussion on the objectives of germination testing, it is very much clear that, the germination test results are relevant at least in three major areas of interest as detailed below.

    (i)   In the determination of field planting value of seed lots,

    (ii)   In the commercial seed trading across the globe and

    (iii)   In the determination of duration and conditions of seed storage.

Germination tests have been developed and fine-tuned over a considerable period of time in such a way that maximum germination percentage can be obtained from any seed lot. Probably, this has been made possible only with the use of standardized, artificial, sterile media together with temperature and humidity controlled in germinators for a sufficiently longer duration. Thereby provide enough scope even for the lame seeds to germinate. Since this test has to be conducted under optimum conditions, needs to be designed specifically for each species, and thereby making the results undoubtedly highly repeatable. Further, the results of this test are reported as percentage of normal seedlings, abnormal seedlings and dead or un-germinable seeds of a seed lot. Theoretically, in the absence of dormancy, a seed lot will be at its peak viability and should have nearly 100 per cent germination. The commercial concept of seed germination as proposed by Wellington (1966), is still remain in force in seed trade even today, i.e., the use of germination test results by merchants to compare the value and fix approximate price for different seed lots for which highest results are desirable and reproducibility being an essential criterion.

2.3.11 Causes of Variation in Germination Test Results

The germination test results are usually prone to some degree of variation due to several causes both between and among germination tests and testing laboratories, which could be due to the use of either of the following.

    (i)   Poor method,

    (ii)   Poor technique,

    (iii)   Poor equipment including variation in temperature within germinator,

    (iv)   Reduced sampling variations,

    (v)   Non -random selection of seeds for testing,

    (vi)   Inaccuracies in counting or recording,

    (vii)   Inconsistency and inability in distinguishing between normal / abnormal seeds,

    (viii) Seeds already treated with chemicals,

    (ix)   Contamination with bacteria and fungi *etc.*

    (x)   In any case, whenever, the variation is arising between replications, it must be tested statistically duly applying the prescribed tolerance limits.

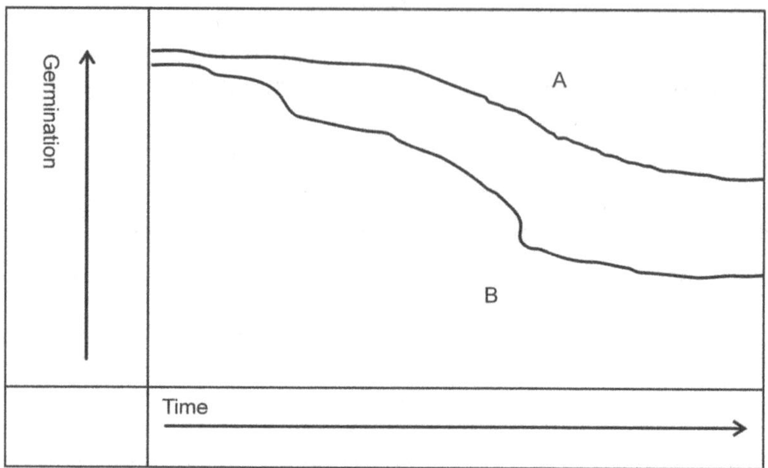

Fig. 2.7 Germination

Despite of all these limitations, even today, the germination test still remains and continues to be the principal and internationally accepted yardstick for seed viability testing. Having identified the problems, it is necessary to initially separate the seed lots based on their results. Standard germination percentage less than 90 per cent automatically indicates that the seed lots are bound to perform very poorly both in the field and in storage. Therefore, this peculiar situation has created a necessity for a more sensitive differentiation of potential performance. This ultimately has led to the development of vigour testing. (Hampton and Cool bear (1990). In the seed survival curve (Fig.2.7) drawn between time on X-axis and percentage germination on Y–axis, the upper line (A) is represented by physiological germination where as the lower line (B) is represented by those seedlings produced beyond physiological germination as per ISTA guidelines.

2.3.12   Limitations of the Germination Test

The standard germination test has failed to provide accurate information pertaining to a seed lot's field performance at least in the following four cases, considered to be the major limitations.

1. The standard germination test, according to the definition of ISTA (1985), emphasizes that the seed analyst is expected to focus only on the emergence and development of **essential structures**. It obviously means that it has nothing to do with the rapidity of growth, considered to be one of the most important and crucial benchmarks for potential plant stand establishment. Therefore, this clearly indicates that the definition is precisely silent on the frequent failure of seed lots and their subsequent performance both in the field as well as in storage. As such, it is only capable of meeting the first objective and hence the germination results continue to be used only as a basis for seed trading, Hampton, (1991).

2. Though it appears that, the methodology for conducting the germination test is standardized, it is only partial and incomplete. Testing for germination is always aimed only at producing uniformly reproducible results both between and within seed testing laboratories duly ensuring the favourable test conditions. As per the stipulations, all germination tests have to be conducted on artificial, standardized and sterile media in

humidified and temperature controlled environment chambers. It simply appears, even to a layman, that these conditions are so artificial and hardly relate to the field conditions anywhere on the globe. Further, since this test has been designed in such a way and conducted under the most favourable and controlled optimum conditions specific to each species, obviously, it establishes maximum plant stand producing ability of the seed lots. It means, the standard germination test precisely predicts the stand production potential of the seed lots only when the seeds are planted in the optimum field conditions. But, in reality, the test is not fully standardized, i.e., ISTA Rules state that, the substrate must always contain sufficient moisture, but never specifies the amount/ quantity of water to be used. Phaneendranath (1980) successfully demonstrated significant decrease in germination of maize and sorghum with increase in water per paper towel. Similarly, ISTA has recommended optimum conditions for obtaining regular, rapid and complete germination, as per definition; temperature or alternating temperature can only be optimal. As a consequence, the results of the germination test always overestimates the actual field emergence. The laboratory germination test result shows 85 or 90 per cent but in reality the actual emergence under field conditions rarely reaches these figures and in majority of cases the field emergence is considerably far less.

3. The germination test, as per the definition, is based on the relative proportion of normal seedlings, and has been designed accordingly to count the normal seedlings in two stages, viz. first count and final count, specific to species. In the first count, all those seedlings that have already germinated from the robust seeds will be removed. Whereas, the final seed count has been so designed as to provide with an ample opportunity in respect of maximum time for the remaining weak and even the lame seeds to be considered as germinable. It means, only dead, badly diseased, and irrevocably lame seedlings are eliminated on one hand and while on the other hand the weak, semi lame and robust seedlings are given same weightage.

4. As per definition, germination is scale less, because, a seed may be either germinable or not. No account is maintained for strong or weak seedlings, as germination percentage is the sum of both strong and weak seedlings. It means, in the absence of distinctions provided to identify strong and weak seedlings, obviously it becomes very difficult to detect, and record the degree of progressive deterioration because even those seeds considered germinated will differ significantly with respect to field emergence as they range from weak, to semi lame to robust and in turn directly exerts a major impact on plant stand establishment.

Under standardized and prescribed optimum germination conditions, the test evaluates individual seedlings and reports the percentage normal seedlings of a given seed sample. Though the germination test provides vital information of seeds, it suffers from two most serious limitations, viz., (i) it does not reflect the potential of a seed lot to grow when exposed to stress conditions and as a consequence requires additional interpretation to avoid any misunderstanding of the results and (ii) it fails to quantify either the location or the distribution of germination response in time.

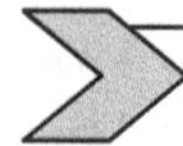 **2.4 Summary**

Plants exhibit vigour throughout their life cycle, but with varying degrees in different growth phases. Heydecker (1972) proposed only four stages of a plant where seedling vigour exerts direct effect. The origin of vigour in respect of seeds is genetical. Genes responsible for producing more vigorous seed was identified. Hardseededness, though an undesirable trait offers protection from ageing. Chemical composition of seed and susceptibility diseases etc., are under genetic control. Increased vigour is due to the accumulation of favourable dominant genes and heterotic response for grain yield was demonstrated through QTL mapping.

The origin of seed vigour is physiological and biochemical too. The term 'triebkraft' introduced by F. Nobbe, virtually means driving force and shooting strength. Seed vigour is a state of good health and natural robustness of seeds which upon planting gives rapid germination, uniform stand establishment coupled with normal growth in a wide range of environmental conditions. Interest on seed vigour is increasing to meet the demand for uniform produce by consumers, transporters and supermarkets and mechanized harvesting. Vigour differences may not always reflect in the yield.

Planting value, as mandated by law, automatically ensures that failure to emerge is not due to the seed alone. Vigour is only an indication of seeds ability to emerge even under suboptimal conditions. Seed vigour is only a relative superiority in performance of seed lots under well-defined conditions. Seed lots with poor emergence potential despite having high laboratory germination are called as low vigour seed. Different vigour parameters were identified. Seed lots when stored under suboptimal conditions show rapid decline in germination.

Viability, germination and vigour are three different but unique parameters of seed quality. Viability curve represents germination under optimal conditions whereas vigour curve expresses germination only under stress conditions. Decline in vigour is much faster than decline in viability. Seeds must remain viable in order to save species from extinction. Farmers across the globe are worried more about the seeds they have planted in the field rather than their genetic potentiality or physical purity. Germination is measured and expressed as percentage of normal seedlings expected to grow into healthy plants. A standard germination test is conducted with 400 seeds in four replicates of 100 each. Germination test results are relevant in determining field planting value, duration of storage and in seed trading across the globe.

There are several causes of variation in germination test results. Similarly germination test also suffers from certain limitations. Vigour can be differentiated from germination in several aspects. Germination and seedling growth are the two major components of seed vigour. Bradford, (1995) proposed three different models to explain germination and field emergence. Tri-phasic seed germination model of Bewley (1997) explains that imbibition in phase -1 occurs both in living and dead cells. Dead cells absorb more water than living cells. Dry seeds will have a very high negative water potential ranging from -100 to -200 MPa due to the colloidal properties of seed coat. Three respiratory pathways, PPP, Glycolysis and citric acid cycle operate both during and after germination to produce key intermediates of metabolism.

The site of initial ATP synthesis during germination is not clearly understood. Cyanide insensitive oxidative pathways operate in the presence of alternative oxidase and divert the flow of electrons from NADH to O_2 and thereby prevent ATP synthesis. Most seeds require oxygen in aerobic respiration for obtaining physiological germination. The pattern of oxygen uptake is similar to water uptake. Synthesis of DNA, including mitochondrial DNA occurs in the cells of the radicle. Pre-existing enzymes like DNA polymerases and Lygases effectively repair single strand breaks. Delay in repair of DNA leads to reduction in the ability to replicate and produce RNA and thereby affects the speed and efficiency of germination. Water surrounding the seed exerts direct pressure on the growth of embryo. High source water potential promotes germination and vice versa. In majority of seeds, germination is prevented at water potential approximately around −1 MPa.

Phase-III, usually occurs in storage organs and coincides with water uptake; associated with embryo growth; marks the completion of germination; and coincides with radicle protrusion due to the greater availability of oxygen. In most cases radicle protrusion is the irreversible commitment of the seed to complete germination. The endosperm cap, due to the initiation of endo-3mannanase activity, allows penetration of radicle into endosperm and leads to radicle emergence. A number of hydrolases were identified to have role in germination. The coordinated action between hydrolases and expansins brings in considerable changes in cell wall structure with respect to stretching under tension by breaking hydrogen bonds between polymers. As the seedlings emerge from soil, they are already making chlorophyll and turn green instantaneously. Germination capacity is the number of seeds that are able to complete germination in a population or seed lot.

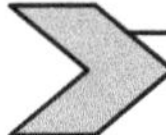

2.5 Exercise

I. Explain the Following

1. Meaning of seed vigour
2. A seed may or may not be germinable, hence may or may not be viable. Why?
3. The relationship between viability and vigour over a period of time with diagram.
4. Longevity.
5. Seed survival curves with the help of a diagram.
6. Why immature seeds require more time to germinate than fully ripened seeds?
7. Imbibitional damage
8. Germination capacity
9. Types of germination

II. Briefly List out

1. List out different stages in the life cycle of a plant where seed or seedling vigour exerts direct effect?
2. List out the technological solutions proposed to tackle the negative effects of seed production and storage environments?
3. List out the micro-environmental conditions to reduce the unproductive plants in the field?
4. List our different vigour parameters?
5. List out the adverse affects of planting low vigour seeds?
6. List out the factors affecting germination inherent to seed?
7. List out the parameters recorded in a standard germination test?

8. List out three basic objectives of a standard germination test?

9. List out the causes of variation in a standard germination test results?

10. List out the limitations of a standard germination test?

11. List out the chief factors for maintaining highest vigour of seed lots?

III. Write short notes on

1. Genetic origin of seed vigour.

2. Physiological and Biochemical origin of seed vigour.

3. Why Seed Technologist's definition of vigour is apt from seed grower's point of view?

4. Pre-requisites of a germination test.

5. Components of seed vigour.

IV. With the help of a diagram explain the process of germination.

V. Differentiate between

1. Germination and vigour.

2. Viability and vigour.

3. Crop Botanist's vigour and Seed Technologist's Vigour.

4. Low and high vigour seed lots.

5. State of being alive and degree of aliveness.

6. Different vigour parameters.

VI. Fill in the blanks with suitable words/ phrases.

1. The scientist who gave 4 different stages in the life cycle of a plant that seed or seedling vigour exerts direct effect --------.

2. The term "*Triebkraft*" was introduced by --------.

3. The term "*Triebkraft*" refers to -------- or --------.

4. Most of the cucurbit seed lots contain high percentage of -------- seeds.

5. Delayed harvesting of cucurbit seeds makes them dead, because seeds have already advanced from physiological maturity to --------.

6. Hard Seededness, an undesirable trait offers protection to seed from --------.

7. Seed germination test together with purity estimate provides -------- of seed lots.

8. The seeds lose their vigour much faster than they lose their --------.

9. The vigour and viability of any seed population immediately after harvesting and processing is --------.

10. Decline in seed vigour is much faster than decline in --------.

11. Seeds will not retain their indefinitely--------.

12. Germination of seeds is measured and represented as -------- of normal seedlings under optimal conditions.

13.   Usually, a standard germination test is conducted with 4 replicates of -------- each.

14.   Seed vigour is considered to be the closest measure of --------.

15.   -------- is the most important pre-requisite for triggering germination.

16.   Too much moisture in germination media deprives off the seed with -------- and causes death of seeds.

17.   The rapid inrush of water in to the seed tissue is referred to as --------.

18.   Water potential of dry seeds ranges between --------.

19.   The highest negative water potential dry seed is due to -------- properties of the seed coat.

20.   -------- attract water osmotically in to the cell.

21.   Reactivation of the existing pre-conserved systems in the seed takes place during -------- phase of germination.

22.   -------- phase of germination occurs in the storage organs of the seed.

23.   Radicle emergence is the -------- commitment of seed during germination.

24.   Phase –II of germination is aimed at the synthesis and supply of -------- required for various synthetic activities.

25.   Proteins produced during germination process are called as --------.

26.   Bradford, 1995 proposed three models to study --------.

27.   Germination percentage is the sum of both -------- and -------- seedlings.

28.   Germination test is usually conducted under most -------- and -------- conditions specific to each species.

Answers

1.	Heydecker, 1972	2.	Fredrich Nobbie in 1876
3.	Shooting strength or driving force	4.	Osmotically distended
5.	Accelerated ageing	6.	ageing
7.	Planting value	8.	ability to germinate
9.	100%	10.	Viability
11.	Viability	12.	Percentage
13.	50 or 100 seeds	14.	Potential field performance
15.	Moisture	16.	Oxygen
17.	Imbibitional damage	18.	-100 to -200 mpa
19.	Colloidal	20.	Selectively permeable membranes
21.	First	22.	Third

23. Ireversible	24. ATP
25. Housekeeping proteins	26. Germination and field emergence
27. Strong,weak	28. Favourable, optimum

 ## 2.6 References

Al-Chaarani, G.R., Gentzbittel,L.,Wedzony,M.,Sarrafi, A., (2005).Identification of QTLs for germination and seedling development in sunflower. Plant Science.169:221-227.

Amuti, K S, and Pollard CJ. (1977). Soluble carbohydrates of dry and developing seeds. Phyto- chemistry. 16 : 529 -532.

An, Y,Q., and Lin, L., (2011).Transcriptional regulatory programmes underlying barley germination and regulatory functions of gibberellin and ascorbic acid. BMC. Plant Biol. 11: 105.

Apel, K., and Hirt, H.,(2004). Reactive oxygen species : Metabolism, oxidative stress and signal transduction. Annu. Rev. Plant.Biol. 55: 373-399.

Bentsink,L.,Jowett,J.,Hanhart,C.J.,Koornneef, M.,(2006). Cloning of DOG1 a QTL controlling seed dormancy in Arabidopsis. Proc. Natl. Acad. Sci. USA. 103 : 17042-17047.

Betty, M., Finch –savage, W.E., King, G.J., Lynn, J.R.(2000).Quantitative genetic analysis of seed vigour and pre-emergence seedling growth traits in *Brassica oleracia,*. New Phytol.148:227- 286

Bernal lugo,I., and Leopold,A.C., (1998). The dynamics of seed mortality. Journal of Experimental. Botany.49:1455-1461.

Bernal lugo,I., and Leopold,A.C., (1995). Seed Stability during storage. Raffinose content and seed glossy state. Seed Science Research.5 : 75 -80.

Bewley, J.D., (1997). Seed Germination and Dormancy. The plant Cell. 9(7): 1055-1066.

Bewley, J.D and Black, M (1985). Physiology and Biochemistry of seeds in relation to Germination. Vol. II. Viability, Dormancy and Environmental Control.Springer- Verlag, New York.

Black, M., and Pitchard, H.W., (1979).(eds.). Desiccation and survival in plants. Drying without dying. CABI, Wallingford, USA.

Bradford, K.J., (1995) Water relations in Seed Germination. In : J . Kigel and G. Galili, (eds.) Seed development and germination. Pp 351-396. Marceel, Dekker, New York.

Bradford, K.J., Tarquis, A.M., Duran, J.M., (1993). A population based threshold model describing the relationship between germination rates and seed deterioration. J. Exptl. Bot. 44: 1225 – 1234.

Bradnock,W.T., andmatthews, S. (1970). Assessing field emergence potential of wrinkled seeded peas. Hort. Res. 10: 50-58.

Brock, A.K., Willmann, R., Kolb, D., Grefen, L., Lajunen, H.M., (2010). The *Arabidopsis* nitrogen activated protein kinase phosphatase PP2C5 affects seed germination, stomata aperture, and abscisic acid inducible gene expression. Plant Physiol.153 :1098-1111.

Buchanan, B.B., andBalmer, Y., (2005). Redox regulation: a broadening horizon. Annu. Rev. Plant Biol. 56: 187 - 220.

Copeland, L. O., and McDonald, M.B. (1995).Principles of Seed Science and Technology.3rd ed. Chapman and Hill, New York.

Copeland, L.O.(1976) Seed and seedling vigour. In Principles of Seed Science and Technology.(Ed). L.O.Copeland. 149 -184, Burgers Publishers. Ltd. Minnesota, USA.

Crowe, J.H. and Crowe, L.M.(1986) Stabilization of membranes in enhydrobiotic organisms. In: Membranes, Metabolism and Dry organisms. Ed. Leopold, A.C. pp 188-209.Ithaca, New York.

Carrera, E.,Holman, T., Medhurst, A., Peer , W,., Schmuths, H., (2007). Gene expression profiling reveals defined functions of the ABC transporter COMATOSE, late in phase–II of germination. Plant Physiol.143 : 1669-16679.

Chiang, G.C., Bartsch, M.,Barua, D., Nakabayashi, K., Debieu,M., (2011). DOG1 expression is predicted by the seed maturation environment and contributes to geographical variation in germination in Arabidopsis thaliana. Mol. Ecol. 20 :336- 349.

Das Gupta, J., and Bewley, J.D., (1982). Desiccation axes of *Phaseolus vulgaris* using development of a switch from a development pattern of protein synthesis to a germination pattern.Plant . Physiol. 70: 1224-1227.

Dave, A., Hernandez, M.L., He,Z., Andriotis, V,M., Vaistij, F.E., (2011). 12 oxo-phyto dienoic acid accumulation during seed development represses seed germination in *Arabidopsis.* Plant Cell. 23 : 583-599.

Delouche, J. C., and Baskin, C.C., (1973). Accelerated ageing techniques for predicting storability of seed lots. Seed Science and Technology. 1 : 427 - 452.

Delouche, J.C. and Caldwell, W.P. (1960). Seed Vigour and Vigour Tests. Proceedings. Of AOSA, 50: 124-129.

Dias, MAN., Pinto, TLF, Mondo, VHV., Cicero,SM.,and Pedrini, LG., (2011). Direct effects of soybean seed vigour on weed competition. Brazilian seed journal. 33(2) 346-351.

Dias, M.B.P., Brunel-Muguet S., Durr, C., Huguet, T., Demilly, D., (2011). QTL analysis of seed germination and preemergencegrowthat extreme temperatures in *Medicago truncatula.* Theor. Appl. Genet. 122 : 429-444.

Dias,MAN., Mondo, VHV., and Cicero, SM., (20100Maize seed vigour and weed competition. Brazilian seed Journal. 32: (2): 93-101.

Dure, L., and Walters,L., (1965). Long lived messenger RNA. – evidence from cotton seed germination. Science. 147: 410-412

Egli, DB., and Rucker,M., (2012). Seed vigour and the uniformity of emergence of corn seedlings. Crop science. 52(6): 2774-2782.

Elder, R.H., Dell"Aquilla,A., Mezzina, M., Sarin, A ., and Osborne, D.J., (1987). DNA ligase in repair and replication in the embryos of rye.(Secale cereal.). Mutation Research. 181 :161 - 171.

Ellis, R. H., and Roberts, E.H., (1980). Towards a rational basis for testing seed quality. In.P D Hebblethwaite (ed.) Seed Production. Pp. 605 - 635.Butterworths, London.

Farrant, J.M., and Moore, J.P.,(2011). Programming desiccation tolerance from plants to seeds to resurrection plants. Curr. Opin. Plant Biology. 14 : 340-345.

Finch savage, W.E., Clay, H.A., Lynn, J.R., Morris, K., (2010).Towards a genetic understanding of seed vigour in small seeded crops using natural variation in *Brassica oleracia*. Plant Sci. 179 : 582- 589.

Fu, H andDooner, H.K. (2002). Interspecific violation of genetic coliniarity and its implications in maize. Proc. Nat. Acad. Sci. 99 : 9573 – 9578.

Graham, G.I., Wolff, D.W., and Stuber, C.W., (1997). Characterization of a yield quantitative trait locus on chromosome five of maize by fine mapping. Crop Science. 37:1601-10.

Galpaz, N and Reymond,M., (2010).Natural variation in Arabidopsis thaliana revealed a genetic network controlling germination under salt stress. PLoS. One 5: e15198.

Gallardo, K., Job,C., Groot, SPC., Puype,M., Demo,H., (2002). Proteomix of Arabidopsis seed germination: a comparative study of wild type and gibberellin-deficient seeds. Plant Physiol. 129: 823-837.

Hampton, J. G., (1992). Prolonging seed quality. Proceedings of 4th Australian Seeds Research Conference. pp 181-194.

Hampton, J. G., (1991). Herbage seed lot vigour- Do problems start with seed production. Journal of Applied Seed Production.9 : 87 – 93.

Hampton, J.G., and Coolbear, P. 1990. Potential versus actual seed performances- Can vigour testing provide an answer? Seed Science and Technology 18: 215-228.

Harrington, J.F., (1972) Seed Storage and longevity.In TT. Kozlowski (ed.) Seed Biology. VI. III. Academic Press. pp145- 245. New york.

Heydecker, W. (1972). Report of the vigour test committee. 1968-1971. Proc. of ISTA.37 : 379- 395.

Holdsworth, M.J., Bentsink,L., Soppe, W.J.J., (2008). Molecular networks regulating Arabidopsis seed maturation. After ripening, dormancy and germination. New. Phytol 179 : 33-54.

Howell, K.A., Narsai, R., Carrol.A.,Ivanova, A., Lohse, M., (2009). Mapping metabolic and transcript temporal switches during germination in rice highlights specific transcription factors and role of RNA instability in the germination process. Plant Physiol. 149 : 961-980.

ISTA, (2009).International Rules for seed Testing. ISTA. Switzerland.

Jendrisak, J.J., and Burgess, R.R., (1975). New method for large scale purification of wheat germ DNA dependent RNA polymerase, -II. Biochemistry. 14 : 4639-1645.

Job, C.,Rajjou,L., Lovigny, Y., Belghazi,M., and Job,D., (2005). Patterns of protein oxidation in Arabidopsis seeds and during germination. Plant Physiol.138: 790-802.

Kendall, S.L., Hellwege, A., Marriot, P., Whalley, C.,Graham, I.A., (2011). Induction of dormancy in Arabidopsis summer annuals requires parallel regulation of DOG1 and hormone metabolism by low temperature and CBF transcription factors. Plant Cell. 23 : 2568-2580.

Kidd, F and Wett. C., (1918). The effects of soaking seeds in water.Ann. Applied Biol. 5: 1-10. Kimura, M., and nambara, E., (2010). Stored and neosynthesized m RNA in Arabidopsis seeds: effects of cycloheximide and controlled deterioration treatment on the resumption of transcription during imbibition. Plant. Mol. Biol. 73 : 19- 129.

Kristal, B.S., and Yu, B.P. , (1992). An emerging hypothesis: Synergistic induction of aging by free radical and Millard reactions. J Gerontol. Biol. Sci. 47: B 107- 114.

Leprince, O and Walters Vertucci, C., (1995). A calorimetric studyof the glass transition behavior in axes of bean seeds with relevance to storage stability. Plant Physiology.109: 1471-1481.

Linkies, A., Muller, K., Morris, K., Tureckova, V., Wenk, M ., (2009). Ethylene interacts with abscisic acid to regulate endosperm rupture during germination: a comparative approach using *Lepidiumsativum*and *Arabidopsis thaliana.* Plant Cell. 21: 3803-3822.

Lopez–Molina, L., Mongrand, S., Chua, N,H., (2001). A post germination developmental arrest checkpoint is mediated by abscisic acid and requires the AB 15 transcription factor in *Arabidopsis.* Proc. Natl. Acad. Sci. USA., 98 : 4782-4787.

Lopez–Molina, L., Mongrand, S., McLachlin,D.T., Chait, B.T., Chua, N.H., (2002). AB 15 acts downstream of AB 13 toexecute an ABA dependent growth arrest during germination. Plant j. 32 : 317-328.

Marcus,A., and Feeley,J., (1964). Activation of protein synthesis in the imbibition phase of seed germination. Proc. Natl. Acad. Sci. USA., 51 : 1075- 1079.

Matthews, S. and Bradnock,W.T. (1967). The detection of seed samples wrinkled peas (*Pisum sativum*) of potentially low planting value. Proc. ISTA. 32: 553-563.

Mondo,VHV., Cicero, SM., Dourado- Neto,D., Pupim,TL., Dias, MAN., (2012a). Seed vigour and plant performance. Brazilian Seed Journal. 34(1): 143-155.

Milosevic, M., Vujakovic,M and Kargi, D., (2010). Vigour tests as Indicators of seed viability. Genetica, Vol 42(1). 103-118.

Mudgett, M.B., and Clarke, S. (1993). Characterization of plant L-isoaspartyl methyltransferases that may be involved in seed survival: Purification, cloning, and sequence analysis of the wheat germ enzyme. Biochemistry32 11100–11111.

Munn, M.T., (1935). Further work with soil for testing the viability of seeds. Proc. ISTA. 30: 369- 380.

Nass, H.G., and Crane, P.L., (1970) Effect of endosperm mutants on germination and early seedling growth in maize. Crop Science. 10 : 139- 140.

Nobbe, Fredric.(1876). Handbuch der SamenKunde, Wiegandt- Hempel, Parey, Berlin. Nonogaki, H., Bassel, G.W., and Bewley, J.D., (2010). Germination. Still a mistery. Plant Sci. 179: 574- 581.

Ogawa, M., Handa, A., Yamauchi, Y., Kuwahara, A., Kumiya, Y., (2003). Gibberellin biosynthesis and response during Arabidopsis seed germination. Plant Cell. 15 :1591-1604.

Osborne, D.J., (1983) Biochemical control systems operating in the early hours of germination. Canadian journal of Botany.61 : 3568 – 3577.

Perry, D. A . (1970). The relation of seed vigour to field establishment of garden pea cultivars. Journal of Agricultural Science Cambridge, 74 : 343- 348.

Potts, H.C., Duangpatra, J.D., Hairston, W.G., Delouche, DC., (1978) Some influence of Hardseededness on soybean quality. Crop Science 18 : 221-224.

Preston, J., Tatematsu, K., Kanno,Y., Hobo,T., Kimura ,M., (2009). Temporal expression patterns of hormone metabolism genes during imbibition of *Arabidopsis thaliana* seeds; A comparative study on dormant and non dormant accessions. Plant Cell Physiol. 50: 1786- 1800.

Rajcan,I., and Swanton,C.J., (2001). Understanding maize weed competition: resource competition light quality and the whole plant. Field crops Research. 71(2): 139-150..,

Rajjou, L., Gallardo, K., Debeaujon, I., Vandekerckhove, J., Job, C., (2004). The effect of amanitin on the Arabidopsis seed proteome highlights the distinct roles of stored and neosynthesized m RNAs during germination.., Plant Physiol. 134 : 1598- 1613.

Ravanel, S., Gakiere, G., Job,D., Douce R., (2008). The specific features of methionine biosynthesis and metabolism in plants.Proc. Natl. Acad. Sci. USA. 95: 7805-7812.

Roberts, E.H., and Ellis, R.H., (1989). Water and seed survival.Annl. Bot. 63 : 39 – 52.

Roberts, E.H. and Smith, R.D. (1977).Dormancy and the pentose phosphate pathway. In: Khan, A.A.(ed.) The Physiology and Biochemistry of Seed Dormancy and Germination. North Holland. Publishing, Amsterdam, pp. 385–411

Schuch, L.O.B., Nedel, J.L., Assis, F.N., and Maia,,M.S.,(2000). Seed vigour and growth analysis of black oats. Scientia Agricola. 57(2) 305-312.

Schuch, LOB., Kolchinski, EM., Finatto, JA., (2009). Seed Physiological quality and individual plants performance in soybean.Brazilian seed Journal. 31(1): 144-149.

Sen, S.,and Osborne, D.J.,(1977). Decline in ribonucleic acid and protein synthesis with loss of viability during the early hours of imbibition of rye embryos. Biochem. J.166: 33-38.

Smith, C.A.D., and BrayC.M., (1982). Intracellular levels of polyadenylated RNA and loss of vigour in germinating wheat embryos.Planta156 : 413-420.

Sreenivasulu, N., Usadel, B., Winter, A., Radchuk, V., Scholz, U., ((2008). Barley grain maturation and germination: Metabolic pathway and regulatory network commonalities and differences highlighted by new Map Man/Page Man profiling tools. Plant Physiol. 146: 1738-1758.

Subbaiah, V., and Reddy, K.J., (2010). Interactions between ethylene, abscisic acid and cytokinen during germination and seedling establishment in Arabidopsis. J. Bio Sci. 35 : 451- 458.

Sun, Q., Wang, J.H., Sun, B.Q., (2007). Advances in seed vigour, Physiological and Genetic Mechanisms. Agril Sci. China. 6 (9): 1060 - 1066.

Sun, W.Q., (1997).Glassy state and seed storage stability. The WLF kinetics of seed viability loss at TTg and the plasticization effect of water on storage stability.Annals of Botany.79:291-297.

Vandecasteele,C.,Teulat–Merah,B.,Morere-Le–Paven,M.C.,Leprince,O.,Vu, B.L., (2011). QTL analysis reveals a correlation between the ratio of sucrose / raffinose family oligosaccharides and seed vigour in Medicagotruncatula. Plant Cell. Environ. 34 : 1473- 1487.

Walters, C. (1998a). Understanding the mechanism and kinetics of seed ageing. Seed science research 8:223-244.

Walters, C. (1998b). Ultra dry technology perspective for National seed storage laboratory, USA. Seed Science research-8. Supplement no.1. pp- 11-14.

Walters, C., Ballesteros, D., Vertucci, D.A., (2010). Structural mechanics of seed deterioration. Standing the test of time. Plant Sci. 179: 565-573.

Willey, R.W., and Hath ,S.B., (1969). The quantitative relationship between plant population and crop yield. Advances in Agronomy.21 : 281 – 321.

Xiao, J., Li,J., Yuan, L.,Tanksley, S.D., (1995). Dominance is the major basis of heterosis in rice as revealed by QTL analysis using molecular markers. Genetics. 140: 745-754.

Zheng, Q.H., Jing X.M and Tao, K.L (1998) Ultradry seed storage cuts costs in seed bank. Nature 393(223-224)

CHAPTER 3

Need, History, Definition and Concepts in Seed Vigour and Testing

(It is better to light one candle than to curse the darkness)

3.1 Introduction

Prediction of seed vigour is as important as estimating the liveliness of a seed or seed lots. Though, testing of seeds for germination is an internationally accepted criterion for viability, at times, even the germination test results may not reveal

enough information as to the real potential status of the seed lot's performance. This is true particularly in case of carry over seed. Estimation of vigour status assumes much significance especially when stored under unknown or unfavourable conditions. Various researchers have clearly demonstrated the evidence that seed exposed to unfavourable environmental conditions might cause serious degenerative changes in metabolic activities of seed and elucidated the mechanisms involved in seed deterioration and ultimately the possible transformations that can provoke differences in the performance of the seeds. Seed physiology, especially seed germination and its influence on plant performance, have been of central interest to humans since the establishment of Agriculture when men discovered the potential use of seeds for the multiplication of plants, around 10,000 B.C.

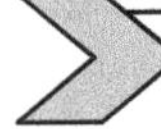

3.2 What is a Seed Vigour Test?

Determination of the degree of aliveness (vigour) of a seed is just as important as determining whether or not the seed is alive. Since inception, vigour tests have been designed to mimic poor seedling and seedbed conditions, usually and predominantly existing under natural field conditions, so as to find out and know how best a particular seed lot will perform under well-defined stress conditions. Seed lots with high germination exhibit differences in vigour. The low vigour ultimately leads to slow, uneven or non-uniform and poor emergence due to ageing. Therefore, vigour tests are expected to reveal differences in seed vigour. Today, a seed vigour test is a sophisticated seed testing methodology designed to reveal the quality of a seed lot aimed at guiding the consumers in avoiding planting under unfavourable conditions. It means, it is exactly the opposite of germination testing where in the seed is usually exposed to the most favourable and optimum conditions of temperature, moisture and light that facilitates the growth of seedlings, but these seedlings may not be capable of continuing their growth anymore and complete their life cycle under a wide range of field conditions, and hence the seeds are classified as viable. Contrary to this, in a vigour test, the seed is exposed to a stressful environment just unfavourable for seedling emergence and development. The environment can be cool, cold or warm or even a combination of either high humidity and high temperatures or high moisture loaded with low temperatures. Since the germination test will almost always have a higher result because the parameters of testing adopted are more forgiving in nature. Whereas, a lower vigour number always represents the result of a vigour test, i.e. lowest possible performance level one would expect to get from a seed lot under stressful conditions. Therefore, a vigour test is one, which has been designed to reveal a seed lot's ability to withstand a variety of different stress factors. In other words, a test or a series of tests that would reflect the potential performance of seed lots encountering stressful situations at planting and still provide a more sensitive measure of seed quality are referred to as vigour tests.

A vigour test does not only measure the percentage of viable seeds in a sample, it also reflects the ability of those seeds to produce normal seedlings under less than optimum or average growing conditions nearly similar to those which are most likely to occur in the field. It is not a test for field response *per se,* because, a particular seed lot shows close correlation with field performance depending on the conditions and nature of field in which the seed is planted. Usually, a vigour test is conducted in conjunction with a germination test because the later is required before the seed can be sold. On the other hand, a seed vigour test conducted in combination with a germination test offers a complete profile for a wide range of field situations and facilitates the end user in taking important seed related decisions like when to go for seeding, or whether is it necessary to wait for some more time till the soil temperature increases, or whether the seed can survive planting even in the existing cooler conditions of the soil, etc. Traditionally, a vigour test is conducted on a seed lot to determine the rate and extent of deterioration. By all means, a seed vigour test is simply, an examination of seed lots under a given set of environmental conditions such that the following can be worked out:

(i)   how do seed lots with similar and high laboratory germination can be compared when the conditions in the field are not favourable.

(ii)   which seed lots will remain good and retain their potential for germination even after storage under unknown conditions.

(iii)   which of the seed lots will germinate and emerge uniformly under a given set of field conditions.

(iv)   which of the seed lots will establish a good plant stand under field conditions.

A vigour test is considered to be worth-while subject to meeting the following requirements.

(i)   When the test is sensible enough and is capable of explaining the background information,

(ii)   when the test is capable of evaluating the commercial seed lots with high germination,

(iii)   when the test results can be related to the expression of vigour in respect of emergence or storage and finally

(iv)   whether the results obtained are repeatable, reproducible and validated.

3.2.1 Essential Characteristics of a Seed Vigour Test

1. The test should have a good theoretical background to ensure that the outcome of the unexpected results can be identified, explained and examined further easily.

2. The test should be simple to complete and hence applicable to wide range of testing environments.

3. The test should be rapid to enable prompt reporting of results at the shortest notice.

4. The vigour test should be repeatable both within and between many seed testing laboratories duly making the results comparable.

5. Positively the result of the vigour tests should correlate with emergence in the field or glasshouse and seed shelf life in storage.

6. The test should provide a quantitative method of assessment duly avoiding subjective assessments in a manner that makes standardization much easier.

7. Finally the test should be economically viable and practical.

 ## 3.3 The Need of Seed Vigour Tests

If the question before us is "why should we have a vigour test at all" when "the nearly perfected germination test is already available at our disposal", may be, the probable answer available to this question is "a standard germination test as carried out by most farmers is basically aimed at determining the seed rate required for planting a crop of their choice. Therefore, since inception, all the germination tests have been designed accordingly with an ultimate target of establishing the highest possible potential for a seed lot and hence the test is typically conducted under the most favourable laboratory conditions. But in reality these optimum favourable conditions are only rarely available under farmer's real field conditions. It means, the germination test do not differentiate between seed lots based on their capabilities to perform better under exact field conditions.

Testing of seeds for germination remains the internationally accepted criteria for seed viability and the test results if less than acceptable standard, say 90 per cent for temperate herbage grasses, is usually reflected as deterioration and hence indicates the performance of seed lot as poor. At the same time, germination test results of high germinating seed lots also may not provide enough information as to potential seed lot performance. Therefore, under these circumstances, testing of vigour status of seed lots not only assume much importance but also a sheer necessity. For the same reason, the testing of carry over seeds for vigour becomes more important especially when the seeds are stored under unknown conditions or unfavourable storage conditions. Therefore, in order to derive maximum economic benefits, a seed vigour test should be used as a supplementary test and not as a substitute for germination test.

One of the major limitations observed while conducting the germination test with soil, i.e., the germination test when conducted under favourable conditions gives good correlation between germination and field emergence, but under poor field conditions, the results of germination test may fail to indicate the precise ability of a seed lot to establish a standard normal crop. It means, in a good majority of cases, it has been observed that, seed lots with high laboratory germination percentage frequently exhibited huge differences in their field emergence capacity. Several researchers have reported this in a wide range of species. Among these only a few are presented in the table 3.1.

Table 3.1 Range in per cent field emergence in different crops

S. No	Crop	Range in emergence %	Reported by
1.	Peas	31-85	Clark and Little (1955)
2.	Soybean	17-90	Oliveria *et al.* (1984)
3.	Field beans	62-85	Hegarty (1977)
4.	*Phaseolus vulgaris*	52-97	Powel *et al.* (1986)

Further, it is very much clear from the following hypothetical example (Table 3.2.) that the germination test has failed to predict differences in field emergence potential of seed lots especially under poor field conditions. This situation obviously suggests the existence of another physiological aspect of seed quality. Accordingly, Frank (1950) proposed that tests related to germination when performed in soil should be called as seedling vigour tests. Consequent to this, considerable attention has been paid in the next three decades towards the development of tests that led to the reliable prediction of the stand production potential of seed lots. This thus made the seed vigour determination as one of the most important methods for assessing the seed quality. The seed vigour tests have been designed to reveal the basic and subtle signs of damage usually caused by moisture and temperature and provide a more reliable indicator in field and greenhouse performance than the standard germination or viability tests.

Table 3.2 Hypothetical example of germination and seedling emergence in different seed lots

Lot	Seedling emergence (%)			
	Per cent Germination	Near ideal Conditions	Stressful Condition	Unfavourable Conditions
A	90	88	80	70
B	90	87	60	40
C	90	85	55	30

According to McDonald (1980), the germination test results on one hand do not correspond to the performance of the seed lots when they are sown under field conditions and on the other hand these tests neither provide a complete evaluation of seed lots deterioration or quality nor differentiate between strong and weak seedlings. This gap in knowledge, thus existing led both Seed Technologists and Seed Growers to search for methods that can offer a still more reliable explanation and accurate information pertaining to the physiological potential of seed lots. It means, the potential difference between the performance of different high germinating seed lots is often referred to as vigour, has obviously created a basic need for the study and understanding the vigour of seed lots.

Basically, the need of seed vigour testing came in to fame initially from the developed countries, that are actually not only facing acute dearth of agricultural labour, but also as a consequence of an absolute necessity of facilitating mechanized harvesting of the forage species in general and agri-horticultural crops in particular. In order to achieve this goal, all plants in the field not only must emerge and grow uniformly but also reach uniform maturity at the same time. In a wide range of situations, may be, if not impossible, at least, it is a very much difficult task to achieve, despite the best management of soil physical conditions, soil fertility, soil moisture and depth of sowing *etc.* under the micro- level field conditions. On the other hand, the inability to control light and heat at macro level often results in the appearance of a good number of unproductive plants, may be obviously a consequence of the use of an abnormally low vigour of seeds. Further, seedling establishment and growth during early stages, determines the subsequent development and establishment of optimum plant population and ultimately the seed yield and its quality, *i.e.,* the seeds thus obtained from such low in vigour plants can never resume a normal growth, give only abnormal plants, which hardly reach maturity stage as per the schedule.

In order to select a best vigour test for use, first of all one has to ascertain the fact that whether vigour testing is really needed or not. The answer to this question is not that simple, because in the first instance, one has to identify whether or not the problem exists with the commercial seed lots. If it is so, whether it is in respect of either emergence or storage or variability in seedlings. One can select a test on the basis of either ISTA Rules or Research Publications or even use commercially developed potential vigour tests on the basis of availability of sound equipment and test kits developed for the purpose.

Establishment of the ISTA Biochemical and Seedling Vigor Committee proposed by Franck at the ISTA Congress 1950 in Washington D.C., USA, provided the much needed impetus for the concept of vigor testing. Over the past 60 years different vigor testing methods have been proposed, studied and used by seed technologists. The seed vigour tests have been designed to reveal the basic and subtle signs of damage caused by moisture and temperature and to provide a more reliable indicator of field and greenhouse performance than the standard germination or viability tests. However, research refinement has demonstrated that the approach of vigor testing has become excessively diversified leading to a trend to redirect the goals since only a few methods are internationally used. Although more than 60 different vigor test procedures have been proposed during the period mentioned, the ISTA Vigor Testing Committee has concentrated its efforts on the standardization of nine vigor tests (Hampton & TeKrony, 1995), while AOSA has suggested or recommended procedures for seven methods since 1983. Methods considered efficient by ISTA and AOSA are almost the same.

However, it becomes essential for both the seed analysts and consumers to understand that vigour tests have not been designed to predict the exact number of seedlings that will emerge and survive in the field, although the vigour test results usually well correlate with field emergence. Since, one of the basic purposes of seed vigour testing is to predict and indicate whether or not any trouble is anticipated from a high germinating seed lot when placed under adverse conditions. According to Ferguson (1993) the vigour test results should provide reliable information to rank seed lots according to seed quality level and eliminate those lots that fall below the prescribed company standards.

Assuming that when the production systems are diverse, the environments with suboptimal germination become a rule rather than exception. In order to maximize yields, the Seed Growers not only need uniform but also rapid plant stand establishment under diversified environmental conditions so as to achieve synchronous emergence and uniform maturity. This in turn demands a still more accurate estimate of field emergence than that provided by a mere germination test.

In addition to this, all most all agri-horticultural crops have specified as well as recommended plant densities for diversified situations so as to realize optimum yields.

Therefore, this situation not only necessitates that, the Seed Growers should have simultaneous access to seed lots with high vigour, but also a reliable vigour information so as to facilitate better predictions than germination test. So that it allows adjustment of seed sowing rates before planting itself so as to achieve the accurate and targeted plant population and ultimately the desired plant stand establishment required per unit area. This in turn assures the yield potential for a given production environment.

The impact of seed vigour on crop yield is tremendous. Since, low vigour seeds lead to poor plant stand establishment, the effect on yield is direct. But, in cereals, when the emergence is slightly reduced, the effect is limited due to the compensatory growth produced by tillers. Hence a drastic reduction in emergence only markedly affects the yield. Contrary to this, the problem of establishment in case of vegetatively propagated crops is very much pronounced because the crops have to be grown at specific densities so as to produce uniform and specific sized crops for the commercial market. Therefore, rate and uniformity in emergence also plays a critical role in the production of uniform sized produce at appropriate stage of maturity (Salter, 1985).

At times, it becomes necessary for the growers to opt for early plantings in cool and wet soils not only to exploit the length of the season but also to capitalize on the maximum available soil moisture. This in turn helps in minimizing the cost of production by way of adopting no-till or minimum tillage methods and also facilitates in catching the market early. TeKrony, Egli and Wickham, (1989) reported a lower emergence and seedling rates in maize due to the lower temperatures in the no-till production system. Contrary to this, in soybean, Oplinger and Phillbrook (1992) reported 14 percent better emergence rates under conventional tillage systems when compared with no-till system.

The plant populations produced by high and low vigour seeds are though not the same, but still the impact of vigour on yield of commercial crops depends largely on the type and nature of the material harvested. In case of crops like lettuce, cabbage, carrot and radish etc., which are harvested at vegetative stage, always requires the use of high vigour seeds. The relationship between vegetative growth and yield clearly points out the effects of vigour, as it is the ability of plant community to accumulate dry matter (TeKrony and Egli, 1991). Contrary to this, in case of cereals and pulses harvested at full reproductive maturity, no relationship between vigour and yield could be established thus far.

No doubt, germination, viability and vigour are the three different and important aspects of quality of a seed lot. Therefore, this clearly shows that, germination data without vigour data and vice versa will be less meaningful than when considered jointly. So, together they describe the quality of a seed lot more precisely. Therefore, improved seed vigour tests, together with accurate germination analysis is the need of the hour, not only to provide/ supply high quality seed but also to facilitate the farmers to exploit the existing production system in more reliable economic terms.

As far as seeds attaining their maximum quality on the mother plant is concerned, despite the debate is going on for several years, this clearly points out one thing that, the quality of seed in terms of both germination and vigour was said to be achieved at or close to physiological maturity, when seeds have reached their maximum dry weight (Harrington, 1977). But, Ellis and Pieto Filho, (1992) on wheat and Barley; Demir and Ellis, (1992) on pepper; Zanakis *et al.,* (1994) on soybean; Sanhewe and Ellis, (1996) in *Phaseolus vulgaris* seeds demonstrated that, seeds do not reach their maximum quality until after the so called physiological maturity. Hence they proposed the time of maximum dry weight as mass maturity. This has been proved with the help of probit analysis of seed survival curve by obtaining the estimates of initial seed viability parameter of seed quality at different stages of development, referred to as K_i. Similarly, TeKrony and Egli (1997) reported that, with the exception of pepper and tomato harvested as fleshy fruits, maximum germination occurred at or before physiological maturity in all crops where dry seeds are harvested.

3.4 History of Seed Vigour Testing

Seed germination and its consequent influence on plant performance in specific and seed physiology in general have greatly attracted the interest to humans since the initiation of Agriculture when men discovered the potential use of seeds for the multiplication of plants around 10,000 years. In the published literature the evaluation of physiological potential of seed is well documented. The observations made by Nobbe, in 1876, in his "Handbuch, der Samenkunde" led to the proposals for conducting a germination test. The term "Germination energy" gained increased acceptance in USA to express speed of germination. The first half of the 20th Century added very little to the knowledge of evaluation of physiological potential of seeds barring the development of TZ Test, for determination of viability, by Lakon in 1940s is one of the most commonly used test, as reported by Franca- Neto *et al.,* (1998). Hiltner andIhssen, 1911, Used the term ' *triebkraft"* to imply "driving force' and 'shooting strength' of germinating seedlings. Highlighted that, seedlings with longer roots in comparison to those from weaker seeds from the same seed lot and is responsible for development of Brick gravel test. Similarly, Fick and Hibbard developed electrical conductivity test in 1925,

The decade of 1930s gained acceptance due to frequent use of the term germination energy to express speed of germination. In the light of changed scenario, after World War-II, during the establishment of the ISTA Biochemical and Seedling Vigour Committee, at the International Seed Testing Association, (ISTA) Congress held in Washington, DC, in1950, Frank focused on seed vigour as a separate quality parameter independent of seed germination as a component of seed physiological potential and defined seed vigour as the plant producing ability of a seed lot in the field and laid much emphasis on the need for vigour testing. This obviously enabled to understand the disparities existing in germination testing philosophies, between American and European seed testing laboratories. Therefore, a new term vigour tests was proposed and accepted to better express the seed physiological potential mainly under suboptimal conditions. Further, the cold test for corn seed became routine use by both seed testing laboratories and private hybrid maize seed production companies for their internal quality control. The two articles published subsequently by Isely (1957), DeLouche and Caldwell (1960) provided the much needed stimulus for the first AOSA Vigour Test Committee, established in 1961, to refocus on the development of various concepts and reorient the advantages and disadvantages of direct and indirect vigour tests. The involvement of diverse points of interest, the more the topic was studied, the more complex and challenging it became.

The decade of 1960s has not only witnessed vigour evaluation and its influence on seed performance including establishment of a definition and a universally accepted concept of seed vigour but also helped the seed industry and seed technologists to recognize seed vigour as an important physiological characteristic and led to (i) improvement of Tetrazolium test, (ii) development of accelerated ageing test, (iii) creation of electrical conductivity test, (iv) a lot of information generated, quickly documented and disseminated and (v) offered training to a new generation of seed research scientists.

Consequent to the understanding of seed maturation process, seed vigour assumed yet another level of importance and led to further studies (i) on ageing process with focus on cell membrane integrity, (ii) lipid peroxidation, (iii) glossy state of cytoplasm and ultimately on how seed deterioration affected the performance of seed lots during the course of their storage. Wilson and McDonald (1986); McDonald (1999). Due to the relentless efforts of AOSA in cooperation with SCST and ISTA and its ability to estimate field performance led to the identification of a number of vigour tests by 1974. Under the able guidance of Woodstock, Seed Vigour Testing Progress Report, considered to be a significant landmark, was published in 1976, as a special edition of newsletter of AOSA, with detailed guidelines for the conduct of eight proposed vigour tests along with historical perspectives. Important contributions to this knowledge were provided by Priestley (1986), Wilson and McDonald (1986), and McDonald (1999).

The accumulated knowledge on seed maturation, seed vigour and deterioration led to identification and synchronization of a series of different themes associated closely with seed physiological potential. Similarly, knowledge generated on repair mechanisms, recalcitrance in seeds, desiccation tolerance led to the development of priming techniques. By 1980, a satisfactory definition for vigour has been evolved by ISTA with active cooperation and acceptance from all concerned organizations involved in seed testing and trade. Accelerated ageing test became the first vigour test to be recommended for soybean after several rounds of refinement in 1987. The number of vigour tests, one or more, conducted by seed testing laboratories increased from 52 per cent in 1978 to 75 per cent in 1990 (McDonald, 1994). Utilization of vigour tests for quality assessment suffered several set backs like lack of methodologies for standardization, subjective interpretation of results and discrepancy in results among seed testing laboratories despite the best testing protocols and education provided to its users. Oliveria *et al.,* (1984); Loeffler *et al.,* (1988); and Blum and Copeland (1995) indicated that Cold Test for Corn is as repeatable as standard germination test. The different facets of seed vigour have been fully reviewed by Heydecker, (1972,1977), Pollock and Ross (1972), McDonald (1975), Cantlife (1981), Halmer and Bewely (1984), and TeKrony and Egli (1991). A comprehensive set of working sheets on methods for 125 agri-horticultural species and 122 tree and shrub species was published by ISTA in 2003, duly describing the minimum extent and location of staining that will lead to production of normal seedlings for each species. Similarly, ISTA has also published 'Handbook of Seedling Evaluation' covering 427 genera based on the lab and field expertise of members of ISTA and AOSA, comprising of identification of characteristics of unacceptable, abnormal seedlings that are not likely to produce satisfactory plants (ISTA, 2003). Some important landmark events in the history of seed vigour are presented in Table 3.3.

Table 3.3 Landmark events in the history of seed vigour testing.

S.No.	Year	Name of the Scientist	Reported
1	1816	Berne	First seed Legislation on seed was passed in Switzerland
2	1876	F. Nobbe	First to distinguish between germination and seed vigour
3	1903	Lundstrom	Isolated empty seeds from filled ones with X-rays
4	1911	Hiltner and Ihssen	Demonstrated seed vigour through the Brick gravel Test
5	1925	Fick and Hibbord	Electrical conductivity
6	1926	Yuasa	Advocated X-ray test for detection of insect damage
7	1926	Kotowsky	Proposed Speed of germination as a concept of seed vigour
8	1927	Alberts	Provided need and practical relevance of seed vigour and first observations on cold test
9	1931	Sthal *et al.,*	Speed of germination
10	1942	Lakon	Dehydrogenase activity Test of Vigour, Tetrazolium Test
11	1950	Frank	First to introduce the word vigour
12	1950	ISTA	Formation of ISTA vigour Test Committee
13	1950	Isley	First to design cold test for corn
14	1953	Clark	Proposed Cool Germination Test for Cotton crop
15	1953	Semac and Gustafson	X- rays to detect abnormalities in seedlings
16	1957	Isely	Concept of vigour based on *per se* performance
17	1959	Linko and Milner	Proposed GADA test as an index of seed quality
18	1960	Delouche and Caldwall	Defined seed vigour in negative sense
19	1960	Germ	Proposed Exhaustion Test
20	1961	Barriga	First to identify lines resistant to mechanical injury
21	1962	Magurie	First to propose the formula for vigour rating
22	1964	Nutice	First proposed and viewed seed vigour in a positive sense
23	1964	Smith and Mollet	Identified 2 tomato varieties with germination capacity at10 ^{o}C
24	1964	Grabe	Developed GADA Test to measure seed deterioration
25	1965	Fritz	Suggested paper piercing test to replace Brick Gravel Test
26	1965	Delouche	Proposed Accelerating ageing Test for Determination of vigour
27	1965	Ader	First to define vigour under suboptimal conditions
28	1965	W J Frank	Vigour, plant producing ability of a seed lot in the field
29	1965	Woodstock	Significant mile stone-first vigour testing progress report
30	1966	Bain and Mercer	Developed Mitochondrial Efficiency Test
31	1968	Palmar and Moore	Proposed Osmotic Stress Test
32	1968	Lin and Cohen	Proposed ATP Level Test
33	1969	Burries *et al.*	Proposed First Count as a measure of early germination
34	1969	Willey and Heath	Explained the relationship between seed vigour and yield.
33	1970	MacKay	Suggested electrical conductivity test
34	1972	Pollock and Roos	Estimated relative vigour of seeds on the basis of leaching
35	1972	Heydecker	Characteristics of seed vigour tests

Table 3.3 *Contd…*

S.No.	Year	Name of the Scientist	Reported
36	1973	Michel & Kaufman	Determination of osmotic potentialities with PEG
37	1973	AbdulBaki and Anderson	Proposed vigour Index Mass and Vigour index length as derived measures of Seed Vigour
38	1973	Delouche and Baskin	Propose AA Test propose evaluation of storage potential
39	1974	Srivastava and Sareen	Proposed Seed Mobilization Efficiency Test
40	1974	Woodstock	Identified a number of Vigour Tests
41	1976	AOSA	Development of Standard Seed Vigour Test Procedures
42	1976	Copeland	Proposed Seedling Growth Rate Test
43	1976	Djavanshir and Pourbeik	Proposed Formulae for Germination Value of Tree Seeds
44	1980	AOSA/ISTA	organizations involved in seed testing and trade approved Vigour definition
45	1980	Matthews	Demonstrated Controlled Deterioration Test
46	1981	Steere *et al.*	Estimation of Conductivity of Single Seeds
47	1983	AOSA	Published Vigour Testing Hand Book
48	1984	Perry	Proposed Membrane Integrity Test
49	1987	ISTA	Revised AA Test significantly which became the first Internationally accepted Vigour Test for Soybean
50	1993	Rao and Sinha	Demonstrated Seed Metabolic Efficiency Test
51	1995	ISTA	Established ISTA Laboratory Accreditation scheme
52	1995	Hampton and TeKrony	Classification of vigour tests
53	1996	Jianhua and McDonald	Suggested modified AA Test for small seeded crops
54	1998	Asiedu and Powell	Demonstrated ageing and imbibition damage interacts only to give still low vigour seeds
55	2001	ISTA Rules	AA test recognized as one of the validated vigour tests
56	2002	AOSA	Classification of Seed Vigour Tests
57	2002	Cui *et al.*	Mapped 31 QTLs for five traits of seedling vigour
58	2002	Miura *et al.*	Reported 3 putative QTLs for seed longevity using BILs
59	2003	ISTA	Working sheets for each species to describe the minimum extent of location and Hand Book of Seedling Evaluation
60	2003	TeKrony	Emphasized precision is the key factor in vigour testing
61	2004	McDonald	Temperature variation should not exceed $\pm 3^{O}C$
62	2004	Fingo *et al.*	Reported 3 QTLs associated with low temperature germination
63	2005	Zhang *et al.*	Reported 34 QTLs for 4 seedling vigour traits
64	2007	ISTA	Standardized AA test for soybean
65	2007	Balesevic *et al.*	Use of Vigour tests for determination of chemical changes in seeds
66	2008	ISTA	1st Edition of Hand Book on Flower Seed Testing
67	2008	Hayashi *et al.*	Reported 28 QTLs for germination rate in Lettuce using RILs

Table 3.3 *Contd…*

S.No.	Year	Name of the Scientist	Reported
68	2010	Zhou-fel-Wang *et al.*	In rice seed vigour during germination is controlled by a limited number of major genes and several minor QTLs
69	2013	ISTA	Finalized Proficiency Test Programme in place of Referee Tests
70	2014	ISTA	Electronic version of Rules for seed testing made available
71	2014	ISTA	Validated Conductivity test for *Kabuli* type *Cicer arietinum*

3.5 Definitions of Seed Vigour and Vigour Testing

It is not an exaggeration to state that we have travelled a long way from 1876, when the concept of vigour was first conceived by Fredrich Nobbe to 1977, nearly after a century only, ISTA came out with a definition on seed vigour, that too for *academic* purpose. It means, the inordinate delay in providing a meaningful and an appropriate definition, for nearly over a century, at least the ability to measure or test this undefined entity became a difficult task, if not impossible. No doubt, several definitions were proposed during the course of dynamic development of the vigour studies. Some important definitions are as follows.

1. **Isley** (1957) defined Seed Vigour as 'the sum total of all seed attributes which favour stand establishment under favourable conditions'.

2. **Delouche and Caldwell** (1960) concluded two aspects from the vigour definition given by Isley (1957). (a). Vigour had an influence only on stand establishment and (b). Differences in vigour among seed lots is usually manifested only under unfavourable conditions for germination and emergence and hence, defined seed vigour as "the sum of all seed attributes which favour rapid and uniform stand establishment". Observe the subtle difference between this definition *vis-a vis* the Isely's (1957) definition. They have deleted the reference "under favourable conditions" and emphasized on "rapid and uniform performance" while "the sum total of all seed attributes" remained still left unresolved. So, they defined seed vigour in a negative sense, "thought of as something not adequately measured or reflected by the standard germination test".

3. **Germ** (1960) defined seed vigour as "the ability of seeds to produce seedlings well capable of increased length and volume while still dependent on their reserves".

4. **Nutile** (1964) viewed seed vigour in a more positive sense as 'the ability of seeds to produce vigorous seedlings'.

5. **Ader** (1965) defined seed vigour as "the percentage of seeds able to produce normal seedlings under suboptimal conditions".

6. **Isley** (1966), in the light of above conclusion by Delouche and Caldwell (1960), modified his earlier definition of 1957 as, "the sum of all seed attributes which favour rapid and uniform stand establishment under field conditions." Non-consideration of the effects of vigour beyond stand establishment is a limitation of this definition.

7.   **Linder** (1967) defined seed vigour as "the efficiency of seeds".

8. **Delouche** (1968) defined seed vigour as "the physiological stamina of seeds".

9.   **Woodstock** (1969) proposed a definition different from earlier ones as "that condition of good health and natural robustness in the seed, which upon planting permits germination to proceed rapidly and to completion under wide variety of environmental conditions".

10.   **Neeb** (1970) defined seed vigour as "the totality of properties contributing to the defense against and successful resistance to biotic and abiotic hazards during germination under suboptimal conditions".

11.   **Pollock and Roos** (1972) considered seed vigour as "the maximum potential for seedling establishment and as a quantum of potential decrease from maximum until the seed is dead, *i.e.,* zero potential for establishment".

12.   **Heydecker** (1972) definite seed vigour as "The overall ability of the seed to perform well when sown in the field and the ability of seedling upon germination to grow rapidly".

13.   **Ching** (1973) defined seed vigour as "Those seed properties, which determine the potential for rapid, uniform germination, emergence and fast seedling growth and development of normal seedlings under wide range of field conditions". By this definition he has identified two chief parameters for the seedling vigour viz., germination and seedling growth,

14. **Perry** (1973) considered seed vigour is a "physiological property determined by the genotype and modified by the environment, which governs the ability of the seeds to produce seedlings rapidly in the soil and the extent to which the seed tolerated the range of environmental factors". It persists throughout lifecycle of plant and affects the yield. Therefore, the emphasis was laid on both genetic and environmental components.

15.   **Mc Daniel** (1973) defined seed vigour as the superior performance of a seed after planting compared to the seed of same genotype or other genotypes under defined experimental conditions. By this definition he has identified the differences prevalent between different varieties and their differential performance under a set of well-defined experimental conditions.

16.   **AOSA** (1975) defines Seed vigour as "the sum total of all those properties of seed, which upon planting result in rapid and uniform production of healthy seedlings under a wide range of environments including both favourable and stress conditions".

17.   **ISTA** (1977) defined Seed vigour is "the sum of those properties of the seed, which determine the potential level of activity and performance of a non-dormant seed or seed lot during germination and seedling emergence". For the first time classified seeds, which performed well as high vigour seeds and those perform poorly are called low vigour seeds.

On the other hand, the above definition adopted by ISTA Congress, (1977) points out the following variations associated with differences in seed vigour.

1. Various biochemical reactions and processes occurring during germination like activities related to respiration, and processes related to enzymatic reactions etc.,

2. Rate and uniformity of seed germination and seedling growth,

3. Rate and uniformity of seedling emergence and growth in the field and

4. The ability of emergence of seedlings under unfavourable environmental conditions. Since this definition identifies and describes seed vigour, it can only be considered **as definition of *academic* importance**.

18. **ISTA, Vigour Committee**, (1979) defined seed vigour as 'those seed properties, which determine the potential for rapid, uniform emergence and development of normal seedlings in a wide range of field conditions'. This definition has not only quantified vigour in terms of rapid and uniform emergence and development of normal seedlings but also focuses on what seed vigour does and hence can be considered as an **operational definition**. According to which vigorous seeds germinate and produce normal seedlings even under adverse soil or environmental conditions.

Thus, the definition of seed vigour has to fulfill the following three essential features;

1. it is the sum total of those properties of non dormant seeds,

2. which led to a rapid and uniform emergence of healthy seedlings in the field,

3. under a range of both favourable and unfavourable environments.

Over a period of time, the definition of seed vigour has undergone several modifications. The differential expression of seed vigour in laboratory and field has obviously created several practical difficulties in agreeing upon a suitable definition of vigour that incorporates all its characteristics. As a consequence, the definition has been modified several times (Heydecker, 1972; Perry, 1973, AOSA, 1983).

Lee *et al.,* (1986) defined seedling vigour as a quality factor that denotes the potential for rapid germination and fast seedling growth under field conditions. Wherein the potential varies in tune with the genetic and environmental conditions and backgrounds. Vigour is the sum total of all seed attributes, which favour rapid and uniform stand establishment in the field (Agarwal, 1993). According to the definition proposed by Perry (1978) and accepted by ISTA and incorporated in "Rules for Seed Testing" (2001) 'Seed Vigour is the sum of those properties that determine the activity and level of performance of seed lots with acceptable germination in a wide range of environments'. This definition clearly points out that, seed vigour is not a single measurable property.

As one goes deep, the subject of vigour appears to be more and more complex because, it is influenced by a number of factors, not only inherent but also due to the wide variation in field and environmental conditions. Hence, the vigour tests thus developed are unable to account for the variability encountered in the field during seedling emergence. Despite the best effort put forth over a considerable period of time to identify a seed vigour test(s) that accurately predicts the emergence potential in the field could not met with significant success. However, when maize was sown under less than ideal conditions in the cold test, the results were more closely related to final emergence than standard germination test or accelerated ageing test. Similarly the GADA test for wheat, seedling root and tetrazolium tests followed by seedling dry weight and Respiratory Quotient tests *etc.,* though could predict the better field emergence, it is not possible to recommend a single best vigour test for all the crops. This can be illustrated by the fact that it took more than two and half decades for the ISTA "Vigour Test Committee" to agree upon a definition of vigour acceptable to all engaged in the seed testing and seed trade. After nearly four decades of intense research and debate, only one test for one species, the "Conductivity Test for Peas", has met the requirements to be accepted internationally.

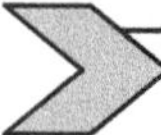 ## 3.6 The Concept of Seed Vigour Testing

The concept of seed vigour is relatively new and very little work was done till the "Cold Test for Corn' was developed. Seed vigour in simple terminology is a measure of the extent damage that has accumulated as seed viability declines. Friedrich Nobbe (1876) first recognized the concept of seed vigour and introduced the term *Triebkraft,* and later on this term was mostly used in agronomic sense rather than physiological. He clearly distinguished between germination and the concept of seed vigour in his "Handbuch der Samenkunde" and also proposed procedures for germination test. However, the real credit of explaining the concept of seed vigour goes to two German pathologists, Hiltner and Ilhssen (1911), who used the term "triebkraft" to imply 'driving force' and shooting strength of germinating seedlings, and demonstrated the same with the help of Brick Gravel Test. and for the first time explained the concept of seed vigour while working with *Fusarium spp.* on cereal seeds and reported that though both infected and non-infected seeds germinated, the seedlings from infected seed could not emerge out of 3-4 cm thick brick grit composed of 2-3 mm particles. Hence, they suggested that both, successful and unsuccessful types of seeds in a "Brick Gravel Test", were alive, but differed in their degree of aliveness, *i.e.,* the infected seeds have a relatively lower aliveness than non-infected seeds and called this ability of seeds as '*Triebkraft*' the 'driving force' to cover the shoot length and subsequently translated into English as 'Vigour.' The original meaning of seedling vigour was the ability that enables a seedling upon germination to grow rapidly. Alberts (1927), during the course of investigation to identify the relationship between injury to pericarp, absorption of moisture, temperature and fungal attack, developed "Seed Corn Viability Test". It was refined and applied to different kinds of seeds, which in turn contributed significantly to the understanding of the need and practical relevance of seed vigour.

During the Biochemical and Seedling vigour committee of ISTA Congress, Frank (1950) demanded (i) definition of seed vigour and (ii) development of standardized vigour tests, because without understanding what seed vigour was, the possibility of creating reliable methodology to assess this quality parameter would be very difficult, if not impossible. This is so because, since inception, seed vigour was never characterized as a single defined physiological process and suggested that vigour could not be defined because it was so complex. Accordingly the first concept considers seed vigour as a "Sum total of seed attributes, which favour stand establishment under favourable conditions", Isley, (1957). Basically, the initial concepts of seed vigour mostly banked on the merits of germination and tetrazolium tests with respect to identification of seed lots that were able to achieve quick and uniform emergence and plant stand establishment under unfavourable or less favourable conditions, Shraf, (1953). Contrary to this, Delouche and Grabe (1958) referred seed vigour as a "seed physiological resistance".

Woodstock (1965) emphasized that seed vigour was a condition of good health and natural robustness associated with rapid and complete germination under wide range of environmental conditions. The term 'Wide range' was used for the first time in place of favourable or unfavourable environmental conditions. Subsequently, Pollock and Roos, (1972) assumed that seed vigour might be considered as a potential for seedling establishment in the field and the same idea was supported by several seed technologists. However, the new approaches and ideas proposed by TeKrony, (2003) became the foundation for the current concepts of seed vigour, *i.e.,* Seed vigour is the sum of all those properties which determine the potential for rapid, uniform emergence and development of normal seedlings under a wide range of field conditions (AOSA, 1983), Baalbaki *et al.,* (2009). A vigorous seed lot, according to ISTA (2014) is the one that is potentially able to perform well under environmental conditions, which are not optimum for the species.

According to ISTA (2014) seed vigour is the sum of those properties that determine the activity and performance of seed lots of acceptable germination in a wide range of environments. Therefore, a vigorous seed lot is one that is potentially able to perform well under environmental conditions which are not optimal for the species. However, Marcos Filho (2015) felt that the current accepted concepts do not include the possible persistence of the effects of seed vigour throughout the plant development cycle persisting until yield. No doubt, emergence of vigorous seedlings can result in high productivity due to rapid and uniform stand establishment. Therefore, it is not appropriate to identify seed vigour as a single seed physiological process. This obviously demands consideration of seed vigour as an abstract characteristic. May be the introduction of this expression is not only aimed at bridging the gap in the absence of parameters to interpret frequent questions pertaining to performance of seed under less favourable conditions.

On one hand, this situation necessitates identification of seed vigour as an interaction product of several independent attributes such as speed of germination, seedling growth, ability to germinate above or below optimal temperatures and other aspects of tolerance to stresses. On the other hand, due to the involvement of several factors, only few scientists could get success in their attempts because they felt that it is easier to understand the main effects of vigour rather than the defining it, Julio Marcos Filho (2015).

3.6.1 Biological Basis for the Concept of Seed Vigour

The causes of variations in vigour are numerous and diverse. Seeds attain their physiological potential close to their maturity. Once seeds pass through this stage, they become prone to deterioration and depends on time of harvesting, prevailing environmental conditions, procedures adopted for seed drying, processing and storage. The physiological potential of seed in turn governs the theoretical capacity of seed and hence exerts control on the level of expression of their vital functions both under favourable and unfavourable conditions such as temperature, moisture availability, physical and chemical properties of soil, plant protection chemicals used and management practices adopted and varies with species, cultivar and stage of plant development. The biological basis of seed vigour concept states that, in addition to the genetic constitution of the seed (seed size, weight and specific gravity), the conditions of seed development, viz. maturation (stage of maturity at harvest, quantity and quality of nutrition provided by the mother plant), deterioration and ageing, pathogens and storage *etc.,* exerts considerable influence on seed vigour. This is so because the seeds are developed under varying levels of stress situations like moisture stress, nutrient deficiency and extreme temperature *etc.,* This situation obviously result in light weight, chaffy and shriveled seed collectively called as poor vigour seed. The pre-harvest environment with high humidity coupled with warm temperature can also cause loss in seed vigour. In addition to these, the genetic factors like hard seededness, resistance to diseases and chemical composition of the seed too exert influence on seed vigour. The mechanical damage caused by harvesting and conditioning equipment together with poor storage conditions too adversely affect vigour. During seed testing in the laboratory, seed dormancy is likely to obscure the germination potential of a seed lot, even then it should not be regarded as a component of vigour if the seedling emergence is not affected in the field sowings. In the earlier days, the concept of vigour was focused and mostly confined to the differences in the stand producing potential of seed lots in the field. Seed vigour is frequently expressed as the level of tolerance of seeds to adverse soil conditions, *i.e.,* when compared to a low vigour seed lot, a high vigour seed lot will produce a greater percentage of seedlings. On the other hand, initial seed ageing result in the decline in germination speed of viable seeds, which is followed by decrease in seedling size, increased number of abnormal seedlings. This low germination rate is associated with early signs of membrane disintegration and the wider occurrence of seedling abnormalities could be attributed to the death of the tissues in different parts of the meristematic tissues.

3.6.2   Modern Basis for the concept of Seed Vigour

Appreciation of seed vigour, in the modern context, probably began with the discussions on the concepts behind the conducting of germination test in USA and Europe. The US researchers believed in conducting the germination test in soil as a media for accurate realization and true expression of the emergence potential of a seed lot. Contrary to this, the European scientists believed in conducting the germination test under the most optimum conditions specified for a species so as to obtain maximum number of normal seedlings that showed the potential for emergence. The objective of seed testing is to provide an accurate estimate of planting value to the grower together with an unbiased guarantee of quality in commercial seed transactions. So, it becomes obvious and essential that the seed testing results thus obtained should be uniform, reliable and reproducible both between and within testing stations.

At the ISTA Congress (1950), ISTA Biochemical and Seedling Vigour Committees were established to define seedling vigour and to standardize the test methods. After several rounds of deliberations between the US and European group of scientists, initially it was decided and agreed upon that germination test should be conducted under standard normal conditions specified for that species. Because, various types of seed require different optima from which deviations can cause variations in germination percentage, more particularly while dealing with low / poor quality seed lots.

3.6.3 The Per-se Concept of Vigour

The wide array of definitions proposed from time to time has changed the concept of vigour. In the positive side, there is no precise definition of vigour and on the negative side, it is something not adequately measured in a standard germination test. The concept remained ambiguous till Isley (1957) explained it based on two important considerations, *viz*, vigour *per se* in terms of germination and rapid growth of seedlings and non- susceptibility of seedlings to unfavourable conditions. So, susceptibility of seed to unfavourable field conditions together with mechanical damage of the seed and presence of short period of post dormancy also needs to be taken into consideration while defining the **per se concept** of vigour as "it is a direct expression of physiological and to some extent the physical condition of the seed."

Accordingly, Isley defined seed vigour as 'the sum total of all seed attributes, which favour stand establishment under unfavourable field conditions, primarily as a function of seed and soil microorganism inter-relationship". The chief drawback of this definition lies in its greater emphasis on the environment rather than the seed itself. This definition can be explained and elaborated with the help of a vigour scale (Fig. 3.1) where, the vigorous seeds relatively occupy a higher position as detailed below.

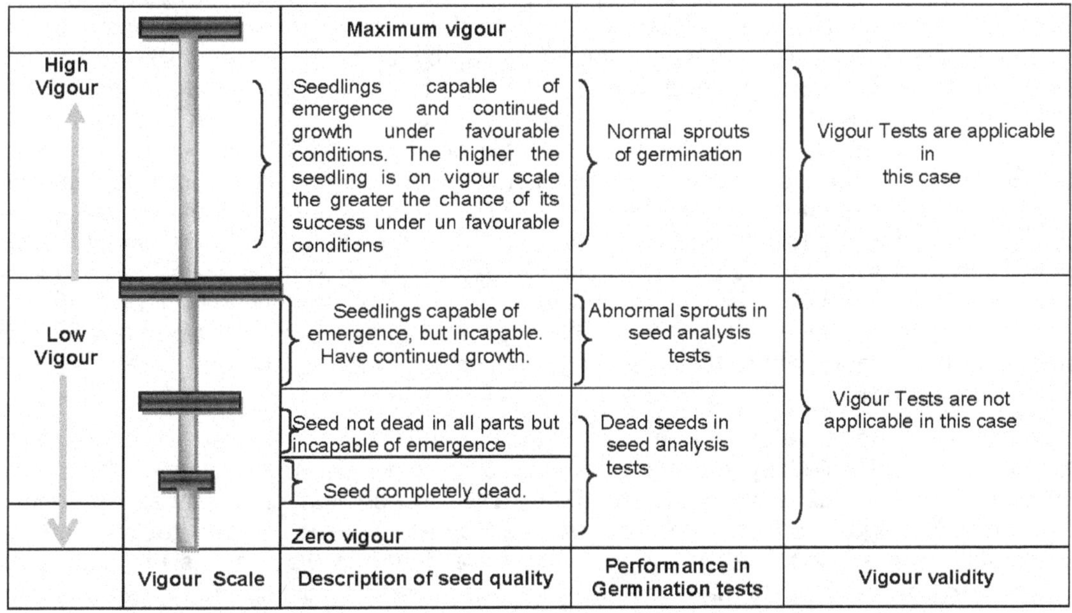

	Vigour Scale	Description of seed quality	Performance in Germination tests	Vigour validity
		Maximum vigour		
High Vigour		Seedlings capable of emergence and continued growth under favourable conditions. The higher the seedling is on vigour scale the greater the chance of its success under un favourable conditions	Normal sprouts of germination	Vigour Tests are applicable in this case
Low Vigour		Seedlings capable of emergence, but incapable. Have continued growth.	Abnormal sprouts in seed analysis tests	Vigour Tests are not applicable in this case
		Seed not dead in all parts but incapable of emergence	Dead seeds in seed analysis tests	
		Seed completely dead.		
		Zero vigour		

Fig. 3.1 Schematic representation of the relationship between germination and vigour.

The definitions of seed vigour provided by Woodstock (1969), Perry (1972) and Heydecker (1972) and others are not only informative but also interesting and contributed significantly to our understanding on the concept of vigour. Despite their merit, none of these definitions were able to provide a satisfactory working definition, which lead, Pollock and Roos (1972) to view the concept of vigour in two different aspects. (i) Vigour is a maximum potential for seedling establishment set by the genetic purity of the plant and (ii) a quantum of potential decrease from the maximum until the seed was dead, *i.e.,* zero potential for establishment. Under poor field conditions, the germination test has failed to predict the differences in field emergence, suggested the existence of one more physiological aspect of seed quality, and largely referred to as Vigour. The foundation for the current concepts of seed vigour, i.e., 'seed vigour is the sum of all those properties which determine the potential for rapid, uniform emergence and development of normal seedlings under a wide range of field conditions' was proposed by TeKrony, (2003). A vigorous seed lot, according to ISTA (2014), is the one that is potentially able to perform well under environmental conditions, which are not optimal for the species.

3.6.4   The Multiple Concept of Seed Vigour

In practice, differences in seed vigour of different seed lots become relevant only when germination test fail to indicate the differences in their seedling emergence as observed in the field. Therefore, it must be unambiguously emphasized that seed vigour, after all, is only one of the several estimates of the seed quality. Since, it is not an absolute quantity and is only a relative measure used for comparing different seed lots.

Greater emphasis in developing and conducting both germination and vigour tests together continued to increase and resulted in more than 50 different tests proposed on the basis of different aspects of seed comprising of stress at physical, physiological, and biochemical levels and among these, only a hand full of tests are less accurate than desired, and are still used actively in seed testing, by both AOSA and ISTA. The wide range of these definitions on seed vigour by AOSA/ISTA has provided a basis for the emergence of two different views, (i) susceptibility of seed to unfavourable conditions (ii) the reflection of seed vigour on the speed of germination and rapidity of growth rate of seedlings. The rate and uniformity of germination and emergence when included as vigour characters, they assume importance especially under protected environments like production of seedlings for transplanting at a specific growth stage. One should always remember that, low vigour seed lots usually have not only a slower mean rate of germination but also a wider distribution of rate of germination of individuals than high vigour seeds. Further, the rapid rate of early seedling growth can also be considered as another parameter of high vigour seeds that are likely to exert influence in enhancing the crop yields at least in some cases.

Further, one should always remember that, Seed vigour is clearly a quantitative character and is controlled by a number of interacting seed traits, and thereby exert direct influence not only on the seed germination but also on seedling emergence. As such the definition of seed vigour, as accepted and adopted by all involved in seed testing and seed trade should comprise of the following three essential features *viz.*, (i) it is the sum total of those properties of non-dormant seeds, (ii) capable of giving rapid and uniform emergence of seedlings in the field, (iii) perform better under a wide range of both favourable and unfavourable environmental conditions. Seed vigour (i) is not an absolute quantity, (ii) is only a relative measure of comparing the seed lots and hence (iii) is just one of the several estimates of seed quality. Aliveness and liveliness of the seed in addition to viability can be termed as vigour. Therefore, all we can say about a vigorous seed lot is one that is potentially able to perform well even under environmental conditions not optimal for the species.

A look at the different and diverse above-mentioned facts, on one hand, seed vigour appears to be a very simple concept rather than a specific property of a seed or seed lot. But, according to the latest definition of ISTA, vigour of a seed lot is not merrily a single measurable property. Because Seed vigour is simultaneously associated with the description of several characteristics of seed lot performance like (i) initial higher capital (food reserves), (ii) rate and uniformity of seed germination (iii) seedling growth and development, (iv) emergence ability of seeds under unfavourable conditions like crusted soil, cold, wet and pathogen infested soil etc., and (v) specific performance both during and after storage with reference to the retention capacity to germinate. Therefore, it becomes a multiple concept, developed by the tireless efforts of several researchers round the globe. Hence, a vigorous seed lot is defined as, "one that is able to perform potentially well even under environmental conditions that are not optimal for the species". In other words, a vigorous seed lot is one, which emerges faster and is likely to survive and succeed under a wide range of fluctuating field conditions. Probably, there is no other component of seed quality, which has been the subject matter of as much controversy and misunderstanding as seed vigour. Even though, the complex mechanisms underlying seed vigour have not been fully described, much progress has been made in understanding this phenomenon besides its practical consequences.

3.6.5 The Molecular basis of Seed Vigour

Seeds with strong vigour significantly improve the speed and uniformity of seed germination and the final per cent of germination and lead to perfect field emergence, good crop performance and even high yield under different conditions (Foolad *et al.,* 2007). Contrary to this Sun *et al.,* (2007) reported that seed vigour is a comprehensive character affected by several characters like genetic makeup and background, environmental factors during seed development and stages of seed storage.

The dominance hypothesis attributes increased vigour either to the accumulation of favourable dominant genes or by masking of deleterious recessives in the hybrid. This hypothesis is found to be consistent with recent genomic evidence of differences in genetic content between maize inbred lines, Fu and Dooner, (2002) and has been demonstrated as the underlying cause of heterotic response for grain yield in a QTL mapping study, Graham *et al.,* (1997). Consequent to the development of techniques of DNA molecular marker and genome graphing, QTL analysis for seed vigour has been reported, and is mainly focused only on a limited number of plant species, which includes barley, Mano and Takeda (1997), rice (Cui *et al.,* 2002), Miura *et al,* (2002), Zhang *et al.,* (2005), Fugino *et al.,* (2008); Arabidopsis, Clerkx *et al.,* (2004); in maize, Hund *et al.,* (2004); in tomato, Foolad *et al.,* (2003), and (2007) and in Lettuce, Hayashi *et al.,* (2008). On the other hand, the Over Dominance hypothesis argues that the heterozygous combination of the alleles at a single locus is superior to either of the homozygous combinations. The origin of seed vigour is physiological and biochemical too. The physiological and biochemical reactions can create and cause the difference in vigour between two seed lots belonging to the same genetic line. Therefore, physiological and biochemical vigour has its basis always lying in genetic vigour because obviously these reactions are under direct genetic control. Hence it is very difficult to determine the cause of the vigour especially in genetically heterogeneous populations. In respect to rice, one of the most important and staple food crops of the world, the farmers especially dealing with direct sown culture system badly felt the necessity of identification of cultivars with strong seed vigour for obtaining optimum plant stand establishment. There are several reports of QTLs for rice seed vigour, duly focusing on seedling growth, seed longevity and tolerance to adversity. Miura *et al.,* (2002) reported three putative QTLs for seed longevity, 1G-2, qLG-4 and qLG-9 using 98 Backcross Inbred Lines (BILs). Where as Figino *et al.,* (2004) detected three putative QTLs, qLTG 3-1, qLTG 3-2 and qLTG-4 associated with low temperature germination using 122 BILs. Contrary to this Cui *et al.,* 2002 mapped 31 QTLs for five traits of seedling vigour using a set of 241 F10 recombinant inbred lines (RILs); Zhang *et al.,* (2005) reported 34 QTLs for four seedling vigour traits using 282 F13 RILs. Fugino *et al.,* (2008) reported map based cloned Qltg 3-1.

Bettey *et al.,* (2000), Finch – Savage *et al.,* (2005) have identified significant QTLs for the three seed vigour components by screening a population of *Brassica oleracia* dihaploid lines. During the evaluation of seed vigour, Hayashi *et al.,* (2008) reported that the speed of germination might be more important than germination rate because the former usually decreases more quickly than the later during seed storage, and hence makes the trait speed of germination more desirable in seed vigour testing when compared to germination rate and germination index.

Further, the reports of Foolad *et al.,* (1999), and (2007) in Tomato and Hayashi *et al.,* 2008 in Lettuce are indicating that QTLs associated with germination speed are important for germination are not specifically affected by stress and hence suggests that rapid seed germination is controlled by similar mechanisms under different conditions, viz., the QTLs for speed of germination coincided with those for ABA (absessic acid) sensitivity and salt tolerance in Arabidopsis, Clerks *et al.,* (2004). The little progress achieved in rice seed vigour through conventional breeding methods like selection for seed size impedes the requirements of consumer preference as increased grain size is likely to reduce the popularity of a cultivar (McKenzie *et al.,* 1994). Therefore, this clearly facilitates use of Marker assisted Selection for bringing improvement in seed vigour through identification of QTLs not linked to seed size. Evaluation and screening of quantitative traits appears to be the most crucial step in any QTL mapping programme. The continuous segregation exhibited as the frequency distribution for seed vigour in the Recombinant Inbred Lines (RILs) suggests that during germination stage, seed vigour is a quantitative character controlled by several genes.

Table 3.6 QTLs for germination speed

Author	Report	Crop / Populations
Foolad *et al.,* (1999)	Mean time to 50% germination under non-stress, cold stress and salt stress conditions	Tomato,119 BC_1S_1 populations
Cui *et al.,* (2002)	Two QTLs for seed vigour and α-amylase not linked to seed weight for use in MAS, 31 QTLS for 5 traits of seedling vigour.	241 F10 RILS
Miura *et al.,* (2002)	Found 3 putative QTLs for seed longevity	Used 98 BILs
Fuzino *et al.,* (2004)	Detected 3 putative QTLs associated with low temperature germination	Using 122 BILs
Al-Charani *et al.,* (2005)	Detected two QTLs for time of 50% germination	Sunflower, 84 RILs
Wan *et al.,* (2005)	The region of major QTL qGP -6 on chromosome 6	34 QTLs
Fuzino *et al.,* (2008)	Identified QTLG-3-1 map based cloned	2 LTG3
Hayashi *et al.,* (2008)	Reported 28 QTLs for germination rate under different temperatures	Lettuce, 131 F_8 RILs
Zhou-fel wang *et al.,* 2010	10 QTLs were mapped on chromosomes 1, 2, 4, 6, 7, 8 and 11. Amount of variation explained by an individual QTL ranged from 7.5 – 68.5%	Rice, 150 RILs

Fait *et al.,* (2006) and Bethke *et al.,* (2007) reported that probably the speed of germination is affected by the genes related to reserve mobilization and endosperm weakening. According to Catusse *et al.,* (2008), the molecular evidence indicates the requirement of de novo transcripts like those involved in gibberellic acid biosynthesis in increasing seed vigour. Zhou-fel wang *et al.,* (2010), by using a Recombinant Inbred Line population derived from a cross between Japonica variety Dagundao and an indica variety IR28 through multiple interval mapping approach proved that indica rice has presented strong vigour during germination stage when compared to japonica rice and further reported that out of the 10 QTLs five are novel and the speed and uniformity of germination under different conditions share the same basic genetic mechanism and in rice seed vigour during germination is controlled by a limited number of major and several minor QTLs. Since seed germination and vigour are closely related to seed weight, seed size and seed coat etc., Zhou-fel wang *et al.,* (2010) reported that major QTLs for seed vigour coincided with QTLs for seed weight, seed size and seed dormancy etc., only suggesting that these are under the partial control of same genetic mechanism. (Table 3.6). Therefore it can be concluded that seed vigour is a very complex physiological process with a comprehensive genetic background and hence needs further study to clarify the functions of major QTLs and for developing new varieties with high level of seed vigour duly using the MAS method.

3.6.6 The Genetic basis of Seed Vigour

Today a number of reviews are available on seed germination and dormancy. But, neither understanding the initial mechanisms determining the vigour of seed in the context of apiculture to nor the consequences of environmental variation are focused sufficiently. No doubt, Arabidopsis adopted to wide range of environmental conditions have emerged as an excellent model for understanding the regulation of germination and seed dormancy. Seed vigour i a quantitative trait under the influence and interaction between genetics and seed vigour is a the environment. Though the role of vigour in successful crop establishment and yield in known beyond doubt long ago, never it was considered as objective in commercial breeding programmes. Seed lots can be quantified for their characters during post maturation period as the mother plant as well as predicting the storage environment factory such as temperature, water potential, soil impedance on seed germination. Today even preemergence seeding, growth can also be predicted. There for by understanding and minimizing there negatively influencing factors further improvement in vigour can be obtained through the targeted genetic inharcement. Since the mechanisms controlling vigour are poorly understood clearly represent a gap in our ability to enhance vigour.

3.7 Summary

The optimum conditions prescribed for conducting a standard germination test are rarely available under real field conditions; the germination test is unable to differentiate between seed lots based on their capabilities to perform under exacting field conditions. Further, uniform maturity to facilitate mechanical harvesting under acute labour shortage conditions coupled with the demand for uniformity in produce to meet the demand of consumers by transporters and supermarket chains of developed countries, necessitated the search for another quality parameter, seed vigour so as to facilitate exploitation of early planting in cool and wet soils; to exploit the length of season; to capitalize on maximum available moisture; to minimize the cost of production.

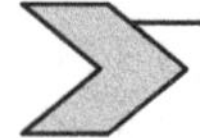 **3.8 Exercise**

I. Answer the following questions.

1. Why should we have a vigour test at all?
2. Explain the basic purpose of seed vigour testing.
3. Define and differentiate between physiological and mass maturity.
4. How rate and uniformity in emergence play role in uniform produce production.
5. Why there is no relationship between yield and vigour in cereals and pulses.
6. Highly vigorous seeds are used in cole crops production. Why?
7. It took nearly more than a century to provide a meaningful definition to vigour. Discuss?
8. List out the variations associated with differences in seed vigour definition of ISTA 1977.
9. List out three essential features operational definition of seed vigour has to fulfill.
10. Seed vigour is not a single measurable property. Why?
11. How can you judge whether a vigour test is worth-while or not?

II. Answer the following questions in detail.

1. Explain the biological and modern basis for the concept of seed vigour.
2. With the help of a diagram explain the *per se* concept of seed vigour.
3. Give a detailed account on the multiple concept of seed vigour.
4. Explain the genetic basis of seed vigour at molecular level with suitable examples.
5. Explain why the subject of vigour testing appears more and more complex.

III. Fill in the Blanks with suitable words/phrases.

1. The need for vigour testing initially came from developed countries facing -------- .
2. The need for vigour testing is an absolute necessity to facilitate mechanical harvesting of -------- species.
3. Appearance of a good number of unproductive plants in the field due to the inability to control light and heat in the microclimate in the field often results in ------- .
4. The low vigour seeds can never resume normal growth and give only --------.
5. As proposed by Frank, 1950, tests related to germination when performed in soil should be referred to as --------.
6. Seed vigour tests have been designed to reveal the damage caused by -------- and --------.
7. The potential difference between the performances of different high germinating seed lots is referred to as --------.

8. In cereals, though seedling emergence is slightly reduced, the effect on yield is limited due to the compensatory growth offered by --------.

9. The real credit of explaining the concept of seed vigour goes to --------.

10. ISTA Biochemical and Seedling vigour committee was established in the year, ------

11. First seed vigour testing progress report was published by --------

12. Hiltner and Ilhssen (1911) demonstrated seed vigour through -------- test.

13. Hiltner and Ilhssen (1911) developed brick gravel test while dealing with -------- species in -------- crops.

14. The causes of variation in vigour are -------- and --------.

15. The pre-harvest environment with -------- and -------- causes loss in seed vigour.

16. The result of seed testing should be --------, -------- and uniform both between and within seed testing stations.

17. The objective of seed testing is to provide an accurate estimate of -------- and -------- to the grower.

18. The per se concept of seed vigour was proposed by --------.

19. In the vigour scale proposed by Isley (1957), zero vigour seeds occupy a --------.

Answers

1.	shortage of labour	2.	Forage
3.	Abnormally low vigour seeds	4.	Abnormal plants
5.	Vigour tests	6.	Moisture and temperature
7.	Vigour	8.	Tillers
9.	Hiltner and Ilhssen (1911)	10.	1950
11.	Woodstck in 1976	12.	Brick – gravel
13.	Fusarium, cereal	14.	Numerous and diverse
15.	High humidity and warm temperature	16.	Reliable, reproducible
17.	Planting value and an un biased guarantee of quality		
18.	Isley, 1950	19.	lower portion

3.9 References

Alberts, H.W. (1927). Effects of pericarp injury on moisture absorption, fungus attack and vitality in corn. Jour. Of American Society of Agronomy.19 : 1021-1030.

Al- Charani, G.R., Gentzbittel, L., Wedzoney, M., Sarrati, A., (2005). Identification of QTLs for germination and seedling development in sunflower. Plant Sci. 169 (1): 221-227.

AOSA (1983). Seed Vigour testing Hand Book. Contribution No. 32 to the Hand Book of Seed Testing. (ed.) Clark, B.E.., McDonald, M.B., and Joo, P.K., pp 88., Lincoln. NE. AOSA. Basavarajappa, B.; Shetty, H.S.; Prakash, H.S. 1991. Membrane deterioration and other bioche-mical changes associated with accelerated ageing of maize seeds. Seed Science and Technology 19: 279-286.

Betty, M., Finch –savage, W.E., King, G.J., Lynn, J.R. (2000). Quantitative genetic analysis of seed vigour and pre-emergence seedling growth traits in *Brassica oleracia,*. New Phytol.148: 227-286.

Bethke,P.C., Libourrel, I.G., Aoyama ,N., Chung, Y.Y., Still, D.W., Jones, R.L., (2007). The Arabidopsis aleurone layer responds to nitric oxide , gibberellin and abscisic acid and is sufficient and necessary for seed dormancy. Plant Physiology. Mar.143(3):1173-88.

Cantlife,D.J.,(1981). Vigour in vegetable seeds. Acta. Horticulture. 11: 219-226.

Cantliffe, D.J., K.D. Shuler and A.C. Guedes. (1981). Overcoming dormancy in heat sensitive romaine lettuce by seed priming. HortScience. 1981. Vol. 15, pp. 196-198.

Catusse,J., Strub,J.M., Job,C., Van Dorsselaer, A., Job, D., (2008). Proteome wide characterization of sugar beet seed vigour and its tissue specific expression. Proc. Natl. Acad. Sci. USA., 105: 10262-10267

Clerkx EJ, El-Lithy ME, Vierling E, Ruys GJ, Blankestijn-De Vries H, Groot SP, Vreugdenhil D, Koornneef M. (1955). Analysis of natural allelic variation of Arabidopsis seed germination and seed longevity traits between the accessions Landsberg *erecta* and Shakdara, using a new recombinant inbred line population. Plant Physiol. 2004;135(1):432–443. doi: 10.1104/pp.103.036814.

Clark, B.E and Little, H. B. (1955). Quality of seeds. New York State Agricultural Experimental Station Bulletin. No. 770.

Cui, K.H., Feng, S.B., Xing, Y.Z., XU, C.G., Yu, S.B. and Zhang, Q., (2002). Molecular dissection of seedling vigour and associated physiological traits in rice. Theor. Appl. Genetics.105: 745 – 753.

Demir I, and Ellis, R.H., (1992). Changes in seed quality during seed development and maturation in tomato. Seed Sci. Res. 2 : 81 -87.

Delouche, J.C. and Caldwell, W.P. (1960). Seed Vigour and Vigour Tests. Proceedings. of AOSA, 50 : 124-129.

Ellis, R.H. and Pieta Filho, C., (1992). The development of seed quality in spring and winter cultivars of barley and wheat. Seed Science research 2 : pp 9 -15.

Fait A, Angelovici R, Less H, Ohad I, Urbanczyk-Wochniak E, Fernie AR, Galili G. (2006).

Arabidopsis seed development and germination is associated with temporally distinct metabolic switches. Plant Physiol. 142(3):839–854 doi: 0.1104/pp.106.086694.

Ferguson, J.M. (1993). AOSA perspective of seed vigour testing. Journal of seed Technology. 17 : 101-104.

Finch savage, W.E., Come,D., Linn, J.R., Corbineau, F., (2005). Sensitivity of *Brassica oleracia* seed germination to hypoxia : a QTL analysis. Plant Sci. 169 : 753- 759.

Frank, W. J., (1950). Introductory remarks concerning modified working of the International Rules for seed testing on the basis of experience gained after the world war. Proc. ISTA. 16: 405-433.

Foolad, M.R., Subbaiah, P., and Zhang, L., (2007). Common QTL affect the rate of tomato seed germination under different stress and non-stress conditions. International Journal of Plant Genomics. PMC2246063.

Foolad, M.R., Zhang, L.P., Subbaiah, P., (2003). Genetics of drought tolerance during seed germination in tomato inheritance and QTL mapping. Genome. 46 (4) : 536 -545.

Foolad, M.R., Lin, G.Y., Chen, F.Q., (1999). Comparison of QTLs for seed germination under non-stress cold stress , salt stress in tomato. Plant Breeding. 118 : 167 – 173.

França-Neto, J.B.; Krzyzanowski, F.C.; Costa, N.P. (1998). The tetrazolium test for soybean seeds. Embrapa Soja, Londrina, PR, Brazil. (Documents, 115).

Fu, H and Dooner, H.K. (2002). Interspecific violation of genetic coliniarity and its implications in maize. Proc. Nat. Acad. Sci. 99 : 9573 – 9578.

Fugino,K., Sekiguchi, H., Matsuda, Y., Sugimoto, K., Ono, K., Yano, M., (2008). Molecular identification of Quantitative Trait Loci – qLTG 3-1 controlling low temperature germinability in rice. PNAS 105 (34) : 12623 – 12628.

Ganguli, S.; Sen-Mandi, S. 1990. Some physiological differences between naturally and artificially aged wheat seeds. Seed Science and Technology 18: 507-514.

Hampton, J. G., and TeKrony 1995. (ed.) Hand Book of vigour Test Methods. 3rd Edition. ISTA Vigour Test Committee.

Harrington,J..F., (1972). Seed Storage and longevity. In TT Kozlowski (ed.) Seed Biology. Academic Press. London. U..K.

Halmer,P and Bewely, J.D. (1984). A physiological perspective on seed vigour testing. Seed Science and Technology 12 : 561-575.

Hayashi E, Aoyama N, Still DW. 2008. Quantitative trait loci associated with lettuce seed germination under different temperature and light environments. Genome: 51(11):928–947. doi: 10.1139/G08-077.

Hegarty, T. W., (1977). Seed Vigour in Field beans (*Vicia faba* L) and its influence on plant stand. Journal of Agricultural Sciences. Cambridge, 88 : 169-17

Heydecker, W. (1977). Stress and seed germination. An agronomic view. In : The physiology and Biochemistry of seed Dormancy and Germination. (ed.) A.A.Khan, pp 237 –276. Elsevier Press. Amsterdam.

Heydecker, W. (1972). Report of vigour test committee. 1968-1971. Proc. ISTA37 : 379-395. Heydecker, W. (1972). In E.H. Roberts (ed.) Vigour and viability of seeds. pp 209-252. Chapman & Hall, London.

Hiltner, L., and Ihssen, G., (1911). Uber das schlchte Auflaufen and die Auswinterung des Getreides infloge Befalls des Saatgutes duresh Fusarium Landirtschafiliches. Jahabuch fur Bayern: 20-60 : 231-278.

Hund, A., Fracheboud, Y., Soldati, A., Frascaroli, E., Salvi, S., Stamp, P., (2004). QTL controlling root and shoot traits of maize under stress. Theor. Appllied , Genetics. 109 (3) : 618 -629.

Isley, D. (1957). Vigour Tests. Proceedings of AOSA. 47 : 176-182.

Isely, D., (1950). The cold test for Corn. Proceedings of ISTA. 16 : 299 - 311.

ISTA, 2003. International Rules for seed Testing, Edtion2003. ISTA, Bassersdorf, Switzerland.

Lee, C.C., Liand, C. C., Sung, J.M., (1980). Physiological and genetic studies on seedling vigour in rice. Inheritance of Alfa- amylase activity and seedling vigour in rice. Jour.. Agric. Assoc. China. 135 : 1724.

Loeffler,T.M., Tekrony,D.M. and Egli D.B., (1988). The bulk conductivity test as an indicator of soybean seed quality. Journal of seed Technology. 12 : 37 - 53.

Matthews, S. (1985). Physiology of seed ageing. Outlook on Agriculture 14: 89-94.

Matthews, S., (1980). Controlled deterioration, a new vigour test for crop seeds. In. P D.

Hebblethwaite (ed.) Seed Production. Pp. 647 - 660. Butterworths. London.

McDonald, M.B. 1999. Seed deterioration: physiology, repair and assessment. Seed Science and Technology 27: 177-237.

McDonald, M. B., (1994). The History of seed vigour testing.   Journal of Seed Technology. 17 : 93 -101.

McDonald, M. B., Jr.(1975). A review and evaluation of seed vigour tests. Proc. of AOSA.65 : 109 –139.

Mc Kenzie, K.S., Johnson, C.W., Tseng, S.T., Oster, J.J., and Brandon, D.M., (1994). Breeding improved rice cultivars for temperate regions. A case study. Austr. J., Expt. Agric. 34 (7).: 897- 905.

Mano, Y., and Takeda, K., (1997). Mappung QTL for salt tolerance at germination and the seedling stage in barley (Hordeum vulgare). Euphytica. 94 (3) : 263 – 272.

Miura, K., Lin, S., Yano, M., Nagamini, T., (2002). Mapping QTL controlling seed longevity in rice.Theoretical and Applied Genetics.104 (6-7)   981-986.

Oliveria, M. De A., Matthews, S., and A.A. Powell, A. A.,(1984). The role of split seed coats in determining seed vigour in commercial seed lots of soybean as measured by Electrical conductivity test. Seed Science and technology 12 : 659 - 668.

Oplinger, E.S., and Phillbrook, B.D., (1992). Soybean planting date, row width, and seedling rate response in three tillage systems. Jour. Prod. Agric. 5 : 94 – 99.

Perry, D.A., (1978). Report of the vigour test committee.(1974-1977). Seed Science and Technology. 6 : 159 -181.

Perry, D.A. (1977). A vigour test for seeds of barley based on measurement of plumule growth. Seed Science and Technology. 5 : 709 - 711.,

Perry, D. A. (1973). Interacting effects of seed vigour and the environment. In W Heydecker ed.) Seed Ecology. Butterworths. London.

Perry, D. A ., (1972). Seed vigour and field establishment. Hort. Abstr. 42 : 334 – 342.

Pollock, B. M. and Roos, E. E., (1972). Seed and seedling vigour, In T.T. Kozlowski (ed.) Seed Biology Vol.I. Academic Press, NY. pp 314 - 388.

Powell, A. A., Oliveira, M. De A and Matthews, S., (1986a). Seed vigour in cultivars of dwarf French bean in relation to the colour of the testa. Journal of Agricultural Sciences. Cambridge. 109 : 419 - 425.

Salter, P., (1985). Crop establishment. Recent research and trends in commercial practice. Scientific Horti. 36 : 32 - 47.

Sanhewe, A. J. and Ellis, R.H. (1996). Seed development and maturation in *Phaseolus vulgaris*. Journal of experimental Bpotany.47 : 949 - 958.

Sun, Q., Wang, J.H., Sun, B.Q., (2007). Advances in seed vigour, Physiological and Genetic Mechanisms. Agril Sci. China. 6 (9): 1060 - 1066.

TeKrony, D. M., Egli, D. B., and Wickham (1989). Seed vigour effect on tillage or no tillage. Field performance. I. Field emergence. Crop Scienc. 29 (6) : 1523 - 1528.

Tekrony, D.M. and D.B. Egli, 1989. Corn seed vigor on No-tillage field performance, Plant growth and grain yield. Crop Sci., 29: 1528-1531

TeKrony, D. M., and Egli, D. B., (1991). Relationship of seed vigour to crop yield, A Review. Crop Science. 31 : 816 - 822.

TeKrony, D. M., and Egli, D. B., (1977). Relationship between laboratory indices of soybean seed vigour and field emergence. Crop Sci. 17 : 573 - 577.

Woodstock, L.W. (1969b). Seedling growth as a measure of seed vigour. Proceedings of International Seed Testing Association. 34 : 273 – 280.

Wilson, D.O.; McDonald, M.B. 1986. The lipid peroxidation model of seed ageing. Seed Science and Technology 14: 269-300.

Zanakis, G.N., Ellis, R.H. and Summerfield, R.J. (1994). Seed quality in relation to seed development and maturation in three genotypes of soybean. (Glycine max). Experimental Agril. 30 : 139-156

Zhang, Z.H., Yu, S.B., Yu, T., (2005). Mapping QTLs for seedling vigour using RILs of rice. Field crop Research. 91 (2-3): 161-170.

Zhou- fei Wang, Jian fei Wang., Yong-mei bao., Fuohua Wang., and Hong- Sheng Zhang., (2010). Quantitative trait loci analysis for rice seed vigour during the germination stage. Jour. Zhejiang Univ Sci B. Dec. 11(12): 958-964. Doi. 10. 1631/ jzus.B.

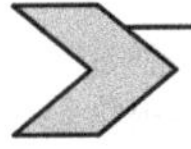

CHAPTER 4

Principles and Elements in Seed Vigour Testing

(Self trust is the first secret of success)

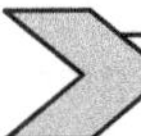

4.1 Introduction

On global basis, irrespective of the economic status of the country whether developed / developing or under developed, the soil and planting conditions are always suboptimal wherever agriculture is important and practiced for deriving social and economic benefits. It is estimated that around 80% of economically important crops are planted with direct or indirect use of seeds. Under such suboptimal field conditions usually the standard germination test results only overestimate the field emergence potential (Delouche and Caldwell, (1960); Johnson and Wax, (1978); Yaklich *et al.*, (1979). Even a moderate to severe stress, imposed by environmental conditions on seed is likely to either delay or prevent germination and thereby result in the establishment of plant stands that are far below the expectation from the results of germination as reported on tags and labels. The inability of seeds of desired cultivar to emerge rapidly and uniformly so as to ensure high plant performance in terms of yield and quality of produce emphasizes the need for production and supply of quality seed to meet the demand in sufficient quantities.

4.2 Aims of Vigour Testing

Vigour tests aim to identify and measure the performance of seed lots at a higher probability both during storage and after sowing even under unfavourable conditions so as

(i) to discriminate between different seed lots for their suitability for storage for varying lengths of time to meet short, medium and long term needs and

(ii) to discriminate between seed lots for planting value in relation to optimum environment, as high, medium or low in vigour, i.e., to promote synchronous emergence or maximum performance under suboptimal seedbed conditions or by adjusting the seed rate.

Vigour of seed lots is tested in the laboratory just "to know and predict their performance in the field well before planting". But in reality, several delicate justifications involved are as follows.

(i)   Seed lot is composed of a population of individual seed units; each one is unique with its own capability to produce a mature plant.

(ii) Vigour test is only a procedure for evaluating seed lots under standardized conditions. It enables a seed producer to determine and compare the vigour of different seed lots before marketing.

(iii) It enables the end user to compare seed lots before making purchasing decisions.

4.3 Goals of Seed Vigour Testing

The basic objective of seed vigour testing is to provide clear cut recognition of key differences in physiological potential between commercial seed lots having similar germination percentage. Therefore, an efficient seed vigour test demands higher sensitivity to differences in physiological potential that could not be detected by any viability tests. Further, a well-organized seed vigour test is also expected to have the capability to position the seed lots in tune with their performance potential. The goal of seed vigour testing can be broadly divided in to common goal and ultimate goal, based on the stakeholders, as detailed below.

4.3.1   The Ultimate Goal

According to Hampton (1991), the ultimate goal of seed vigour testing, from the view point of both Seed Technologists and Seed Physiologists is to understand enough about the variables, which may limit any aspect of seed performance, so as to be able to avoid vigour loss through the production systems.

4.3.2   The Common Goal

Keeping in view the necessity, the common goal is fixed from seeds men and seed growers' point of view, i.e., to produce and also to have access to high quality seed material for planting purpose with a better index than germination percentage so as to identify high and low vigour seed lots in a comparable way to have better field seedling emergence.

4.4 Basis of Vigour Testing

The basis of vigour testing lies in the assumption that 'Almost all seeds undergo a sequential loss in cell function, that ultimately culminates in the loss of germinability'. Therefore, all those events, that precedes loss of germinability could also serve as a basis for the loss of vigour and ultimately the vigour tests. So, the earlier the parameter is measured during the loss of germination, the more sensitive will be the index of seed vigour. Since, the degradation of membranes precedes loss of germination, it is always better if one selects a test that is capable of dealing with and consistently monitors the integrity of the membranes.

 ## 4.5 Principal Objectives of Seed Vigour Testing

1. The basic objective of seed vigour testing is basically to provide a more reliable identification of important differences in physiological potential among seed lots having similar germination percent-tage and commercial value and sensitive index of seed quality than the standard germination test.

2. The principal objective of seed vigour tests is to differentiate the quality of seed lot's with consistent ranking as high, medium and low vigour seeds in terms of potential performance. The seed lots thus identified with high vigour can be used for special purposes like: (i) in crops where uniform emergence is required, and (ii) for use in mechanized vegetable production systems.

3. Farmers need the information on seed vigour to make timely decisions on date and quantity of seed to be sown and expected uniformity of plant stand. Selection of high vigour seed lots reduces the chances of crop failure at seedling emergence stage itself when planted in suboptimal soil conditions and extremes of moisture and temperature.

4. Seeds men require this information because they are aware that loss in vigour precedes loss in viability and that vigour tests could aid in monitoring the quality of seed during production and conditioning.

5. Determination of seed vigour also helps to know whether the loss in vigour is occurring either when the seed is still on the mother plant intact or at harvesting, or during transportation or during storage, etc.

6. Estimation of seed vigour leads to adoption of improved marketing strategies by the seed companies that ultimately benefit the consumers.

7. Seed companies could also initiate certain preventive measures by either improving the storage conditions or disposing off the low vigour seed lots on top priority basis.

8. The vigour tests also provide an opportunity to evaluate physiological seed treatments aimed at enhancing the performance level of low vigour seed lots.

4.5.1 Requirements of Vigour Test

The chief requirements of an efficient vigour test should obviously include higher sensitivity to differences in physiological potential which could not be detected by viability of germination tests and the ability to rank lots according to their performance potential.

4.5.2 Key Elements of Seed Vigour

The success of seed germination and subsequent establishment of normal seedlings, indicating both economic and ecological significance, largely govern the capacity of propagation of a plant species. Germination is considered to be the most crucial phase in the life cycle of a plant due to its high vulnerability to injury, diseases and various kinds of stresses. The process of germination incorporates and gets initiated with the uptake of water by the dry matured seed and terminates with the protrusion of the radicle through various layers of seed coat and there after leading to seedling growth. Hence this process is very much complex because the imbibed mature seed has to quickly change from maturation to germination driven programme of development and prepare for the seedling growth Nonogaki *et al.,* (2010). The extensive field crop studies conducted recently are oriented at the development of 'Predictive Seedling Establishment Models' by Finch-Savage *et al.* (1993, 1998, 2001, 2005); Rowse and Finch-Savage (2002), and Whaley *et al.* (1999) led to the identification of three key features, a seed must possess in order to establish a seedling well across a wide range of field / seed bed conditions. They are

(i)   Rapidity in germination, Rapid initial downward growth to maintain contact with moisture in a drying seed, and high potential for upward shoot growth in soil of increasing impedance as the soil dries.

(ii) Basically, all these three features are aimed at reducing the time gap between sowing and seedling emergence particularly when seedbed deteriorates negatively and thereby shows its negative impact on seedling establishment significantly.

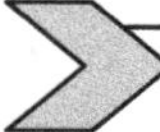

4.6 Criteria for Vigour Testing

Since seed vigour can be considered as a combination of characteristics or independent attributes of physiological potential such as (i) speed of germination, (ii) seedling growth, (iii) ability to germinate beyond optimum conditions of temperature, and other aspects of stress tolerance that are likely to resolve the high latent performance consequent to sowing in the field. No doubt, this particular situation is bound to hinder the establishment of a precise definition as several interlinked factors are involved in the composition and manifestation of seed vigour. As a result, several criteria and techniques have been developed for determining its assessment both directly and indirectly so as to establish a relationship with seedling emergence and storability through the evaluation of current metabolic state of the seed.   Contrary to this, some other tests have been designed to ascertain the level of tolerance to stresses.

According to ISTA Rules, all vigour tests must meet the following two basic criteria, which along with other things depend on plant species being tested.

1. Repeatability of the applied method i.e., sufficient uniformity of the terms under which the test is conducted in order to obtain results, which are approximately equal under these terms.

2. The relation between the result of vigour test in the laboratory and the results of seedling growth in the field should be approximately the same.

4.7 Predominant Views in Seed Vigour Testing

1. Some vigour tests are designed to identify seed performance (germination) after exposure to different kinds of stresses. Eg: Cold test (low temperature and high moist substratum); Accelerated ageing test (high temperature and high relative humidity); and controlled deterioration test (high seed water content), *etc.*

2. *Per-se* performance can also be evaluated either through estimation of enzyme activity, or changes in cell membrane structure and organization and / or seedling growth rate etc

 ## 4.8 Importance of Seed Vigour and Vigour Testing

Possession of high vigour status is essential for seed lots, as it determines the success of a crop by giving high quality seeds. The performance of any crop is determined, to a large extent, by the interaction of the genotype with its environment and simultaneously is under constant influence of vigour component of the seed quality. No doubt, improved varieties are highly essential for the best exploitation of full genetic potential. But, the use of seeds with low vigour as planting material not only slows down and lowers the growth at initial stages of crop establishment but also causes suboptimal plant population on one hand and ultimately lowers performance on the other hand. Therefore, use of highly vigourous seed only at a little extra cost per unit area is advocated in F_1 hybrids of vegetable Cole crops like carrot, lettuce, etc., that offers early emergence coupled with earliness and uniformity in produce and ultimately economic gains.

4.8.1   Importance of Seed Vigour

Even the high germinating seed lots also exhibit substantial differences in their field emergence capacity when sown at the same time and in the same field. Further, the seed lots may also differ in performance after storage in the same environment for sometime or transport to the same destination. Probably, either the germination results are wrong or else it appears that the germination test is not sensitive enough to indicate the subtle but significant quality differences, because these differences are caused by another parameter of quality referred to as vigour.

Vigour testing assumes much importance in crop production in general and seed technology in specific for the following reasons.

1. The sole purpose of conducting seed vigour tests is to make a distinction between different seed lots that are vulnerable to suboptimal conditions either at soil level or in the form of extreme moisture and temperature.

2. Basically, the vigour tests, when conducted on low vigour seed lots, extend an opportunity for evaluation of physiological seed treatments aimed at enhancing shelf life and performance.

3. Identification and use of high vigour seed lots minimizes the chances of crop failure at emergence itself.

4. Use of high vigour seed lots in suboptimal conditions promotes the reduced application of seed dressing chemicals and thereby reduced soil and water pollution with chemicals.

5. Identification of high vigour seed lots facilitates their use in high end and special purpose mechanized vegetable production systems where achieving synchronized emergence, flowering, maturity are high priority

6. Vigour test results can be used as a means of alerting seed growers who have stored the seeds or seed users regarding the probable rapid decline in seed vigour and possible failure of germination well in advance.

4.8.2   Importance of Seed Vigour Testing in Agriculture

The result of a germination test, which provides optimum conditions to the growing seedlings, though classify seeds as viable, but they may not be capable of continuing their lifecycle under a wide range of field conditions. Therefore, a vigour test is designed to measure the percentage of viable seeds in a sample. This test also reflects the ability of these seeds to provide normal seedlings under less than optimum or adverse growing conditions similar to those occurring in the field.

Vigour testing assumes importance for at least two obvious reasons as follows.

(i)   As a component of seed testing it provides a still more sensitive and better index than a standard germination test and

(ii)   Generally, seeds begin to lose their vigour well before they lose their ability to germination and hence vigour testing becomes an important practice in all seed production programmes.

Of late, seed vigour has been recognized as one of the most important components of seed quality encompassing potential seed performance both in the field and in storage. It means, in order to have a complete assessment of this quality trait of seed, the following three criterions should be taken into consideration carefully.

(i)   The capacity of seed to germinate and produce normal seedlings,

(ii)   The expected uniformity in field emergence and

(iii)   Potential storability

Seed lots with low seed vigour often fail to emerge rapidly and uniformly and produce a plant population just above the critical threshold level and hence indirectly responsible for the reduced yields. At the same time, one has to remember that germination test alone cannot give the exact prediction of plant stand establishment in the field. In other words, laboratory germination test is not a true indicative of field emergence potential of a seed lot. This is mainly due to the fact that germination test is being conducted in the laboratory under most favourable and artificial media/substrate conditions which are seldom available in the field. Further, under laboratory conditions, opportunity will be provided even to the weakest seed to germinate since it receives moisture, temperature and substrate not loaded with microorganisms, herbicide, fertilizer, fungicide and systemic insecticides. Further, a standard germination test is designed with a purpose, in the first count, most of the strong seedlings that have germinated will be removed while the final count is designed to provide a sufficiently longer duration of time to give a chance to germinate even the weakest seeds.

Further, the estimated laboratory germination percentage being the sum of strong and weak seedlings, and during germination analysis, since we consider only whether a seed is germinated or not, whether the seed has produced a visible radicle or not, the most frequently used physiological criterion of germination appears to be unsatisfactory as a measure of germination especially in the context of seed testing, because, the process of germination becomes scale less and even those germinable seeds may differ in their robustness in terms of field emergence. This is so because the seedlings that have emerged subsequently may be abnormal in their structure and hence are incapable of establishing a normal plant in the field. Therefore, it simply depends on our inability to measure or document the progressive nature of deterioration, because it exerts a major and direct impact on the stand establishment potential. As a result, the standard germination test values, in most of the cases are nothing but the over-estimates of actual field emergence potential of seed lots. These higher results, most probably, could also be due to the use of inert media in the quest for uniformity in test results.

International Rules for Seed Testing (ISTA) clearly specifies the adaptation of optimum conditions like planting media, duration of evaluation, temperature and criteria of classification for conducting the germination test on one hand, so as to determine the maximum germination potentiality of a given seed lot. During planting time when the field conditions are nearly optimum, the results of the standard germination test usually correlates with field emergence Perry, (1972); Egli and TeKrony, (1979). On the other hand, to make these results more reliable and reproducible, the seed testing laboratories adopt these conditions and rules too rigidly and in toto. Contrary to this, a farmer is interested only in knowing how the seed lot performs in his field rather than how they have performed in/or between the seed-testing laboratories.

 ## 4.9 Factors Limiting Seed Performance

Usually experiments are conducted in seed testing laboratories on Petri dishes or other appropriate media under all favourable and controlled conditions. Contrary to this, an agricultural seed bed environment is encased in a soil matrix wherein the planted seeds are exposed to and experience a variety of different stresses. Therefore, in order to understand and improve seed vigour and establishment, it becomes mandatory to understand the limiting factors based at field level for all species not only for percentage germination, seedling growth but also for speed and uniformity in emergence especially in respect of small seeded vegetable crops that too with epigeal germination, because it is the key for optimizing the seed rates.

Seed germination is said to have been occurred when the growth of radicle bursts through the seed and protrudes as a young root. **Koning, (1994)**. Whereas, a plant is considered established once the seedling becomes autotrophic and is no longer reliant on its endosperm,

Accordingly **Bouaziz and Bruckler (1998ab)** divided the physiology of establishment process in to three broad stages namely imbibitions, germination and seedling gro wth, and seedling emergence.

1. **Imbibition:** It is the first stage in the process of germination, where in water penetrates the seed coat and softens the dry and hard tissues inside the seed, activates the GA3 inside the embryo. The rate of water uptake is primarily controlled by hydraulic conductivities of soil, seed epidermis and the extent of seed soil contact. At times, despite normal imbibitions, subsequent stage of germination is controlled due to wet, nearly saturated conditions where in oxygen diffusion may also become a limiting factor. The Gibberellic acid triggers cells of aleurone layer surrounding the endosperm and facilitate secretion of amylase enzyme that acts as a catalyst for the hydrolysis of starch in to its components of sugars. Seedling growth and emergence involves development of radicle and coleoptiles.

2. **Seed related factors affecting germination and emergence:** These factors include seed size, dormancy, parent crop nutrition and autotoxicity.

   a. **Seed Size:** Growth and vigour of several species are affected by seed size and weight. Increasing seed size had a small but positive effect on final germination percentage of wheat. **Naylor (1993).**

b. **Dormancy:** A small degree of dormancy is of great value in cereals that prevent pre-harvest sprouting in high rainfall areas. Seeds are shed from their respective mother plants and remain in /on the soil till a time and place best suited are available for establishment of a new plant. In most species, seeds adapted to natural conditions are shed in dormant condition, **Baskin and Baskin, (1998); Finch-Savage and Leubner- Metzer (2006).** The depth of dormancy is not fixed at shedding and is determined by the genotype and maternal environment and is altered further by environmental conditions prevailing at shedding **Fotitt *et al.*, (2011, 2013).** The dormancy mechanisms adapted in different species result in germination at different times of a year and that too under different environmental conditions. Dormancy can be overcome by sorting seeds/ grains for extended periods or warm drying and warm storage is known to accelerate decline in dormancy.

c. **Parent crop management:** Increased nitrogen application to parental plant in early stages may result in proliferation of shoots and large quantities of partially filled grains at the end of the season. Contrary to this, application of nitrogen in the later stages of crop development or deposition of storage proteins.

d. **seed vigour:** Differential partitioning of resources to roots and shoots between cultivars will result in the better emergence of fast rooting lines. The ability to grow fast and strong in fact results from the capacity to generate greater mechanical forces. Therefore, the ability to unconditionally generate this foe across a wide range of stress conditions can be defined as the vigour of a seed. Hence, vigour can be considered as a mechanical driven crop trait.

e. **Allelopathy and Autotoxicity:** The former refers to beneficial and detrimental biochemical interaction between plants. **Molisch (1937),** whereas, the later refers to an inter-specific type of allelopathy that occurs when a plant species releases some chemical substances that inhibit / delay germination and growth of same plant species, **Putman (1985).** By releasing characteristic allelochemicals a plant species prevents the growing of other seeds and seedlings in its own proximity and compete for resources.

3. **Germination and Pre-emergence growth:** Germination is a mechanically driven process usually manifested by opposing forces of intercellular turgor pressure, considered to be the driving force for expansion by cell walls of germinating embryo and depends on the availability of water within the heterogeneous soil matrix. Cellular turgor promoting cell expansion depends on the available water and how tightly it is bound by soil. The relationship between turgor and seed vigour is poorly understood and is partly limited by the inability to measure using a pressure probe. A decrease in water and osmotic potentials to maintain the turgor was reported by **Cavalieri and Boyer, (1985).** Availability of alternative mechanisms to alter turgor in seeds and seedlings facilitates generation of mechanical force necessary for increasing seed vigour. Further, genetically coded enzymes, such as expansion, xyloglucan, and endo-trans-glucosylase **(Rose *et al.*, (2002)** and pectin methyl esterase, **Peaucelle *et al,* (2015)** secreted by cells offer a link between cellular signaling and plant growth and modifies the biophysical properties of the cellwall. The role of cellwall modification in the control of germination of Arabidopsis seed was demonstrated by **Muller *et al.*, (2013).** The role of cell wall remodeling enzymes that drive embryonic growth, seed germination and vigour was poorly understood. **Li *et al.*, (2011, 2013)** demonstrated that germination and seedling growth in Arabidopsis and tobacco which were under the control

of physical properties (osmotic stress) of cellwall germination and seedling establishment stages only represent the downstream targets of seed quality and vigour.

4. **Abiotic factors:** The soil seed bed environment is so complex wherein the seeds get exposed to multiple stresses, **Hadas, 2004;** Seedlings are extremely sensitive to the physical stresses such as available water content, mechanical impedance, oxygen and temperature etc., that interact with each other and impose restrictions during germination and seedling expansion. **Whalley and Finch- Savage, (2006).** Neither the seed nor the seedling is directly sensitive to water content.

Hence, in order to obtain a robust seedling, it is necessary to overcome the above mentioned interactions, either by wetting or drying the seedbeds as the case may be. Water stress and mechanical impedance are the two stresses limiting the germination and emergence, **Whalley and Finch-Savage, (2010).** The initial uptake of water causes imbibitional damage in grain legumes resulting in loss of membrane integrity, either due to lack of intact testa or seeds with low vigour due to ageing. **Powell, (1985).**

(a) **The Soil climate:** Reduced oxygen diffusion to seeds and seedlings in nearly saturated soils result in poor germination due to the presence of thick water films. **Milthrope and Moorby (1974).** Similar to Critical seed water content **(Richard and Guerif,1988).** Critical quantity of oxygen must also be consumed by seed in order to reach germination as the same is required for respiration in all biochemically active cells during seedling growth. Soil compaction and water logging result in reduced oxygen availability and directly affect both seed germination and establishment.

(b) **Available / Excess water:** Seeds and seedlings are not sensitive to the water content of soil per se. Availability of water, the sum of matric potential and osmotic potential is measured as water potential (MPa). It affects early plant growth by limiting aerobic respiration. **Shepherd (2003)** reported poor emergence and high seedling losses associated with November and December sowings could be attributed to low temperatures resulting in slow growth process.

(c) **Soil water potential and Seed – soil contact**: Usually germination and emergence are more affected by soil moisture in several parts of the world than any other factor. Soil water retention characteristics differ between various soil types. Soil moisture fluctuates rapidly in surface layers where seeds are sown. Complete germination will occur only when there is adequate moisture in the soil for subsequent growth. In saline soils, osmotic potential is sufficient to affect water uptake by seeds.

But in majority of cases matric potential will determine the availability of water to seeds in the soil. Critical seed moisture content must be achieved in order to occur imbibition **Bruckler (1983)**, which in-turn demands a degree of seed –soil contact to facilitate water movement in to the seed and is determined by the aggregate size of soil particles distributed in the seed bed. **Dexter (1988)** reported increasing seed soil contact achieved through using light rollers/ press wheels enhances the proportion of imbibition in liquid phase, yet this excessive pressure result in soil compaction of the seed bed and in turn may delay germination by reducing oxygen availability. **Bouaziz and Bruckler (1989b)** reported that negative water potentials only increase the time taken to emerge but not the final percentage emergence between -0.02 to 1.31MPa. The mucilage produced by mixospermic seeds with hydrogel properties hold water around the seed and facilitate water uptake during germination especially under salt stress, **Western (2012).** A strong evidence is available regarding effective uptake of water in the vapour phase, **Wuest et al,. (1999).**

(d) **Depth of sowing**: Failure of seeds to emerge was attributed by some researchers to the depletion of seed reserves before emergence. Large seeds are more capable of emergence from deep sowings, **Gan et al., (1992).** The experiments by **Bouaziz and Hicks(1990)** too revealed that poor plants in wheat are not due to the depletion of seed reserves but could be due to the length of coleoptiles, mechanical resistance imposed by soil surface, soil moisture content and even lack of oxygen for respiration under waterlogged conditions. Whereas**, Kerby (1993)** reported non-significant effect of establishment due to depth of sowing of wheat in the range of 2.3 to 8.3 cm. Contrary to this **Mahdi et al ., (1998)** reported negative effects of deep sowings on establishment and concluded that final emergence and seedling vigour will reduced by deeper sowings and 6 cm depth was optimum and is in agreement with most other works.

(e) **Temperature:** Temperature is a key variable in germination and seedling growth. Solar radiation determines the temperature of seedbed to a great extent. Evaporative cooling and soil water content are the two factors that influence soil temperature. On the other hand, soil temperature partly exerts its effect on the rate of water uptake through water temperature and its viscosity during imbibitions. The seed becomes physiologically active consequent to imbibitions. At this stage temperature affects the kinetics of seed germination by controlling the rate of chemical reactions inside the swollen seed. A clear relationship between temperature and rate of emergence in wheat was established and concluded that high soil temperatures are associated with low soil moisture **Lindstrom (1976).** The soil water content is in fact acts as a buffer in summer and prevents fluctuations. Hence, using soil temperature as an indicator of establishment is inappropriate particularly at higher soil temperature range.

(f) **Oxygen:** Temperature together with soil moisture greatly influences the activity of soil microbes which determine the supply of oxygen in the seed bed. In hot wet conditions oxygen stress will be very high. In general oxygen sensitivity varies between species. Seeds of monocot species with high starch content are less sensitive to oxygen than dicot species seeds with high lipid content.

(g) **Soil colour:** Dark coloured soils are warmer than light coloured soils as they absorb and retain more of solar radiation and hence can be directly linked to soil moisture.

(h) **Soil strength:** It is due to capillary pressure of water in the pores holding the soil particles together and hence unlikely to affect the germination of seeds. But increasing strength has a negative impact on the rate of elongation of roots, Jin et al., (2013). Decrease in root and shoot elongation is a linear function of water stress rather than soil strength, **Whalley and Finch- Savage, (2010).** The strength of soil increases rapidly as the soil dries and renders the expansive growth of seedlings more sensitive.

(i) **Soil structure / type and Tillage**: The term soil structure refers to the arrangement of soil particles and pores, **Braunack and Dexter (1989).** Depending on the distribution of aggregate sizes in seed bed crop establishment differs. Clay soils with greater capillary pressure have high degree of saturation at a given matric potential compared to sandy soils. Depending on the soil moisture content, tillage alters particle size. Timely tillage can alleviate soil compaction. However, repeated tillage can cause deterioration in soil structure **(Greenland (1972).** Differences in soil structure exert direct impact on the rate at which physical stresses change with water content. Moreover, all types of soils are not appropriate for direct drilling. A well structured soil provides a relatively weak seed bed when dry and strong when wet and under ideal conditions soil structure helps to minimize water loss by evaporation and hence acts as a weak growth environment. Contrary to this, impeding soil crust following heavy rain fall in case of poorly drained soils, direct drilling causes the walls of the drill slits to smear and restricts oxygen diffusion to seed. **Lynch _et al._, (1981).** Using a plough and press is considered more appropriate form of tillage on sandy, light loams, light and medium silts and peat soils. Shallow tillage is more suitable under dry conditions on medium loams and clays, unless the problem of weeds and drainage is present, **Davis (1988).**

(j) **Mulching and straw residues**: Mulching and straw residues are expected to exert a direct impact on seedling establishment not only by causing mechanical problems at drilling **(Cannel, 1983)** but also release allelopathic chemicals capable of affecting seedling growth. **Guenzi, (1967)** reported that the degree of compaction of straw exerts the allelopathic response on succeeding crop. Hence, maintenance of adequate time gap between straw decomposition and succeeding crop sowing appears to be a prerequisite for the healthy growth and development of succeeding crop.

(k) **Mechanical impediment:** It is also known as soil crust formation and is a consequence of increased soil strength that results in seed bed deterioration. In order to overcome this problem, seedlings have to generate greater force, which in turn result in decreased extensibility of the cell wall. Application of physical impedence has resulted in 36% increase in hypocotyls cross sectional area in snap beans, **Taylor and Tan Broeck, (1988);** In onion and carrot, **Whalley _et al_ ., (1999)** reported redistribution of seed reserves and consumption more seed reserves in both roots and hypocotyls per unit length of thickening. Whereas increased stiffening of cellwall of growing pea roots was reported **Croser et al.,** (2000).

4.10   Why Vigour Testing?

Vigour testing not only measures the percentage of viable seed present in a sample drawn from a seed lot, it also reflects the ability of these seeds to produce normal seedlings under less than optimum growing conditions usually prevailing under field conditions. It means, when the seedbed conditions are ideal and favourable, the vigour of seed lots will be closer to germination and hence the vigour test result will be closer to the germination and ultimately the result is more likely to be consistent with the field emergence. It means, the test information helps the end users directly by indicating how their seed will perform if subjected to stressful field conditions. In addition to the above, the activity of enzymes in immature and dormant seeds, rate of respiration of seeds during harvesting and in storage, genetic damage and accumulation of toxic metabolites etc., are also likely to speed up the process of deterioration. Therefore, all these parameters make vigour testing a very complicated, interesting and a sophisticated tool in the kit of Seed Technologists to reveal the nearly true quality of seed lots and thereby help the end users as a guide for avoiding planting under unfavourable seed-bed conditions.

Therefore, in order to predict seedling emergence under a wide range of field conditions in a better way, two or more tests of germination and vigour are combined to construct vigour test indices. To estimate the seedling emergence, per cent vigour rating is calculated as an average of standard germination percentage; shoot length and seedling fresh weight (Woodstock, 1969). However, not even one best combination could be identified so far for all agronomic situations, as the predictive accuracy varied between 50 - 60 %.

Subsequently, statistical methods were also employed to detect the best combination of vigour tests for a better estimate of wheat seedling emergence. For accurately predicting the field emergence, a combination of seedling growth tests showed some promise. Over a period of time, seed technologists have proposed several tests. Of which, only a few tests were able to gain the acceptance of both, by the seed analysts and seed testing organizations alike (AOSA, 1983). Cold test for *Zea mays L.* and Conductivity test for *Pisum sativum* only could gain wider acceptance, while others have been proved to be of value confined to a local context only. Hence, the vigour tests are therefore exclusively crop specific, and hence a combination of methods may be preferred in specific cases as compared to a single test method.

The vigour tests thus developed are unable to account for the variability encountered in the field emergence. Despite considerable effort put forth over a period of time to identify a seed vigour test(s) that accurately predict(s) the emergence potential in the field could not be successful. Therefore, any seed vigour test must be able to provide a more reasonable and sensitive index of seed quality than the standard germination test. Therefore, it is natural and understandable that the evaluation of seed vigour obviously demands more rigorous control of test variables as well as interpretation criteria so as to obtain consistent results in the first instance. On the other hand, it is also highly essential that seed analysts, seed technologists, consumers and all those connected with seed vigour testing should understand that seed vigour tests are neither designed for predicting the exact number of seedlings that are likely to emerge in plant stand in the field after sowing nor for how long a seed lot will be in a position to maintain the required percentage germination 'x' (90 or 95%) during storage for "n' (required period of storage) months / years.

Precise interpretation of vigour test results are expected to provide information that not only facilitates selection of better seed lots but also determines the required quantity of seeds, their distribution during storage and a clear cut estimation of probable commercial success under diversified field conditions. As, seeds begin to lose vigour much before they lose their ability to germinate, this obviously makes vigour testing an important practice in all seed production programmes. In respect of carry over seeds, testing for vigour becomes more important particularly when seeds were stored under unknown conditions or unavoidable storage conditions. Further, seed vigour testing is also used as an indicator of the storage potential of a seed lot and also in ranking vigourous seed lots with different qualities.

One should always remember that vigour tests will only be relevant where there is a problem of seed vigour within a species. Not all species show differences in vigour for a variety of reasons. Therefore, it is not relevant to apply vigour tests to all species without a problem of emergence or storage potential having been described in that species.

4.11 Principles of Seed Vigour Testing

All seeds, being living entities, deteriorate in quality with time in accordance with the prevailing environmental conditions. Both the length of storage as well as existing environmental conditions will determine the rate of deterioration during storage. The basic principle involved in vigour testing therefore is, if the seed is really vigourous, it will withstand one or all of the prevailing stresses and gives rise to plants as if it were on stimulants. Contrary to this, in case of seeds with moderate to low vigour, the stress applied will automatically suppress the seedling growth. If the seedbed conditions are favourable, the vigour will be closer to germination.

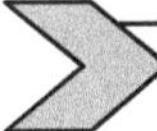

4.12 Nature of Vigour Tests

Seed ageing is one of the major and specific causes for differences in seed vigour, the theoretical background of most of the vigour tests obviously lies in seed ageing and hence involves the application of one or more consequences of ageing.

Vigour testing, no doubt, is a challenging task and hence the success lies in the identification of more than one quantifiable parameters that are common to seed deterioration and ageing which usually starts as soon as the seed attains physiological maturity. Therefore, it is imperative, that all seeds must be handled carefully to prevent the accelerated reduction in performance through the physical damage to the cells and cell membranes. This is particularly true especially in case of large seeded legumes. Beyond doubt one can say that, all those changes that are occurring during seed deterioration are not yet understood fully, but one can only and probably speculate the sequence of events.

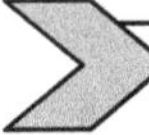

4.13 Characteristics of a Seed Vigour Test

The moment the seed(s) attains physiological maturity, even when they are still intact on the mother plant itself; the process of ageing also begins instantaneously and is likely to continue till the seed is sown in the field in the ensuing season. So, in-between physiological maturity and sowing, including the period of storage and testing, the seed will be in a state of continuous / progressive deterioration. This particular situation thus makes the ability to quantify and detect changes in seed vigour not only difficult but also a challenging task, since vigour test (s) is / are designed to provide a more sensitive index of seed quality than standard germination test and is also supposed to consistently rank the seed lots in terms of potential performance in the field and or in storage. Accordingly, a vigour test should possess certain essential characteristics so as to make it more reliable and useful to both seed producer as well as the seed end user. The characteristics of a seed vigour test are as follows.

1. **Theoretical Background**

   First of all, ensure that the proposed seed vigour test should be based on sound theoretical background such that the outcome of any test could be explained with ease and even the unexpected results could be identified more easily, and further examined critically (Mathews, 1980). But, at the same time more and more emphasis should be laid on the technical aspects and requirements.

2. Simple (Uncomplicated)

Since seed vigour testing assumes prime importance in respect of both farmers and seed marketing point of view, as far as practically possible the vigour test procedure should be too simple to complete and report by the existing seed testing laboratories' staff. It should neither demand any additional staff with special background and / or specialized training and hence should be applicable in a wide range of seed testing environment.

3. Relatively Inexpensive

A vigour test should neither demand too much labor, nor too costly equipment and consumables because all these add only to the cost of the test. It is mandatory to fix the price of the test at a reasonable level duly keeping in view the limited budgets provided for the purpose of seed testing and at the same time, the test should be economically practical and feasible too.

4. Rapid

In the era of season bound and location specific varieties, growing of crops within a specified time frame to catch the early market, taking up of sowing operations have become so rigid and confines to a specific period of season only. Therefore, as far as practically possible, the vigour test should be completed within the shortest possible time for the following reasons.

(i)   The analysis of a large number of samples in a rapid succession during peak season saves the analysts' time,

(ii)   Helps in efficient utilization of germination (germinator's) space in the laboratory,

(iii)   Automatically renders the seed producers who desire a quick turnaround time for a competitive marketing advantage and

(iv)   Enables prompt reporting of results by seed testing officials.

5. Repeatability and Reproducibility

The test is to be used in many laboratories and the results from different laboratories would be comparable. Since all the vigour tests are based on sound scientific protocols and principles, give the same results if the test has to be repeated or if a second analyst does the assessment. If the results cannot be easily reproducible because of intricate procedures or even when the interpretation is subjective, it makes the comparison of results both within and between seed testing laboratories more meaningless. Therefore, the success of any seed vigour test always depends on its accurate reproducibility and repeatability and hence more care should be exercised in this regard.

6. Correlated with field performance

The ultimate value of any vigour test lies in its ability to precisely predict the field emergence or glass house emergence and storage house performance. Most of the definitions of seed vigour, hence, emphasize only on the relationship between vigour and field performance, and this was confirmed by several studies. The results of laboratory test should be nearly equal to its field performance.

7. Objective

The vigour test should be precise in its objective, in order to be easily standardized. A seed analyst, preferably and accordingly should use more quantitative method of assessment or numerical index of quality that always avoids subjective interpretations.

8. Unique

The test should be unique in order to be performed by already employed staff of seed testing laboratory. It should be crop specific as far as possible, and a combination of methods may be preferred in specific cases only.

 **4.14   Significance of Seed Vigour**

In order to exploit the full genetic potential of improved varieties, obviously the genotypes should be (i) pure genetically and (ii) essentially have high seed vigour component of seed quality. The genotypes interact with their environment and exert influence during germination and plant stand establishment. At field level, a reduction in seed viability can be compensated theoretically by using a higher seed rate, which in turn leads to practical difficulties due to strong non linear correlation between viability and vigour parameters. This could be attributed to the continuous decline in vigour of each seed before its ultimate death because the life span of seeds in a population exhibits normal distribution.

4.14.1 Significance of Seed Vigour Tests

The significance of vigour tests can be briefly summarized as follows.

1. The lable of "Vigiour Tested"

Provides only an assurance of seed quality. It does not neither necessarily mean higher yields. No one should not assume that automatically means the result of a sound vigour testing programme or implies higher vigour than other seed which is not promoted as vigour tested. Therefore, for the ultimate end user, the farmer, it would always be advantageous to know and assess the vigour status of different high germinating seed lots before making any decision as to which one to buy and which lot will be more likely to perform better under his own sowing conditions with one or the other forms of stress. The label vigour tested usually will not provide any information on expected field emergence value, but will only indicate whether a seed lot is of high, medium or low in vigour. Therefore, it is always better to select seed lots with high vigour, as they are more likely to perform better under sowing conditions without stress.

2. Seed Production Companies

Vigour tests are commonly employed by seed production companies to establish their own "in house quality standards", both on the (i) standing crop in the field by monitoring seed quality during different stages of production, and (ii) at different stages of post harvest processing operations like cleaning, upgrading and treatment etc., including storage.

This ultimately helps the seed production companies in identifying probable causes / areas where exactly the loss in seed vigour is occurring.   Further it also helps in initiating and suggesting suitable practices and control measures, which could if not improve seed vigour subsequently but could arrest further deterioration.

3. Seed Marketing

In the years of excessive production, seed companies can take a decision regarding which lot should be marketed first (high vigour lots can be stored for later use).

4. Seed storage managers

Similarly, a seed store manager can utilize this information on seed lot's vigour for taking better decisions pertaining to the (i) suitability of seed lot(s) for storage, (ii) the possible length of storage time and (iii) the conditions required and to be maintained for storing the seed lots.

5. Seed Exporters

The seed exporters can use this information to decide which seed lot will withstand the rigors of transportation without further impairment of the seed quality before it reaches the importing country within the stipulated time. It facilitates the exporters to take appropriate decisions regarding the mode of transport, duration required for transporting and the conditions required during transporting the material and how best the material could be delivered at the import point.

6. Revalidation and inventory management

Vigour tests are important particularly in inventory management, revalidation of certified seed lots and seed carryover decisions.

7. Quality assurance

Vigour tests in combination with standard germination tests can provide a greater degree of quality assurance.

 ## 4.15   Summary

Consequent to the initiation of concepts in the year 1950, comprehensive studies started a decade later and became popular with the development of several vigour tests. These vigour tests were constantly improved against standard efficiency and sensitivity giving rise to the procedures of standardization. Since then, it is the most researched subject in seed technology. As a consequence of progress made in standardization protocols, some tests have already been recommended with greater levels of confidence. The suboptimal planting conditions prevailing across the globe either prevent or delay germination and cause uneven establishment far below the expectation printed on labels and necessitates determination of the degree of aliveness just as important as whether or not the seed is alive. Initially the vigour tests were designed to mimic the existing natural stress conditions. Vigour testing is exactly an opposite of germination testing. Germination percentage is always higher as the test conditions are more forgiving.

Contrary to this vigour score, a lower number always, represents the lowest possible performance of seed lot under stressful conditions and offers a more sensitive measure of seed quality, an ability of seeds to give normal seedlings in addition to the percentage of viable seeds. Vigour test when conducted in conjunction with germination test offers a complete profile of quality of the seed lot and renders the end user in taking several seeding decisions.

The principle of vigour testing is very simple. If the seed is really vigourous, they will withstand the stress and grows as if on stimulants, or else stress applied on low vigour seeds will suppress the seedling growth. The goal of vigour testing can be determined from seed technologists or seed growers view. Whereas, discriminating different seed lots either for their suitability to varying lengths of storage or classifying them for planting value into high, medium and low vigour forms the basic aim of vigour testing. All seeds undergo a sequential cell loss and germinability. So all events that precede loss of germinability lead to loss of vigour and form the basis of vigour testing. In order to have a complete assessment of this quality trait of seed (i) capacity of seed to produce normal seedlings, (ii) expected uniformity in emergence and (iii) potential storability needs to be taken in to consideration.

A seed lot is composed of individual seed units capable of producing unique plants, a vigour test is only a standardized procedure for evaluating seed lots which enables the seed producer to determine and compare different seed lots before marketing and enables the user to compare seed lots before making purchasing decisions.

A combination of vigour and germination tests clubbed to construct vigour indices to predict the seedling emergence in a range of field conditions could only explain the predictive accuracy to the tune of 50-60 per cent. Though several tests were proposed, most of them are of local value and is confined to crop specific only.

4.16 Exercise

I. Answer the following questions?

1. List out the characteristics of a seed vigour test.

2. What is vigour test and explain why it is needed.

3. List out the sequence of changes occurring during seed deterioration with a flow chart?

4. Why vigour testing should become an important practice in all seed testing programmes?

5. What are the criteria required for complete assessment of seed quality?

6. List out the three key features required by a seed to establish seedling across wide range of seedbed conditions.

II. Write short notes on the following

1. Goals of vigour testing

2. Aims of vigour testing

3. Principal objectives of vigour testing

4. Criteria for vigour testing

5. Importance of vigour

6. Importance of seed vigour tests

7. Importance of vigour testing

8. Significance of vigour tests

9. Key elements for seed vigour

10. Basis of vigour testing

III. Fill in the blanks

1. A vigour test is designed to reveal a seed lot's ability to withstand different --------.
2. A vigour test is conducted on seed lot to determine the -------- and -------- of deterioration.
3. Vigour tests are designed to mimic poor seedling and seed bed conditions existing under -------- conditions.
4. In germination test seed is exposed to -------- and -------- conditions of temperature, moisture and light.
5. In a vigour test seed is exposed to stressful environment unfavourable for seedling -------- and --------.
6. If the seed is really vigourous, it withstands all -------- and grows as if on stimulants.
7. In case of low vigour seeds, the stress applied will -------- the seedling growth.
8. The common goal for seed vigour testing is fixed from -------- and -------- point of view.
9. Vigour tests discriminate between different seed lots for their suitability to varying periods of --------.
10. Vigour tests discriminate between different seed lots for -------- to promote synchronous emergence.
11. The earlier the vigour parameter is estimated, the better will be the -------- of seed vigour.
12. Laboratory germination test is not a true indicator of -------- of seed lots.
13. Standard germination test results often -------- of field emergence potential of seed lots
14. A vigour test enables seed producer to compare seed lots before --------.
15. A vigour test enables farmer/end user to compare seed lots before making --------.
16. Under favourable seed bed conditions, vigour test results are closer to -------- results.
17. Determination of seed vigour helps us to know where and when -------- is occurring.
18. Use of high vigour seed lots minimizes the chances of -------- at emergence.
19. Use of high vigour seed promotes the reduced use of -------- that ultimately control soil and water pollution
20. Reduction in seed viability theoretically can be overcome by --------.

▶ Answers

1.	Stress factors,	2.	rate and extent,
3.	Natural,	4.	favourable and optimum,
5.	emergence and establishment,	6.	stressful situations,

7. Suppress,
8. Seeds men and seed growers
9. Storage.
10. Planting value,
11. Index,
12. emergence potential,
13. Overestimates,
14. Marketing.
15. purchasing decisions,
16. germination test,
17. deterioration,
18. Crop failure,
19. Seed treatment chemicals,
20. higher seed rate.

4.17   References

AOSA (1983). Seed Vigour testing Hand Book. Contribution No. 32 to the Hand Book of Seed Testing. Ed. Clark, B.E.., McDonald, M.B., and Joo, P.K., pp 88., Lincoln. NE. AOSA.

Baskin CC, Baskin JM. 1998. *Seeds – ecology, biogeography, and evolution of dormancy and germination*. San Diego: Academic Press.

Bouaziz, A and Bruckler (1989a) Modelling of wheat imbibition and germination as influenced by soil physical properties. In Soil Sci. Soc. Am J. , Vol 53, Jan-Feb. pp 219-227.

Bouaziz, A and Hicks, D. (1990) Consumption of wheat seed reserves during germination and early growth as affected by soil water potential. Plant and soil 128, 161-165.

Bouaziz, A and Bruckler (1989b) Modelling wheat seedling Growth and Emergence: 1.seedling growth affected by soil water potential. In Soil Sci. Soc. Am J. Vol 53, Nov-Dec..pp-1832-1838.

Braunack MV, Dexter AR. 1989. Soil aggregation in the seedbed: a review. II. Effect of aggregate sizes on plant growth. *Soil and Tillage Research* 14, 281-298. 39

Bruckler, L. (1983) Role des proprietes physiques du lit de semences sur l'imbibition et la germination. II Controle experimental d'un modele d'imbition des semences et possibilities d'applications. Agronomie (Paris) 3(3): 223-232.

Cavalieri AJ, Boyer JS. 1982. Water potentials induced by growth in soybean hyopcotyls. *Plant Physiology* 69, 492-496

Delouche, J. C., and Baskin, C.C., (1973). Accelerated ageing techniques for predicting storability of seed lots. Seed Science and Technology. 1: 427 - 452.

Delouche, J.C. and Caldwell, W.P. (1960). Seed Vigour and Vigour Tests. Proceedings. Of AOSA, 50: 124-129.

Dexter, A.R. (1988) Advances in characterization of soil structure, Soil & Tillage Research, Volume: 11, Number: 3/4, 199-238.

Davies, B. (1988) Reduced cultivations for cereals, HGCA research review, No 5.

Egli, D. B., and TeKrony,D.M., (1979). Relationship between soybean seed vigour and yield. Agronomy Journal, 71: 755 - 759.

Finch savage, W.E., Come,D., Linn, J.R., Corbineau, F., (2005). Sensitivity of *Brassica oleracia* seed germination to hypoxia: a QTL analysis. Plant Sci. 169: 753- 759.

Footitt S, Douterelo-Soler I, Clay H, Finch-Savage WE. 2011. Dormancy cycling in Arabidopsis seeds is controlled by seasonally distinct hormone signalling pathways. *Proceedings of the National Academy of Science* 108: 20236-20241

Footitt S, Huang Z, Clay H, Mead A, Finch-Savage WE. 2013. Temperature,light and nitrate sensing coordinate Arabidopsis seed dormancy cycling resulting in winter

Finch-Savage WE, Leubner-Metzger G. 2006. Seed dormancy and the control of germination. *New Phytologist* 171, 501-523.

Finch savage, W.E, Phelps, K., Steekel, J.R.A., Whalley,W.R., and Rowse, H.R., (2001). Seed reserve dependent growth response to temperature, water potential in carrot (*Dacus carota* L.). J. Exptl. Bot. 58: 2187 -2197.

Finch savage, W.E., Steekel, J.R.A., and Phelps, K., (1998). Germination and post germination growth to carrot seedling emergence. Predictive technology models and sources of variation between sowing occasions. New Phytologist. 139: 506-516.

Finch-Savage, W.E., Grange, RI., Hendry GAF., AthertonNm., (1993). Embryo water status and loss of viability during desiccation in the recalcitrant species *Quercus robur* L. In: 4th International workshop on seeds. Basic and pplied aspectsof seed Biology (ed.).Come, D and Carbineau, F., pp 723-730. ASFIS, Paris.

Greenland, D.J. (1977) Soil damage by intensive arable cultivation:temporary or permanent? Philosophical Transactions of the Royal Society. London, (B) 281, 193-208.

Guenzi, W.D., McCalla, T.M. and Norstadt, F.A. (1967) Presence and persistence of phytotoxic substances in wheat, oats, corn and sorghum residues. Agronomy Journal 59: 163-165.

Hadas A. 2004. Seedbed preparation – The soil physical environment of germinating seeds. In: Benech-Arnold RL, Sánchez RA, eds *Handbook of Seed physiology: Applications to Agriculture*, New York: Haworth Press 3-49.

Hampton, J. G., (1991). Herbage seed lot vigour. Do problems start with seed production. Journal of Applied Seed Production. 9: 87-93.

Li F, Xing SC, Guo QF, Zhao MR, Zhang J, Gao Q, Wang GP, Wang W. 2011. Drought tolerance through over-expression of the expansin gene TaEXPB23 in transgenic tobacco. *Journal of Plant Physiology* 168, 960-966.

Lu P, Kang M, Jiang X, Dai F, Gao J, Zhang C. 2013. RhEXPA4, a rose expansin gene, modulates leaf growth and confers drought and salt tolerance to Arabidopsis. *Planta* 237, 1547-1559.

Lindstrom, M.J., Papendick, R.I., Koehler, F.E. (1976) A model to predict winter wheat emergence as affected by soil temperature, water potential and depth of planting. Agronomy Journal, 68, 137-140.

Lynch, J.M., Harper, S.H.T and Sladdin, M. (1981) Alleviation by a formulation containing calcium peroxide and lime of microbial inhibition of cereal seedling establishment. Current Microbiology 5, 27-30.

Johnson, R.R. and Wax, L.M. (1978) Relationship of soybean germination and vigour tests to field performance. Agronomy Journal. Vol. 70 (2): 273-278.

Julio Marcos Filho., (2015). Seed vigour Testing: An Overview of the Past, Present and future perspective. Sci. Agric. v.72, n.4, p.363-374.

Koning, R.E. 1994. Home Page for Ross Koning. Plant Physiology Information Website. http://plantphys.info/index.html (4-30-2003).

Matthews, S., (1980). Controlled deterioration, a new vigour test for crop seeds. In. P D. Hebblethwaite (ed.) Seed Production. Pp. 647 - 660. Butterworths. London.

Molisch, H., (1937) Der Einfluss einer Pflanze auf die andere – Alelopathie. Jena, Germany. Fischer (Jena.

Muller K, Levesque-Tremblay G, Bartels S, Weitbrecht K, Wormit A, Usadel B, Haughn G, Kermode AR. (2013). Demethylesterification of cell wall pectins in Arabidopsis plays a role in seed germination. *Plant Physiology* 161, 305-316.

Milthorpe, F.L. and Moorby, J. (1974) An introduction to crop physiology. Cambridge university press, Cambridge.

Naylor, R.E.L. (1993) The effect of parent plant nutrition on seed size, viability and vigour and on germination of wheat and triticale at different temperatures. Annals of Applied Biology 123, 379-390.

Nonogaki,H., Bassel, G.W., Bewley, J.D., (2010). Germination- still a mystery. Plant Sci. 17: 9574-581.

Perry, D. A ., (1972). Seed vigour and field establishment. Hort. Abstr. 24: 334-342.

Powell AA. (1985). Impaired membrane integrity – a fundamental cause of seed quality differences in peas. In *The pea crop*, P.D. Hebblethwaite, M.C. Heath and T.C.K. Dawkins (eds). London: Butterworths, pp. 383-395.

Putnam A.R. (1985) Allelopathic research in agriculture: Past highlights and potential. In The Chemistry of Allelopathy: Biochemical interactions among plants. American Chemical Society Symposium Series No 268, pp.1-8. Ed. AC Thompson, Washington DC, USA: American Chemical Society.

Rose JK, Braam J, Fry SC, Nishitani K. 2002. The XTH family of enzymes involved in xyloglucan endotransglucosylation and endohydrolysis: current perspectives and a new unifying nomenclature. *Plant and Cell Physiology* 43, 1421-1435.

Richard, G and Guerif, J., (1988) Modelisation des transferts gazeux dans le lit de semence: application au diagnostic des conditions d'hypoxie des semences de betterave sucriere (Beta Vulgaris L.) pendant la germination. I. Presentation du modele. Agronomie 8, 539-547.

Rowse, H.R., and Finch –Savage, W.E., (2002). Hydrothermal threshold models can describe the germination response of carrot and onion seed populatios across both sub and supra optimal temperatures. New Physiologist. 158: 101-108.

Shepherd, M.A. (2003) A review of the impact of the wet autumn of 2000 on the main agricultural and horticultural enterprises   in England and Wales Review for the Department for Environment, Food and Rural Affairs (Project CC0372)

Whalley, W.R., Finch –savage, W.E., Cope, R.E., Rowse, H.R., and Bird, NRA., (1999). The response of carrot and onion seedlings to mechanical impedence and water stress at suboptimal temperatures. Plant Cell and Environment. 22: 229-242.

Whalley WR, Finch-Savage WE, Cope RE, Rowse HR, Bird NRA. 1999. The response of carrot (*Daucus carrota* L.) and onion (*Allium cepa* L.) seedlings to mechanical impedance and water stress at sub-optimal temperatures. *Plant Cell and Environme*nt 22, 229-242.

Whalley WR, Finch-Savage WE. 2006. Seedbed environment, In Black M, Bewley JD, Halmer P, eds *The Encyclopedia of Seeds: Science, Technology and Uses*. Wallingford: CAB International 599-602.

Whalley WR, Finch-Savage WE. 2010. Crop emergence, the impact of mechanical impedance. In: Glinski J, Horabik J, Lipiec J. eds. *Encyclopedia of Agrophysics*, Berlin: Springer-Verlag 163-167

Woodstock, L.W., (1969). Seedling growth as a measure of seed vigour. Proceedings of International SeedTesting Association. 34: 273-280.

Western TL. 2012. The sticky tale of seed coat mucilages: production, genetics, and role in seed germination and dispersal. *Seed Science Research* 22, 1-25.

Wuest SB, Albrecht SL, Skirvin. 1999. Vapout transport vs. seed-soil contact in wheat germination. *Agronomy Journal* 91, 783-787. , 367-378.

Taylor AG, Ten Broeck CW. 1988. Seedling emergence forces of vegetable crops. *HortScience* 23, 367-369.

Yaklich, R. W. and Kulilk, M.M. and Anderson, J.D., (1979). Evaluation of vigour tests in soybean seeds; Relationship of ATP, conductivity and radio-active tracer multiple criteria tests to field performance. Crop Science. 19: 806-810

CHAPTER 5

Differences and Manifestation of Seed Vigour

(Knowing others is wisdom, knowing yourself is enlightenment)

 **5.1 Introduction**

A vast majority of agricultural crop seeds, being desiccation tolerant, facilitate storage and transportation with minimal loss to their ability to grow. These seeds have been designed and loaded with full genetic component accompanied by viability, purity, and health, freedom from mechanical damage besides vigour to cater to the food and nutritional security needs of human as well as their domesticated animal populations in tune with the dynamic changes in the climatic conditions. Depending on the varieties within a species and species within a genera differ significantly in represent of their ability to loose their assigned physiological function within a specific time frame widely known as ageing which is initiated immediately after obtaining physiological majority and is likely to continue till sown in the field. inThe seed having planted in soil interacts with several biotic and abiotic factors before giving rise to a small radicle. Successful establishment of rapid, uniform and robust seedlings in the field not only determines the resource use

LEARNING OBJECTIVES

- How the vigour level of a genotype interacts with external factors to cause direct and indirect differences in vigour,
- Conditions governing the seed maturity,
- Various genetic parameters responsible for differences in variability of seed vigour,
- Physiological basis of deterioration of seed vigour,
- Impact of cumulative effect of free radicles on seed vigour,
- Role of field and storage fungi in reduction of seed vigour,
- Deterioration in vigour level due to ageing, imbibition damage and their interaction,
- Reasons for poor plant density establishment in the field by low vigour seed,
- Factors governing the field emergence of a seed lot and
- Morphological abnormalities during germination, establishment of low vigour seed lots.

efficiency but also determines the productivity and profitability in a given environment. Despite a better understanding in our knowledge on seed germination, still lot of gap in our knowledge exists in respect of manifestation of seed vigour. Therefore, understanding the inherent factors together with the external factors contributing directly and indirectly towards creation of differences in seed vigour, a complex trait is the need of the hour not only to have wisdom but also to enlighten our knowledge. A brief attempt is made in this chapter to understand the causes and effects of manifestation of seed vigour.

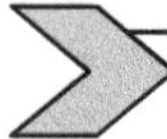 ## 5.2 Differences in Seed Vigour

The causes of variation in vigour are several and diverse, but still, it is the vigour level of a genotype which is of great importance, because of its interaction with several other external factors like moisture, temperature, soil fertility etc., and contribute both directly and indirectly towards differences in vigour.

5.2.1 Morphological Maturity Indices

Seed, during the course of its development, right starting immediately after fertilization, passes through a series of ontogenetic phases like cell division, differentiation, accumulation of nutrients and ultimately attains a maximum dry weight. Each seed nearly approaches an arbitrary stage during the course of development called 'physiological maturity' and could be defined as "accumulation of maximum dry weight." This stage usually coincides with the onset of maternal plant senescence. The physiological maturity is characterized variously in different crops. The morphological maturity indices together with approximate moisture content in some important crops are presented in table 5.1.

Table 5.1 Morphological maturity indices in some important crops.

S.No.	Crop	Maturity indices	Moisture content (%)
1.	Rice	Neither too ripe nor too green full grains, yellow hulls, panicles bend on their own weight	22-28
2.	Wheat	Loss of green colour from glumes and peduncle, development of dark layer at the crease of the kernels.	20-40
3.	Sorghum	Development of dark layer at the base of the seed.	25-30
4.	Bajra	Development of dark layer at the bottom of the kernels.	25-35
5.	Maize	Development of dark layer at the base of the kernels and development of milk line on the smooth side of kernels.	23-28
6.	Soybean	Leaf and Pod colour changes from green to yellow.	30-35
7.	Beans	Pods ripen and turn yellowish. Kernel skin becomes easily detachable.	30-35
8.	Groundnut	Older leaves dry and fall, top leaves turn yellow, kernels attain pink colour, inner side of the shells turns darker than outer shells	30-35
9.	Sunflower	Back side of the head turns yellow from green, bracts turn brown and a dark closing develops in seeds.	35-38
10.	Safflower	Formation of a dark layer at the base of the seeds.	30-35

The presence of high moisture content in the seed at physiological maturity neither permits mechanical harvesting, threshing and conditioning of the crop nor is safe for storage. In order to minimize pre and post harvest losses including processing and storage, one has to wait till the moisture content is reduced to a safer level, say 18% from 55% in maize. The time gap between physiological maturity and harvest maturity is often referred to as maturation. By the end of this maturation period, seeds attain their highest potential vigour in the absence of interaction from dormancy and other related factors. From this stage onwards, depending upon the ambient temperature and relative humidity, the vigour of seeds starts declining, albeit slowly in the beginning. At times, depending on the species, for one reason or the other, this may even start while the seed is still intact on the mother plant itself. This may also be due to the delayed harvesting and is likely to continue till the seed is planted in the ensuing season.

The changes taking place within the developing seed during the above mentioned phases are likely to exert influence, in one way or the other, on the potential performance of the seed. The intrinsic variation, considered to be the most important, is caused by the genotype and its interaction with its external environment and the nutrition provided by the mother plant. These intrinsic factors bring in variation in several internal morphological characters of the seed especially with respect to seed size, seed weight, density, development and stage of maturity at harvest, and so on. On the other hand, the extrinsic variation could be due to either mechanical damage, or microbial activity, or even deterioration due to ageing both while still on the plant and /or even during storage. Further, this could also be due to imbibitional damage caused to seed upon sowing in the field, or the interaction between ageing and imbibition etc. All these factors, which cause loss to viability, also play an important role in determining the vigour of the seed. The loss of viability, which is represented by the presence of a few living tissues in the seed, leads to loss of vigour and this loss of vigour ultimately, may be regarded as a precursor to the death of the seed.

Minimum Seed Certification Standards have already been prescribed for the maintenance of quality of seed in general and germination in particular, below which the seed cannot be offered for sale commercially. Seed germination test together with purity estimation provides the planting value of seed lots, which need to be presented on seed labels as mandated by law. This automatically ensures that, the failure of seed to emerge is not due to the inability of the seed alone. Despite this, differences in emergence of germinable seeds do occur when sown in poor field conditions and this was termed as seed vigour by Perry (1970). From this, it is very much clear that germination and vigour are not the same, vigour is only an indication of the ability of seed / seed lots to germinate and establish seedlings even under suboptimal conditions. Therefore vigour gives only an assurance as to the reliability of emergence especially for costly seeds of new varieties even in less than ideal sowing conditions. It means, different seed lots of a genotype or variety, harvested from the same area on the same date and time and registering similar high laboratory germination percentage are bound to differ significantly with respect to the levels of vigour and ultimately the plant population and yielding ability.

Vigour pertaining to seed or a seed lot is simply, the ability of germination, emergence and seedling growth rate controlled by a co-adaptive genotype in a given environment. In other words, seed vigour is simply a relative superiority in performance of a seed lot of a particular genotype over the seeds of the same or other genotypes in a given environment under well-defined experimental conditions. The meaning of seed germination from a 'Crop Botanists' or 'Plant Physiologist's' view is simply related to the emergence of radicle, contrary to this, a 'Seed Technologist views it is as whether or not the development of all essential structures that lead to the production of the seedling, which in turn will produce a normal plant. So, the difference between these two definitions can be called as vigour. In other words, the failure of the germination test to precisely predict the differences in the field emergence potential, particularly under poor field conditions, reveals the existence of another physiological aspect of seed quality and this has came to be referred to as seed vigour. Therefore, seed lots having poor emergence percentage in the field despite high laboratory germination are referred to as low vigour seeds, where as those giving good emergence are termed as high vigour seeds.

The progressive reduction in the performance capabilities of seeds could be attributed either due to (i) the changes occurring in the integrity of cell membrane(s), or (ii) the level of activity of various enzymes and ultimately the synthesis of appropriate proteins. Further, the changes at biochemical level may also occur either very rapidly in few days, or more slowly and may take even years, depending again up on several factors like (a) the genetic makeup of the species, (b) production environment and (c) the interaction of these two, which, of course, are not yet fully understood. Generally, the end point of this deterioration ultimately leads either to the death of seed or ends up in complete loss of germination. Usually, the seeds lose their vigour much before they lose their ability to germinate. For this reason, the seed lots that have registered high similar germination values differ significantly with respect to the level and extent of deterioration. Hence, the seed lots differ significantly in their seed vigour and therefore, differ in their ability to perform ultimately.

5.3 Factors Responsible for Differences in Seed Vigour

Seed vigour, an impressive term, encompasses around a number of quality traits. Seed vigour being a complex quantitative trait, it will be difficult to find out one or more genes that will have a dramatic impact on seed vigour. The maximum permissible vigour is determined by the genotype and is modified by the environment. This results in the production of immature, under sized and shriveled seed during maturation. The seed may be subjected to mechanical damage during mechanical harvesting, handling and storage and even the conditions of soils after sowing. Further, during transport and storage also the seed is likely to deteriorate due to the invasion of seed born / storage fungi. Therefore, factors responsible for loss of vigour can be broadly divided into (i) genetic (ii) environmental and (iii) Storage related factors.

5.3.1 Genetic Factors

The selection unknowingly practiced by our ancestors for thousands of years in the past and the deliberate selections practiced by the plant breeders of the recent past are only aimed at achieving higher yields either directly or indirectly. During their inadvertent attempts they have also improved characters like seed size, hardseededness, hybrid vigour, resistance to diseases, and chemical composition including protein content. The recent genetic studies in several crops have entitled a number of Quantitative Trait Loci that contribute to seed vigour Foolad *et al.*, (1995); Betty *et al.*, (2000); and Finch- Savage *et al.*, (2005). However, development of molecular markers associated with seed vigour should allow plant breeders to select for improved quality along with other desirable traits. The genetic factors responsible for differences in seed vigour can be further subdivided into three groups viz., variation within a genotype, variation at individual seed level and variation at population level, each of which can be further divided into several categories, and described in detail as follows.

5.3.1.1 Within a Genotype

Most of the variations in seed germination and emergence of seedlings in the field can be attributed to the climatic and edaphic conditions experienced by the mother plant during seed development or maturity at harvesting time. The environmental conditions prevailing at the time of seed filling may also exert affect on the supply of assimilates and other materials to seed situated in different positions in the head. The seeds being sensitive to desiccation can tolerate heat but not the humidity, because major constituents present in the seeds like starch, celluloses, pectin, and proteins together with mucilage are highly hydrophilic in nature and hence readily react with water. The various genetic parameters (a) within a genotype like

(i) Seed size, (ii) Nutrition provided by the mother plant, (iii) Position of seed on the inflorescence and (iv) Stage of maturity at harvest are described in detail hereunder.

5.3.1.1.1 Seed Size

The size of seed and amount of nutrition stored in it are controlled genetically under a high genotype **x** environment interaction. The influence of seed size on seed vigour has been variously reported by several workers. In his "Hand Book of Seed Testing" Nobbe (1876) reported that larger seeds always performed better in terms of production of vigorous seedlings. Similar results were also reported in carrot seeds, Mackey (1970); Ching (1973) on cotton; Scott *et al*, (1974) on sugar beet; Dhillon *et al.* (1978) on Triticale. It is widely accepted that large sized seeds, by virtue of being fully loaded with nutrients, are capable of giving rise to healthy and robust seedlings and ultimately a plant that can establish well and yield better. Contrary to this, Abdullahi and Vanderlip (1972) on sorghum, and Sahoo (1988) in groundnut reported that medium sized seeds recorded more germination in the field. Whereas, Taylor (1971) on clover seeds and Takeda (1972) on rice reported that small sized seeds registered better seedling growth. Hong *et al.* (1982) reported more rapid germination of larger maize seeds. In cowpea also large seeds produced larger seedlings, Lush and Wien, (1980). In soybean, large sized seeds have higher germination, vigour and yield potential than small seeds (Burris, 1973). Town Send (1972) observed only slight differences among the seed sizes for seedling emergence in legumes. Whereas Singh *et al,* (1972) could not found any effect of seed size. Hunter and Kannenberg, (1972) in maize hybrids and Shieh and McDonald, (1982) on maize inbreds reported very little effect of seed size on germination. Rajan (1978) could not find any influence of seed size in French bean. But, in large seeded varieties/ kinds like beans, the smaller seeds germinated faster than larger seed, making them appear to be more vigorous in barley, black gram, green gram, maize and soybean as reported by Haskins and Gorez (1975).

The differences in seed size could be due to the size of cotyledon, total protein and amino acid composition *etc.*, and since there is a source x size interaction, it can be concluded that the effect was not due to seed size alone. But in a wide range of crops, the association between small or large seeds and either high or low vigour is not consistent (Powell, 1988). Only under specific circumstances and in certain species, seed size relates positively to the rate of germination or emergence. But a consistent association could not be demonstrated between seed size and seed vigour over a wide range of crop species. Variation in seed size could occur either within a genotype or at individual seed level.

5.3.1.1.2 The Nutrition Provided by the Mother Plant

Environmental conditions, especially the nitrogen content, affect the seed quality not only during post-fertilization seed formation stage but also at seedling establishment in the subsequent growing season, Zakaria, (2009). The nutrient nitrogen added to the soil affect seed's nitrogen content and can increase or decrease yield or yield components. Nitrogen plays a key role in seed filling (Green, 1984) and may increase dry matter due to increased light absorption through leaves (Willhelm, 1998). This is more important because the nutrient reserves are accumulated, which is the direct effect of nutrition applied to the crop. Hence, more nutrients applied to the plant bring in greater seedling vigour, which is the potential for survival. Perry (1971) attached more importance to the nutrition of the mother plant to get more number of vigorous seeds for future sowing. Contrary to this, Pollock and Ross (1972) considered a set of conditions as a potential factor in seedling vigour, which help to accumulate such nutrients, which control the rate of seedling development. Therefore, nutrition forms the basis of management of a seed crop with judicious application. Holzman (1972) working on different doses of fertilizers on the quality of winter wheat cultivars reported indirect and direct effects. The former being the frequency distribution of different seed sizes while the later was independent of seed or embryo effects. Linear relationship between seedling growth and weight of embryo was observed. Knowles, (1991) reported that nitrogen fertilizer increased seed protein content in wheat, a good index of seed quality and vigour and results in higher final germination percentages while the time to 50 per cent of seeds germinated and mean germination time (MGT) significantly reduced, Warraich,(2002). Application of nitrogen fertilizers at the rate of 120 kg/ha had more vigour when compared to application of 0, 60 and 180 kg/ha in an electrical conductivity test. Similarly, Sawan (1985) also reported that by application of nitrogen @120 kg/ha high germination can be achieved in sunflower. Whereas Osechas and Torres (2002) reported that application of 0, 66, 132 and 600 kg /ha nitrogen had no effect on seed vigour and germination percentage. Contrary to this, in peas high-level application of nitrogen fertilizer resulted in decreased yield and low germination rate, Pollock (1972).

Similarly, seeds treated with various chemicals, obtained from mother plants showed increased height and dry weight of 10 days old seedlings, positively correlated with fertilizer dose despite maintaining the uniformity in seed size. Lowe *et al.* (1972) collected 17 wheat seed lots and reported that seed protein and amino acid composition was related to seedling vigour and yield. Seeds produced under low nitrogen condition showed protein depletion. Increase in protein by application of nitrogen was found to increase in seed vigour as well as yield. Dharmalingam and Vijay Kumar (1990) reported that collection of seeds from acid lime mother plants aged above seven years gives high vigour seeds. In the absence of dormancy, the process of deterioration gets initiated while the seed is still intact on the mother plant that causes delayed harvesting due to inclement weather conditions and continues even during storage.

5.3.1.1.3   Position on the Inflorescence

Position of seed on the plant is one of the components of within plant variation that may account for part of physical variation such as seed weight and shape as well as physiological variation like vigour and viability of attributes of the seeds. Variations in field emergence due to poor management could be attributed to the differences in seed vigour. Therefore, seeds with low vigour obviously emerge either late or produce abnormal seedlings and thereby result in poor yields. The seeding vigour of seedlings raised from the seed is also affected by the position of that seed on the mother plant. Variation in respect of moisture content of seeds in head inflorescence is likely to cause variation in chemical composition of the seeds and hence becomes a source of variation between seeds in storability and ultimately the seed vigour. *Triplasis purpurea,* the annual dune grass, exhibits a position dependent seed heteromorphism in respect of seed number, mass and dormancy characteristics, Cheplick and Sung, (1998). The position of seed on the mother plants exerts influence via seed size and seed composition in many crops, which in turn affect the performance in the ensuing generation. The first formed seeds will have competitive advantage. This fact has been established by Borthwick (1931) in Carrot; Hathron *et al.* (1962) in maize and chillies. The first formed seeds gave 73 per cent germination followed by 67% and 54 % in subsequent harvests. A number of species belonging to the members of Asteraceae with ray and disc flowers are the best examples of how the position within an inflorescence exerts influence on the germination of progeny. Cotton seeds from the lower bolls that opened first, and exposed to longer period of field environment before harvest were consistently low in seed vigour than seed from bolls located in the upper half portion of the plant.

The position of seed or fruit present on its mother plant exerts affect on morphology, mass and dormancy or germination characteristics, Escalante *et al.*, (1993). This type of response is usually referred to as Position dependent effects, Moravcova *et al.*, (2005). Variations in germination may occur in seeds from subterranean versus aerial flowers of amphicarpic plants, Baskina and Baskin, (2001) or from flowers produced at different parts of the same inflorescence Dutta *et al.*, (1970) or heights on the stem or position in a bur (Baskin and Baskin 2001) or fruit Maun *et al.*, (1989). Seeds from the top of plants were heavier than seeds from the bottom and showed faster seedling growth and a higher germination, but there were generally no differences in conductivity. Adam *et al.*, (1989). The seeds of soybean collected from upper strata are usually heavier than those from the lower strata, however the relations between position and physiological quality are not clear.

Some grass species of *Aegilops ovate, A. neglecta, A. geniculata* of Poaceae family, the basal caryopsis in a spikelet is larger and less dormant than the upper one, Maranon (1987). Early pollination and seed set particularly in primary branches resulted in obtaining carbohydrates from the sources at a much faster rate because of greater sink effect was reported by Tashiro *et al* .,(1988). Soybean seeds developed in the upper one-fourth portion of the plant were found to have greater concentration of protein and lower concentration of oil when compared to seeds from the lower one-fourth portion of the plant. Both determinate and indeterminate varieties of soybean contained more oil in the seeds developed on lower nodes (Collins, 1956). Mehdi Babaeian *et al.*, (2012) in wheat reported that tiller order, seed position, genotype and seed nutrient content accumulation changed significantly with tiller over seed position and genotype. In case of wheat, seeds on main tiller had higher nutrient content than those on secondary tillers. Similarly, the inner and mid florets had lower amount of nutrients than florets in the upper spikelets and outer florets in the middle spikelets. Therefore, Hampton *et al.*, (1996) concluded that seeds from the lower third of the plant compared to seeds from medium or upper third tended to have lower 100 seed weight, inferior in germination and higher electrical conductivity after incubation indicating lower seed vigour. They attributed that; these differences could be due to the unequal allocation of resources to all seeds or seeds produced at one position developing under different environmental conditions than those produced at another position because of differences in physiological age of mother plant when seeds are produced, Baskin and Baskin, (2001).

5.3.1.1.4 Stage of Maturity at Harvest

Seed development is the period between fertilization and maximum fresh weight accumulation and seed maturation begins at the end of seed development and continues till harvested, Mehta *et al.*, (1993). Vigour of seeds is controlled by the appropriate stage of harvest, immature or over mature seeds will have less vigour. It means, seeds will have their maximum dry weight or in other words maximum vigour attained at or around physiological maturity stage. Therefore, seed development as well as physiological maturity studies assume immense importance, as the seeds have to be harvested at proper time to ensure their quality in respect of both vigour and germinability. The seeds, if retained on the mother plant after physiological maturity, for one reason or the other, may lead to formation of hard seeds or off coloured seeds in pulse crops. Early harvested seeds will be immature and poorly developed and hence are poor storers when compared to seed harvested at physiological maturity, Singh and Lachanna, (1995); Deshpande *et al.*, (1991). Therefore, harvesting of seed crop at optimum stage of seed maturation is essential for obtaining better seed quality. Harvesting stage influences the quality of seed in respect of germination, vigour, viability and storability.

The recalcitrant seeds usually require relatively high moisture content for longevity, hence, at low moisture contents loose vigour rapidly. Contrary to this, Orthodox seeds maintain their vigour for a long time as they can be dry stored at low moisture level or in a cool storage condition. Seed maturity is governed by variety, nutrition status, moisture and temperature and all these factors put together exert influence in turn on shelf life of the seed. Storability of seeds, though genetical in nature, is influenced by both pre-storage history of seed, seed maturation and environmental factors during pre and post harvest stages, Mahesha *et al.,* (2001). No doubt, the maximum storage potential is attained in any seed around physiological maturity for individual seeds. Contrary to this, seeds of grasses and carrot etc., where indeterminate flowering is the rule. Most of the mature flowers grow at the base of the inflorescence and more number of immature flowers formed on the new branches. This automatically facilitates the occurrence of a range of newly fertilized ovules to varying degrees of developing seed to mature seed on the same plant. Obviously, when the seeds were harvested from these plants, the seeds will be with varying degrees of maturity and ultimately with differential storage potential and vigour. A number of reports by Rics (1971), Lopez and Grabe (1973) have clearly demonstrated that small and immature seeds are inferior to mature seeds with respect to viability and vigour. Khatun *et al.,* (2009) reported that harvesting stage of lentil had significant effect on some parameters.

In case of cultivated groundnut, where seed is the most expensive unit, the rich oil renders it more perishable and prone to rapid deterioration in respect of quality and viability in storage, Perez and Arguello, (1995). The cultivars being indeterminate in growth, at harvest it comprises of seeds of varying maturity stages. As a result around 20-30 per cent of seeds either do not germinate or fail to develop in to healthy seedlings and hence only gives patchy crop stand and consequently yield loss, Nautial *et al.,* (1990). Poor seed quality in groundnut, especially under suboptimal conditions, delays the onset of germination, adversely affects seedling vigour and ultimately crop stand and yield. Nautial *et al* (2010) reported that mutual relations between the physiological processes associated with seed maturity in groundnut and attributes of seed and seedling vigour are yet to be fully understood to enhance the emergence and uniform crop stand.

5.3.1.2 At Individual Seed Level

The various genetic parameters at individual seed level like (i) differences in the activity of enzyme system, (ii) rate and uniformity of seedling growth are presented hereunder.

5.3.1.2.1 Differences in activity of enzyme system, which in turn control O_2 uptake

In order to survive under stress conditions, plants are equipped with oxygen radical detoxifying enzymes such as superoxide dismutase, peroxidase, catalase and antioxidant molecules like ascorbic acid and reduced glutathione, Prochazkova *et al.,* (2001). Generation of ROS causes rapid cell damage by triggering off a chain reaction, Imlay (2003). Therefore, the plants produce some defense mechanisms to protect themselves against abiotic stresses, Vranova *et al.,* (2002). An integrated system of non-enzymatically reduced molecules like enzymatic antioxidants, ascorbate, and glutathione offer the detoxification mechanism for ROS scavenging, Prochazkova *et al.,* (2001).

High temperature and relative humidity during storage of seeds contribute to their deterioration by promoting degenerative changes such as destabilization in the activities of enzymes and destructing and eventual loss of integrity of the cell membranes system, caused mainly by lipid peroxidation due to increased reactive oxygen species Alscher *et al.,* (2002).

Copeland and McDonald, (2001) reported that the most sensitive evaluations to detect the start of deterioration of the seeds are related to the activity of enzymes associated with the biosynthesis of new tissue, since with the process of deterioration, the enzyme becomes less efficient in exerting their catalytic activity. During early stages of germination, α-amylase activity was higher in bold seeded cultivars Deshmukh *et al.*,(1990). Whereas Krishna Sámi and Seshu (1990) reported that α-amylase content increased rapidly in fast germinating rice cultivar, IR- 50.

5.3.1.2.2 Rate and uniformity of seedling growth

AOSA (1983) has considered germination rate as an indicator of seed vigour. Germination rate can be measured by different methods. Allmost all methods are based on time to radical protrusion as parameter of germination. The hybrid seeds of barley and maize germinate more uniformly and faster than the inbreds from which they have been developed, Mc Danniel (1969). Geneve and Kester (2001) and Sako *et al.*, (2001) developed systems that capture images using a flat bed scanner for evaluating seedling growth, growth uniformity and germination percentage. In the absence of published evidence that seedling size or growth rate measured by computer correlates with other standard measures of vigour for small seeded crops. Further there is no evidence that a vigour index value provides more information about seed lot vigour compared to simple growth analysis.

5.3.1.3 At Seed Population Level

The various genetic parameters at seed population level like (i) seed weight, (ii) seed density, (iii) seed coat colour and (iv) Hard seededness, (v) Chemical composition of seed, (vi) Hybrid vigour and (vii) Susceptibility to mechanical damage are described in detail.

5.3.1.3.1 Seed Weight

Positive correlation between seed weight and relative force exerted during seedling emergence was reported in alfalfa and strawberry by Jenssen *et al.*, (1972) and confirmed by Town Send (1972) in *Cicer spp*. Positive correlation between seed weight and seedling weight was reported in rice by Takeda (1972). The effect of seed weight on germination and seedling vigour in maize and soybean was reported by Katiyar *et al.*, (1990) as bold seeds of maize and small sized seeds of soybean were superior in vigour than their graded counterparts. Adibisi *et al.*, (2010) in Nerica genotypes of rice for Africa reported that 100 seed weight had significant correlation with plant height. They further reported that standard germination; energy of germination Mehdi *et al.*, (2012) in wheat reported that seed weight varied significantly and the interactions of tiller order **x** seed position and genotype **x** seed position on the head. Most seed position on the head of main tillers had heavier seed than their counterparts on secondary tillers, except the basal spikelet, which had similar seed weight on both tillers. Irrespective of genotype, seed from outer florets of middle spikelet and basal spikelet's were heavier than inner floret of middle spikelet and upper spikelets. Ali *et al.*, (1992) found positive genetic correlation of seed weight with fresh shoot length, fresh root length, fresh shoot weight and fresh root weight and dry root weight. Jafri *et al.*, (1992) also found positive correlation between 100 seed weight and germination percentage at both genotypic and phenotypic levels.

5.3.1.3.2   Seed Density

Variable performance of seeds in a given seed lot is a consistent problem in rice production, often exhibited in the form of failure to establish a desired crop stand coupled with non-uniform growth of seedlings. Chester (1938), Arndt (1945) observed positive correlation with germination and emergence. Whereas, Wiliams (1956) reported positive relationship between density and seedling emergence forces. Wanjura *et al.,* (1969) demonstrated that majority of the yield was produced by those plants originating from seedlings that had emerged most quickly from the soil. Bartee and Kreig (1974), Gajabee *et al,* (1977) reported that organic and inorganic materials increased the seed density and helped indirectly in better performance. Tupper *et al.,* (1971) reported that seed density was directly related to earliness of germination. Similarly, Ferguson and Turner (1971) reported that degree of fill was positively related to both emergence and seedling vigour. Krieg and Bartee (1975) evaluated the influence of seed density on various aspects of germination and emergence and concluded that high-density seed has bigger embryos and greater lipid content. Whereas Kreig and Caroll, (1978) demonstrated that seed density and seed weight determined seedling growth rate. They strongly recommended seed separation on density basis to improve germination and seedling emergence. Hussaini *et al.,* (1984) reported that larger seed were superior to medium and small seed in germination percentage and seedling vigour. Senthil Kumar *et al.* (1991) based on density grading by flotation technique reported that cotton cultivar LRA 5166 and Hybrid CHB were best. Ali *et al.,* (1991) reported significant genetic variability among rice genotypes for all seedling traits with the exception of fresh root length, dry root weight, and relative root weight. Significant genetic variance and ratio of genotypic and phenotypic variance observed in rice clearly indicate that vigour can be improved through phenotypic selection.

5.3.1.3.3   Colour of Seed Coat

Normal seed coat colour is indicative of high vigour of seeds while off coloured seeds were of poor quality irrespective of seed size classes as reported by Dharma lingam (1978) and is associated with physiological and pathological disorders. The emergence ability of *Phaseolus vulgaris* seeds with pigmented testa resisted the attack by *Rhizactonia solani* and same is true with pigmented testa in groundnut which is found to tolerate fungal invasion in soil and hence have exhibited better emergence potential when compared to non-pigmented seeds. Alison *et al.,* (1990) reported poor emergence of legume seeds of different species with unpigmented seed coats.

5.3.1.3.4   Hard Seededness

The dry conditions prevailing during seed production enhance the proportion of hard seeds, which in turn results in slow, extreme variations in germination and emergence leading to poor plant stand establishment. Though, it is an undesirable genetic trait, efforts are being made to reintroduce this trait into cultivars to prevent ageing and protect against leakage of nutrients during imbibition (Potts *et al.,* 1978). This trait can be eliminated relatively easily by breeding.

5.3.1.3.5   Chemical Composition of Seed

Breeding for nutritional quality resulted in more seed quality problems often resulting in small, shriveled, shrunken or low vigour seeds as in case of high lysine maize. The parameters of selection like increased protein and oil content, resistance to diseases and pests, mechanical integrity or hard seededness are only the physical manifestations of seed vigour aimed at increasing the plant stand establishment potential of the seed in the field which ultimately is supposed to result in higher yields. High protein wheat seed results in increased germination, Fox and Albert (1957), seed vigour, Lowe *et al.,* (1972).

5.3.1.3.6 Hybrid Vigour

In addition to these physical manifestations of vigour, plant breeders were able to introduce and exploit successfully the concept of hybrid vigour, the superiority in performance exhibited by the progeny over and above that of parents from which the seed was developed. The superiority of the hybrid is usually more under stress conditions than under optimal conditions.

5.3.1.3.7 Susceptibility to Mechanical Damage

A general rule is that, high germination capacity is associated with high vigour. Be it, the differential level of tolerances to mechanical abuse in navy beans, Barriga, (1961); high level of mechanical resistance in snap bean coloured cultivars compared to white varieties, Wester, (1970); thickness of lime bean seeds to seed coat cracking, Kannenberg and Allard, (1964); quantitatively inherited trait of visible seed defects of soybean, Green and Pinnell, (1968) are under genetic control. Mechanical injury is one of the chief causes of low seed vigour caused to the embryo or seed coat during harvesting, threshing, cleaning, handling and planting operations. Discrepancies between germination capacity and field performance are not equal in different species. For example, discrepancies in pulses occur more commonly than cereals Burg, (1986).

5.3.2 Environmental Factors

It appears that the seed production conditions tend to influence more on the seed yield rather than seed quality. However, the effect of environment on seed vigour as reported in literature can be divided in to four broad groups viz. (a) during seed production on the mother plant, (b) during harvesting, processing and conditioning (c) during storage and (d) During sowing of the seed in the field. The environmental factors responsible for differences in seed vigour are presented in Table. 5.3.

Seed production in the field on the mother plant can be further sub divided into (i) seed development stage, (ii) Seed filling stage, (iii) Seed maturation stage. Further, operations during harvesting and post harvest processing comprising of (i) physical and mechanical damage, (ii) Harvesting, (iii) Threshing, (iv) Conditioning, (v) Transport (vi) Drying, (vii) Improper handling, (Viii) Excessive drying. During storage the seed is likely to get exposed to physiological and pathological damage and ageing. The fourth group comprising of pre-sowing operations like effect of tillage and fertilizer, soil temperature and moisture, damage caused by imbibition and the interaction of imbibition and ageing are described in detail hereunder.

5.3.2.1 On Mother Plant

The degree of pre-harvest deterioration is dependent on the climatic factors like temperature and moisture. The importance of these factors can be described by using the term weathering. No doubt, the seed vigour declines with age but the rate of decline in vigour depends on the moisture content and the temperature.

5.3.2.1.1   Factors affecting during seed development

An individual seed in a population acquire vigour as it grows and develops. The vegetative growth of mother plant has little or no effect on vigour as neither the cell structures are not yet built nor the storage compounds are deposited. The process of cell division and differentiation leading to the formation of seed directly affects the maximum attainable vigour level. Because at this juncture, it is quite possible that due to some conducive conditions prevailing upon may cause abnormal seedling development and this ultimately leads to loss of vigour. Probably, this could be attributed to the fact that during these early stages of development several biochemical structures like mitochondria, Golgi apparatus, DNA, RNA and various metabolic proteins that are essential for vigorous cellular activity at micro level and germination activity at macro level accumulates gradually.

Soil moisture stress initially exerts direct impact on the development of pollen and ovules. Rowe and Andrew (1964) reported a decrease in the number of kernel rows in maize grown in soils with low moisture content. Contrary to this, Campbell *et al.,* (1969) in wheat reported that excessive soil moisture has reduced the seed set and ultimately yield. Soil moisture content also exerts direct impact on the post fertilization stage, wherein the seeds get accumulated with nutrient reserves considered to bring greater vigour and potential for survival of seeds. Moisture stress during seed development often results in lightweight, chaffy and shriveled seeds, which in turn result in poor-vigour seed. When a production environment is constrained with poor fertility, probably a decline in the production of both number and weight of seeds can be observed. But, the seeds thus produced are as vigorous and germinable as those produced under fertile conditions. Exceptions do occur. Boron deficient pea seeds produce abnormal seedlings. Such abnormalities could only be corrected by application of additional doses of borax. Similarly when groundnuts grown in soils low in boron and calcium produce seeds that exhibit a discolouration of cotyledons associated with boron deficiency and a watery hypocotyl and physiological root breakdown associated with calcium deficiency. Drought stress imposed on soybean plants during flowering stage failed to exhibit any effect on either germination percentage, accelerated ageing percentage, Smiciklas *et al.,* (1989) or reduced seedling axis dry weight, Smiciklas *et al.,* (1992). The association between seed quality and nutrition reported in a number of studies is an exception than the rule. Foliar application of nitrogen in wheat, Bulsani and Warner, (1980) and in sunflower, Szirtes and Szirtes, (1980) reported improved seed vigour while, Austin and Longden (1965) found no influence of nitrogen levels on seed vigour.

5.3.2.1.2   Factors affecting during seed filling stage

At this stage several storage compounds like carbohydrates, proteins and lipids are very rapidly deposited into the developing seed so as to ensure supply of energy during the process of germination. So, any plant growth-inhibiting event occurring at this stage is likely to substantially reduce the subsequent seed vigour. Reduction in germination ability and vigour in groundnut (Ketring, 1984) and soybean (Dornhos and Mullen 1991) were as a consequence of exposure of the crops to drought and heat during seed filling stage.

Soybean plants when exposed to drought even at different stages of seed filling period responded differently, i.e., when exposed to drought in the beginning of seed filling period resulted in reduced germination percentage over the control. Similarly, stress during rapid seed filling period did not reduce germinability. Whereas stressing during seed formation and seed rapid filling stages could neither decrease seedling dry weight, nor electrolyte conductivity. Contrary to this increased leachate conductivity from the stress imposed at rapid filling stage is an indication of reduction in seed vigour. *Phaseolus vulgaris,* when exposed to high temperature during seed filling stage lead to the production of small and poor quality seeds (Siddique and Goodwin,1980). *Pisum sativum,* when exposed to high temperature during wrinkled pod stage (70-80 % moisture) resulted in highest incidence of hallow heart and hence low vigour (Mayers, 1948). Reduced water availability interrupts seed development. Severe drought leads to the production of light, shriveled and misshapen soybean seeds (Delouche, 1980). Vieira *et al.* (1992) could not detect any effect of drought stress during development on germination and vigour of soybean. So, it can be concluded that seed filling period comparatively exerts less influence on vigour and germination of seeds.

5.3.2.1.3   Factors affecting seed vigour during maturation

At physiological maturity, the point at which the funiculus, the connecting link between developing ovules and the maternal tissues gets degenerated leading to disruption in the transfer of nutrients and moisture. Exactly by this time, the seed attains maximum vigour and hence is ready for harvesting. But in reality, one has to wait till the moisture levels are sufficiently brought down to safer levels, which permit mechanical harvesting without causing undue damage to the seed, known as harvest maturity. The time gap between physiological maturity and harvest maturity, is defined as maturation, during which the seed is dependent on the mother plant only for its physical support and some level of protection from husk and pod tissues. At this juncture, exposure of the seed to fluctuations of moisture, temperature, coupled with invasion by a number of pathogens make the seed most vulnerable to rapid loss of vigour despite of precautionary measures taken up by the seed growers.

During this stage of maturation, seed undergoes desiccation, the most obvious change, the extent of which varies with the crop and variety. Several physiological and biochemical processes taking place during this period involve reduction in biochemical composition, viz. decrease in sugar content, increase in oil content due to accumulation of saturated and unsaturated fatty acids. Dornbos and McDonald (1986) reported a change in the concentration of proteins in specific electrophoretic bands despite the gross protein content remained constant. At this stage, the highly organized fluid lamina structure develops into organized cellular lipid membranes during desiccation (Abdul-Baki and Baker, (1972), and associated with dry metabolically inactive seeds in storage at 12-14% moisture contents (Edwards, 1976). Hence, the disruption of normal sequence of physiological and biochemical events associated maturation process could compromise with seed vigour.

5.3.2.1.4   Factors affecting seed vigour during harvesting and processing

Between harvesting and planting in the ensuing season, the seed lots must be conditioned, stored, shipped and bagged to be ready for planting. Hence considerable scope exists for the occurrence of vigour loss due to deterioration in the form of either physical / mechanical damage or physiological decline. Usually, the seeds will be kept under temporary storage during and between different stages, at which ageing also occurs. Soybean seeds undergo severe pressure during threshing and results in broken seeds, seed coat cracks and invisible internal damage. The extent of damage depends on designing and effectiveness of threshing machine and threshing conditions and properties of cultivars. Sosnowski *et al* 1992, Khazaei *et al.*, 2007.

5.3.2.1.4.1   Physical / Mechanical damage to the seed

The physical and mechanical damage caused to the seed is considered to be one of the chief causes of low seed vigour, occurs at different post harvest stages are as follows. The loss of storage and meristematic tissues automatically bring down the seedling vigour. Mechanical damage, which may be external or internal or even microscopic renders the seed susceptible to the invasion of fungal pathogens and thereby reduces the shelf life of the seed. The external damages are visible to the naked eye but the internal damages are visible only after germination whereas the microscopic breaks readily renders the seed susceptible for micro-organism attack.

5.3.2.1.4.2   Harvesting

The problem of mechanical injury during harvesting was reported way back in corn by (Albert, 1927) and is increasing gradually since introduction of new farm machinery. The extent of damage depends on the type of combine, cylinder speed, variety, orientation and % moisture.

5.3.2.1.4.3   Threshing

Methods of threshing influence seed quality in terms of germination and vigour. Kausal *et al* 1992. Copper sulphate ($CuSO_4$) injury was observed in the embryos of seeds threshed with machines was reported, in contrast to hand threshing. When compared to a single or double rotary threshing machine, the percentage splits are significantly greater in a conventional cylinder. Irrespective of cylinder size, reduced cylinder speed decreases the mechanical damage, Newbery, Paulsen and Nave, (1980). As the crop becomes drier and brittle, the problem of threshing injuries too become more and more prominent. As the threshability of a crop changes in tune with ambient conditions, gentle threshing can be achieved by employing skilled and experienced persons so as to obtain good quality seed. In order to avoid mechanical injuries during threshing, it is always better to thresh the crop when the seed moisture content is reasonably low. Sorghum panicles threshed at low moisture content (13-14 %) to obtain high seed quality. Langa *et al.*, (2016).

5.3.2.1.4.4   Conditioning

This is a post harvest process, wherein all the contaminants are removed from the seed without imposing any additional physical damage, McDonald, (1985). But in reality, whenever the seed comes in contact with hard surface, increases the possibility of injuries. It improved germination by 3 per cent in soybean. Misra, (1982) reported that certain devices are more efficient than others. Gravity separators and air screen cleaners are more efficient conditioners and respectively improved germination by 2.1 and 1.5 per cent. Contrary to this, spiral separators increased physical damage by 0.6 per cent and hence could not increase germination.

5.3.2.1.4.5   Transport

Considering the volume of seed material of various classes and varieties in different crop species to be handled, the Indian seed industry has nearly perfected the exercise of maintaining the quality control of seeds during production. But when it comes to the maintenance of quality during transportation of the seed material from one corner to another, be it from south to north or east to west and vice versa, the information available either on transportation environment or its impact on seed vigour is scanty, Kulkarni (2002). In the light of published reports stating that around 26 per cent of the seeds get damaged during transport. It is always better to opt for cooling and refrigerated transport especially in case of high value seed and drying to safe moisture content during bulk transportation. Gill (1983) conducted one simulated transportation environment study on maize and revealed significant signs of weakening of seeds as evidenced by reduction in first count and slower seedling growth in early stages. Seeds, that are very sensitive to high humidity and temperatures, are likely to lose their viability quickly during transit, especially when large consignments of field crop seeds could be killed due to overheating in railway wagons or holds of ships. When compared to seeds with medium and high moisture content, seeds with low moisture content are relatively better in tolerating high temperatures. Therefore, drying the seeds and transporting them in sealed containers, though prevents deterioration in most seeds, still it is a problem with a large number of seeds, which cannot tolerate desiccation tolerance. Large seeded legumes are more susceptible to damage than caryopsis types because the damage occurs between cotyledons and more on the embryonic axis and thereby increases the splits. During long distance transport care should be taken to see that the seed should not be exposed to either extremes of moisture and temperature.

5.3.2.1.4.6   Drying

Optimum drying of each type of seed minimizes the mechanical damage, at less than 10 % moisture, soybean seeds are more resistant to damage than at 14 % moisture.

5.3.2.1.4.7   Improper Handling

Even minimal drop of 5 feet can reduce the vigour in soybean seeds. The vigour tests provide extremely useful information to seed processors as tools to critically evaluate the effectiveness of the material. Damage to seed coat is the most common form of physical damage. The physically damaged seed lots cause substantial reduction in field emergence, hence the yields. The damaged seed coat promotes rapid leakage of cellular contents upon imbibition. On one hand there is a considerable loss of nutrients that are required for deriving vigorous seedlings and on the other hand these nutrients sustain the growth of microorganisms capable of causing pathogenic infection that ultimately renders the seedling as an abnormal one.

5.3.2.1.4.8   Excessive dryness of seed

This is a problem usually associated with grain legumes like soybean, lima bean and *Phaseolus* beans leading to damage during harvesting and processing. Production of cracks on dry cotyledons leads to an increase in the proportion of abnormal seedlings. Higher moisture content in the seed lessens the chances of mechanical injury.

5.3.2.1.4.9   Slow Imbibition

The slow imbibition of non-hard seed leads to reduction in leakage, not necessarily associated with the condition of embryo.

5.3.3 Differences in Seed Vigour due to Storage Related Factors

After harvesting, threshing and conditioning, the seed material will be kept in storage before being planted in the soil in the ensuing season. During storage, deterioration in the level of vigour continues depending on the activity of physiological and pathological factors, whether further sub-divided into various factors as presented below.

5.3.3.1 Physiological Damage

Impairment of normal physiological processes during seed development directly contributes to non-vigorous germination. According to the physiological basis of deterioration proposed by Delouche and Baskin, (1973), membrane degradation is followed by ATP synthesis, reduced respiration, and automatically the rates of biosynthesis leading to poor shelf life and ultimately poor emergence and development of abnormal seedlings. Membrane degradation is associated with low vigour seed. The membrane lipid peroxidation in dry seeds during storage results in the formation of hydroperoxides, oxygenated fatty acids and free radicles (Wilson and McDonald (1986). The free radicles denature DNA, hinder protein translation and transcription, oxidise certain amino acids. The cumulative effect of which ultimately reduces the vigour (Priestly, 1986). Peroxidation causes changes in membrane lipid composition, impair membrane fluidity and hence functionality. Non-functional membranes inhibit ATP, RNA, protein and synthesis.

5.3.3.2 Pathological Damage

Various types of pathogens, fungal, bacterial, viral either individually or in combination with others invade the vegetative as well as reproductive parts of the plant in the field, during transit and in storage, and cause loss to the viability and vigour of seed directly as well as indirectly. The direct loss could be due to various mechanisms like degradation of enzymes, production of toxins, etc., that ultimately limits or regulates the growth. Whereas the indirect reduction in seed vigour could be due to the infection caused by the pathogens that automatically limits the ability of seed to develop into a normal seedling and plant subsequently. A good number of pathogens directly attack the seed and cause infection in the range of complete death of seed to nearly nil effect on seed vigour.

5.3.3.2.1 Field Fungi (*Alternaria species*)

Usually, the field fungi cause infection and damage between seed development and maturation but in any case before harvesting. Several species of fungi are not only crop specific but also crop growth stage specific too and mostly depend on the weather conditions. By weakening the parent plant, the seed borne fungi causes reduction in the seed vigour due to the loss of nutrients into the ambient environment through the same processes of leaching or exudation called differently by plant physiologists and plant pathologists respectively. Since, the process is likely to encourage microbial activity and deteriorates the quality of seed further. Therefore, for several years in the beginning, this principle has been extensively utilized by various researches for detecting the differences in the vigour of seed lots of Maize (Tatum, 1954); Castor beans (Thomas, 1960); cotton (Anderson *et al.,* 1964); Peas and Beans (Matthews and Bradnock, 1965); Lima beans (Pollock and Toole, 1966); and rape (Takayanagi and Murakama, 1968). *Aspergillus flavus* infects maize kernels when the silks are yellow brown rather than brown. Glume blotch infects wheat during late growth stages rather than early. *Claviceps pururea,* responsible for ergot of rye, makes the plant more susceptible to infection especially during cold and rainy weather which delays fertilization and thereby enhances the length of time to invade the plants. Similarly, *Anthrocnoses and bacterial blights* are promoted by high humidity and warm temperature.

5.3.3.2.2 Storage fungi (*Aspergillus species*)

According to Christensen and Kauffman (1969), storage fungi play a vital role in the destruction of seed during storage. The storage fungi are capable of initiating infection after maturation as a function of weather and storage conditions, that is, they require suitable moisture and temperature to infect the seed. Most of the pathogens are well adapted to soil saprophytes and colonize on a wide variety of plant and tissues. They are weakly pathogenic, infection/ attack is not possible without optimum conditions. Contrary to this, *Phomopsis longicola*, is strongly pathogenic. Presence of moisture in the seed surface could be due to either too humid storage conditions or from rainy weather. High moisture coupled with warm temperature promotes the growth of pathogens like *Alternaria, Helminthosporium* and *Fusarium* and cause discolouration, produces mycotoxins, generates heat, develops mustiness, deterioration, decrease in germinability and ultimately total decay. The deterioration contributes to loss of vigour.

5.3.4 Differences in Seed Vigour due to Ageing of Seeds

Seed ageing, also known as deterioration, one of the major causes of differences in seed vigour, defined as the gradual accumulation of deleterious changes within the seed till its ability to germination is lost (Powell, *et al* 1984). One of the first effects of ageing is an increase in mean germination time (MGT) (Guy and Black, 1998; Bailly *et al.,* 2002). Mean germination time (MGT), calculated from many counts of germination, and is the average delay or lag period from the start of imbibition to radicle emergence and describes the germination curve. MGT is highly indicative of the emergence (rate and final level) as well as seedling size and variation of commercial seed lots.

Under Indian conditions, prevalence of high humidity and high temperature are common, the process of deterioration especially in the stored seed is very rapid because of the phenomenon of accelerated ageing. In the absence of dormancy causing factors, the process of deterioration commences while the seed is still intact on the mother plant itself and continues even during the extended or prolonged harvesting due to unfavourable weather conditions and even during the period of storage. Seed ageing, being both a pre and post harvest factor determines the germination and vigour of a seed lot. How a seed lot loses its germination capacity over a period of time can be best explained with the help of a seed survival curve. Seed lots usually show a very slow decline because at this stage, the seeds are considered to be physiologically very young and hence highly vigorous. During the initial stages the decline in germination being very slow, maybe it is very difficult to differentiate between the samples located at different points (say A, B and C) on the survival curve. At the same time, it is also not possible to detect the minor differences in germination only on the basis of just 400 seeds sample drawn for conducting this germination test because all samples will have high germination as they are physiologically very young and have not been subjected to ageing. This slow decline is followed by a rapid fall and ultimately only a few seeds retain their ability to germinate.

However, with the lapse of time, probably due to the accumulation of deleterious changes, the samples, A, B and C will be slowly moving on the survival curve. It means, there is a gradual increase in the number of seeds of a sample that have failed to germinate, by this time the seeds might have become physiologically old and hence have lost their vigour. According to Matthews 1980, germination after a precise period of deterioration is related to seed vigour.

The physiologically young seed lots when exposed to high temperature and moisture for a specified period, though they move right on the seed survival curve, they retain their high germination as they would be still on the initial slow decline stage. Contrary to this, the physiologically older seed lots (seed lots with low vigour) tend to exhibit a drastic reduction in germination because they happen to be on the curve when the germination declines rapidly.

So, the initial extent of seed deterioration that cannot be detected in the germination test could be detected by controlled deterioration test. It means, germination after precise period of controlled deterioration indicates the vigour both in terms of field emergence (Matthews, 1980) and storage potential (Powell and Matthews, 1984). The levels of deterioration in commercial seed lots, measured using accelerated ageing (AA) for Glycine max and control-led deterioration (CD) for Brassica spp, (ISTA 2011), have been consistently and significantly correlated with MGT in a large number of species. The more deteriorated the seed lots, the higher the MGT, that is, the greater the lag period from the start of imbibition and radicle emergence. This association between deterioration and the length of the lag period has been explained by the need for more time for metabolic repair in the more deteriorated seeds before germination processes can begin (Matthews and Khajeh Hosseini, 2007; Matthews *et al.,* 2011 a,b).

5.3.5 Differences in Seed Vigour due to Imbibition Damage and Vigour Loss

During imbibition, water enters into the cotyledons rapidly and causes damage within two minutes from the beginning of imbibition process. It is a physiological process and hence is initiated as a consequence of development of matrix potential generated by long chain molecules in the embryonic cells of dry seed when come in contact with free water. Rapid imbibition is a common feature of soils with high moisture coupled with low temperature or when the seeds are excessively dried or damage caused to the seed during pre and post harvest processing operations have substantially injured the embryo. In all these cases physical damage occurs in the form of disruption and disorganization of cell membranes and ultimately leads to cell death and high solute leakage from the seeds Powell and Matthews, (1978). The extensive loss of cellular material including enzymes from the seeds is an indication of extensive membrane disruption (Duke and Kokifuda, 1981). At low permeability, the imbibition damage is greater as the components of membrane are held more tightly and rigidly and hence are more sensitive to physical damage Powell and Matthews, (1978).

The influence of imbibition damage on seed vigour was successfully and clearly demonstrated in both temperate and tropical grain legumes (Powell, Matthews and Oliveria 1984) and in small seeded dicotyledonous crops (Thronton and Powell, 1992). Imbibition damage has not only highlighted the role and importance of testa in protecting cotyledons from the damaging effects of water uptake but also disproved the hypothesis of Simon and Raja Harun, (1972) by providing evidence that the decline in leakage was due to the formation of a bi-lamellar membrane. Therefore, Matthews *et al.,* (2011a) suggested that initial high leakage resulted from the death of outer cells of the cotyledons due to imbibition damage and decline in leakage represented loss of solutes from within the seed.

Powell and Matthews (1979) and Oliveria *et al.,* (1984) demonstrated in pea and soybean species that the extent of imbibition damage and vigour of seeds has reduced in the presence of intact testa. It means, the seed lots with extensively damaged testa imbibe rapidly and thereby exhibit high imbibition damage and hence emerge poorly in the field and therefore have low vigour. Contrary to this, seed lots with low testa damage imbibe slowly, show little damage and exhibit field emergence indicating that they are highly vigorous. Therefore, this clearly emphasizes the need to minimize the damage to the testa during harvesting, threshing, transportation and storage because the integrity of the testa determines the vigour of seeds in at least some legume species.

Seeds of legume species are found to be susceptible to damage caused by imbibition due to partially or unpigmented testa which imbibe more rapidly when compared with seeds having pigmented testa and hence show greater levels of imbibition damage thereby the vigour of the seeds is reduced and hence poor field emergence than pigmented cultivars Powell, Matthews and Oliveria, (1984). In case of chickpea, and cowpea, Legesse and Powell, (1996) reported that close adherence of the pigmented testa to the cotyledons further limits the rate of entry of water into the seeds in addition to the permeability role played by the testa. Since it is not possible to reduce the proportion of low vigour seeds in the final seed lot, selection of appropriate genotype is the only available solution to minimize or avoiding the losses caused due to rapid imbibition.

5.3.6 Differences in Seed Vigour due to Interaction of Ageing and Imbibition Damage

Ageing and imbibition damage, the two most important causes for reduction in vigour, interact simultaneously and make the aged seeds more susceptible to imbibition damage. Pea seeds with testa intact aged at high temperature and moisture exhibited complete vital staining, whereas pea seeds with testa scarified before imbibition exhibited more water uptake, markedly declined living tissues, clearly indicating the damage caused by imbibition. Similarly, cowpea cultivars with white and brown testae when exposed to high relative humidity and temperature during storage exhibited greater reduction in staining following imbibition when compared to cultivars with pigmented testae. The differences in moisture uptake, be it from water or humid atmosphere, cultivars' differing in testa pigmentation also contribute to the decline in seed quality. Therefore the uptake of moisture by seeds plays a more crucial role in determining the seed quality because it exerts influence at a number of stages during seed production as detailed in the following table 5.5.

Table 5.5 Factors influencing seed vigour

State of seed production	Conditions influencing vigour	Physiological Effect	Vigour level	Interaction of ageing and imbibition damage causes further reduction in vigour.
Harvest maturity	Delayed harvest + Poor weather	Ageing	Low	
	Good harvest Conditions	Little ageing	High	
Harvesting stage	Testa damage	Imbibition Damage	Low	
	Little/no testa damage	No imbibition Damage	High	
Storage	High seed moisture/ And or temperature	Ageing	Low	
	Low seed moisture/ And or temperature	Little ageing	High	
	Mechanical damage		Low	
Pre-sowing	Rapid imbibition		Low	

5.3.7 Other Factors

Selecting seed lots with high germination and vigor would be beneficial for early plantings in no-tillage systems.

5.3.7.1 Effect of Tillage

Loss in seed vigour continues even after sowing in the field depending on the level of tillage, level of chemicals and fertilizers applied, soil moisture and temperature. Tekrony *et al* ., (1988) in case of maize seed reported that delayed germination and slow seedling growth occur commonly following no-tillage planting. Low and medium vigour seed lots recorded consistently lower emergence than high vigour seed lots in all tillage systems and plantings tested. And hence recommended that selecting seedlots with high germination and vigour would be beneficial for early plantings in no tillage systems. Das *et al.,* (1990) reported that proper tillage operation and adequate levels of NPK enhanced the seed vigour and seed yield ultimately. By adopting ample seed spacing while planting, seed vigour can be modified substantially as it eliminates thinning operation and saves soil moisture and up to 40 per cent of labour in celery crop, Zink (1964). In order to achieve space planting in small seeded crops Smith (1961) and Chancellor (1969) adopted seed singulation methods on plastic and paper tapes.

5.3.7.2 In the Field after Sowing

The pre- harvest environment brings changes in the internal structure of the seed and thereby affects the seed quality. Contrary to this, the nature of environment prevailing at the time of sowing also exerts considerable pressure on the viability and vigour of seeds. Selection of seed through density grading before sowing enhances the seed rate to be used and thereby exerts influence on the market cost of seed, which is expected to modify the seed vigour to some extent. The level of normal germination obtained in any seed lot broadly gives an indication of the vigour. It is a common observation that viable seeds obtained from seed lots with low germination will have low vigour than when the viable seeds obtained from lots of high germination.

5.2.7.3 Effect of Chemicals and Fertilizers

Application of fertilizers is one of the primary methods for improving the availability of soil nutrients to plants. Fertilizers can change the rates of plant growth maturity time, size of plant parts, phytochemical content of plants and capabilities of seed. Mevi- Schutz *et al*, 2003. Heavy use of chemical fertilizers have created a variety of economic, social, ecological and environmental problems besides increasing in cost of inputs results in decreasing in seed quality, fall in commodity prices and ultimately reduced farm Income. Tung and Fernandez. 2007. Application of nitrogen fertilizers upto 600 kg/ha has no effect on seed vigour and germination percentage. Osechas *et al*, 2002. Application of organic fertilizers compared to inorganic were better in respect of vigour and seed moisture content in Mungbean. Aquino and Fernandez, 2001. Soybean seeds rich in molybdenum content fared well in molybdenum deficit soils (Gurney and Giddens, 1969). Application of nitrogen to wheat increased the protein content and vigour of seed (Schweizer and Ries 1969). Contrary to this, Hawthorn and Pollard (1966) reported that application of nitrogen to pea decreased germinability and yield. Treatment of seeds with chemicals like Zinc (Abraham *et al.,* 1969) not only eliminates soil deficiency but also enhances seed vigour. Similarly treatment of seeds of maize, bajra and ragi with ascorbic acid has increased both the germination and yield. In respect of winter wheat cultivars, indirect effect of different doses of fertilizers on the quality of seed and linear relationship between seedling growth and weight of grain or embryo. On the basis of a study conducted with 17 lots of wheat seed collected from different locations, fertilizer treatments and growing seasons, Lowe *et al.,* (1972) reported that (a) seedling vigour and yield are determined by the seed protein content and amino acid composition, (b) high and positive correlation exists between seed protein content and seedling dry matter, (c) increased nitrogen application increased protein content and ultimately an increase in the vigour and yield. In case of wheat, Das *et al.,* 1990 reported that use of adequate levels of fertilizers like nitrogen, phosphorus and potash have exerted greater influence on seed yield and seed vigour. Trace elements like Manganese and Boron often exhibit deficiency symptoms in seedlings.

5.3.7.4 Effect of Soil Moisture and Temperature

Moisture and oxygen present in the soil, as a medium for germination, compete for same physical space. Hence, the presence of excess of water in the soil can be considered as a condition of stress that limits the supply of oxygen to the germinating seeds and emerging seedlings. On the other hand, low moisture content present in the soil, leads to dryness, soil compaction and crusting due to mechanization and thereby creates a stress situation. As a consequence, it enhances the energy required for emergence and therefore the force required by the emerging seedling will be more. It is a common knowledge that stresses related to soil never occur individually, i.e., the stresses occurring at soil level are a product of interactions of two or more factors. Low moisture, wet soil and poor aeration on one hand and high temperature, dry soil and adequate aeration on the other hand are just the two extreme cases.

For a moment, assume a situation where in the soil is wet at root zone level and dry at emergence point and vice versa. From this it is possible to conclude that different genera and species of plants respond differently to same stress situation. It appears to be true with different seed lots of same variety. Therefore, standardization of this could be used as a measure for evaluating the vigour performance of the seed lots. By the time the seeds attain physiological maturity, they will reach their peak viability and vigour in the absence of dormancy and declines with ageing. The extent of decline is usually governed by two factors viz. the temperature and the moisture content of the seed. Dexter (1966) observed enhanced seedling emergence from 39 to 78 per cent by increasing the moisture content of bean seeds from 11% to 16% before sowing.

5.4 Manifestation of Differences in Seed Vigour

Seed vigour differences are, expressed throughout the life cycle of a plant, more apparent in some crops under some specified circumstances. It is expressed in terms of percentage and speed of germination, uniformity of population, morphological abnormalities, longevity of seeds in storage, growth and yield as detailed below.

5.4.1 Germination

Seed lots differ in their percentage of germination between laboratory testing and field emergence because reduction in vigour is expressed in the form of differences in speed of germination, (slower and more variable germination); growth rate of seedlings (giving rise to greater proportion of abnormal seedlings); and decreased ability to emerge due to the presence of suboptimal conditions in the field. Germination is a complex interacting phenomenon and is the culmination of three distinct but overlapping phases, *viz.* (i). Reactivation of pre-existing system, (ii). Synthesis of enzymes necessary for the breakdown of stored food reserves and their mobilization into the organelles and (iii). Synthesis of new cellular compounds. Similarly, the seedling growth that sharply accompanies the germination can also be divided into three distinct interacting phases *viz.* (i) cell division, (ii) differentiation and (iii) development of functionally and structurally diverse tissues and organs. The suboptimal conditions present in the field in more than one way are likely to hamper the different phases of germination and seedling growth.

5.4.2   Emergence

The seed lots with low vigour take more time for seedling emergence and the final emergence was less than with high vigour seeds. (Powell *et al.*, 1991), and this ultimately leads to the vulnerability of seedlings either to physical stress or pathogen infection and hence enhances the chances of mortality resulting in low plant population in the field. On the other hand the seedlings produced from the seeds of low vigour lots were also less vigorous and more variable in height at first leaf stage, produced less fresh and dry weight of seedlings. Total emergence determines plant density and there exists a strong relationship between plant density and ultimately the yield (Willey and Heath, 1969).

In the laboratory, under optimum conditions of moisture, temperature and in the absence of constraints like micro-organisms, chemicals etc., it is very difficult to distinguish between the differences in germination and seedling growth of viable seeds at different levels of vigour because of the differences in the rate of field emergence of these seedlings. It means, both biotic and abiotic stress situations existing in the field soil would exert influence on one or more than one phases of germinating seedlings and allow only those seeds that have the capacity to overcome these stresses to emerge from the soil and this phenomenon of progressively increasing rate of germination has been clearly demonstrated by Ghosh and Sen (1981) in size graded jute seeds of JRO-632 cultivar is presented the following table 5.6.

Table 5.6 Germination in size graded seeds of jute variety JRO-632
(Ghosh and Sen (1981)

Seed Size	Germination (Percentage)			Abnormal seed lings/plants (%)	
	5th day on glass plate	10th day on glass plate	10th day in the field	10th day in the field	At harvesting stage in the field
Large	100	100	85	0	0
Medium	98	100	80	2	1
Small	32	85	15	68	60
Ungraded	80	95	65	20	15

Differences in emergence of highly germinable seed lots have been shown to be a problem in grain legumes (Powel *et al.* 1984), Soybean (Oliveria *et al,* 1984), green bean (Powel *et al.* 1986a), maize (Bekendam *et al.* 1987). Seed lots having high germination, but poor emergence are referred to as low vigour seeds, where as those giving good emergence are called as high vigour seeds. A comparison of lab germination percentage and field emergence percentage of different lots of commercial species are presented in the table 5.7.

Table 5.7 Ranges for percentage germination in laboratory and percentage field emergence in commercially sown seed lots of various crops.

S.No.	Crop	No. of seed lots compared	Range among seed lots	
			Germination (%)	Field emergence (%)
1	*Pisum sativum*	80	80-100	8-85
2	*Glycine max*	18	83-96	22-90
3	*Phaseolus vulgaris*	30	75-100	34-93
4	*Allium cepa*	12	77-92	46-72
5	*Brassica oleracia*	15	71-98	25-72
6	*Brassica campestris*	30	88-99	36-76

In the field, the competitive advantage of sorghum, over weeds could be due to the rapid germination and early seedling vigour, which plays a potential role in weed control. It is a well established fact that cereal seeds compete with weeds under adverse moisture conditions, where as weeds compete more effectively under adequate moisture conditions.

5.4.3   Establishment

The field performance of a seed lot is governed by environmental factors like physical condition of the soil, depth of planting, moisture and fertility status of soil, etc., and plant stand establishment is possible only when the seedlings maintain vigour sufficient enough to maintain rapid growth and establish contact with water supply as early as possible because the same will be receding rapidly in the growing season of arid regions. Therefore, under such natural circumstances, in order to compete with the neighbouring plants for light, moisture and other nutrients also vigour is highly essential. An observation of any field crop with a number of barren plants is simply an indication of the consequence of the low vigour seeds. Therefore, elimination or reducing the number of non-productive plants per unit area with a little additional cost in the form of using vigorous seed could permit higher yields (Pollock and Roose, 1972).

The competitive advantage between plants of the same variety is considered to be one of the major considerations of seed vigour and has been successfully demonstrated in several crops by Whalley *et al.*, (1966) where seedling vigour is highly correlated with seed size. Assuming, large size seeds produce high vigour seedlings than low vigour seed, Black (1958) size graded the clover seeds and planted uniform sized seeds in equal densities and found a reduction in the number of plants surviving due to elimination of certain plants throughout the growing season. But, when the large and small seeds are planted together, the number of plants from large seeds eliminated by competition was more or less constant while the number of plants eliminated from the small seeds was reduced by two thirds, and he attributed this elimination to the shading where the seedlings from small seeds received less sunlight at the time of foliage canopy development.

5.4.4   Growth

The size advantage of larger seeds producing larger seedlings than small seeds is often lost during the subsequent crop growth when interplant competition becomes a major limiting factor in realizing crop yield in barley and wheat. Contrary to this, several genera of large seeded legumes have shown improvement in the crop yield.

5.4.5 Morphological Abnormalities

Years of seed testing experience gained both in the field and laboratories by the members of ISTA and AOSA, has resulted in the documentation of a detailed list of descriptions of unacceptable forms of seedling abnormalities like split or missing coleoptiles, stunted or broken roots etc., which are not likely to produce satisfactory seedlings under favourable conditions in the field (Hand Book of Seedling Evaluation, ISTA, 2003). In case of low vigour seeds, loss of storage tissue, injury to meristematic tissue of plumule and radicle, damage to vascular connections, presence of abnormal embryos, transvers breaks in cotyledons are the common morphological abnormalities observed. The abnormalities could be due to mechanical damage during post harvest handling including processing, drying with excessive heat, heat generated by fungi and insects in storage due to the presence of high moisture, use of over dose of seed treatment chemicals and seed ageing ultimately. In all these cases, production of high proportion of abnormal seedlings is accompanied by an increase in non-germinable seed. In addition to these causes, imbibition damage after sowing the seed in the field coupled with the unscheduled DNA synthesis/ repair mechanism in the early stages of imbibition in combination with ageing further adds to the production of more abnormal seedlings.

Poor field emergence in temperate (pea, beans) and tropical soybean and bengalgram is associated with rapid uptake of moisture from the soil leading to high solute leakage and death of cells due to imbibition damage. The problem of rapid uptake of water by soybean and pea seed lots has been attributed to a large proportion of seeds having cracks in their testa. (Powell *et al.,*1993). Contrary to this, (Tomer *et al.,* 1998), reported synchronous emergence, uniform and quick growth of Okra seeds with high vigour when compared to low vigour seeds.

5.4.6 Longevity of Seeds

Seed longevity is the relationship between viability and vigour. The duration of time the embryos retain their viability varies considerably among species and ranges from few days to several thousand years. The influence of maternal genotype is significant in the expression of seed longevity than their genetic constitution. During the time that the seed is held in storage, a gradual decline in germination and vigour can be observed. A high vigour seed lot has good storage potential and hence retains high germination during storage. Whereas, the seed lots with low vigour, exhibit rapid decline during storage due to poor storage potential. The deterioration will be more pronounced when the seeds are stored under suboptimal conditions, especially when the relative humidity permits growth of storage moulds. But, when the seeds take longer time to germinate than usual, the seedling thus obtained is usually smaller, and often malformed. If one is familiar with the general relationship between the viability and vigour, then from the germination data, the vigour of a variety can be partially inferred because of the existence of three distinct phases in this relationship as detailed below. More than 80 per cent germination in the first phase indicates the seed is both viable and vigorous. In the second phase, deterioration progresses very rapidly. In the third phase deterioration slows at approximately 20 per cent or even below, all seeds will slowly die.

5.5 Summary

Despite several and diverse causes, level of vigour of a genotype is more important. Each seed during the course of its development reaches an arbitrary phase coinciding with the mother plant senescence, called physiological maturity and defined as the accumulation of maximum dry weight. Morphological maturity indices vary with crop. The potential performance of a seed is influenced by intrinsic and extrinsic factors. Genetic studies revealed a number of QTLs to influence seed vigour. Development of molecular markers for seed vigour should facilitate breeders select for improvement of desirable traits. In addition to seed size, nutrition provided by the mother plant, position on the inflorescence, stage of maturity at harvest at genotype level; rate and uniformity of seedling growth and differences in activity of enzyme system at individual seed level; seed weight, density and seed coat colour, chemical composition and susceptibility to mechanical damage at population level exert influence on vigour.

On the other hand prevailing seed production conditions to exert influence more on seed yield than quality at four different stages viz., on the mother plant during seed production (comprising of factors at seed development, seed filling, seed maturation stages); during post harvest handling (consisting of mechanical damage, harvesting threshing, conditioning, transport, drying, improper handling, excessive dryness and slow imbibition); storage and sowing of seed in the field (involving pathological and physiological factors); and the impact of seed ageing, damage due to imbibition and impact of the interaction of these two determines the level of vigour of seed lots. Further, differences in seed vigour are manifested through germination, emergence, establishment, morphological abnormalities, growth and ultimately the longevity of seeds.

The small sized seeds of some leguminous crops emerged faster than the large sized seeds and appear to be more vigorous. Seed maturity is governed by the variety, nutrition status, moisture and temperature and hence appropriate stage of harvest determines the level of vigour. Indeterminate flowering facilitates the occurrence of a range of newly fertilized ovules in various degrees of development and maturity and hence differential storage potential and ultimately the vigour. Selection for improved protein and oil content, Hard-seededness, resistance to diseases and pests along with mechanical integrity are only physical manifestations of seed vigour meant for plant population establishment. Soil moisture content directly exerts influence at post fertilization (light weight, shrivelled seed) as well as germination (death of seedling). At maturation, seed is dependent on mother plant for physical support and protection. Seed ageing, also known as deterioration is one of major causes of differences in vigour loss. Imbibition damage arising as a consequence of development of matrix potential generated by long chain molecules of embryonic cells of dry seed coming in contact with water. Rapid imbibition is a common phenomenon in soils with high moisture coupled with low temperature or when the seeds are excessively dried.

Exposure of seed to biotic and a biotic stresses during maturation causes rapid loss of vigour. Drying, transport processing and conditioning, improper handling also lead to rapid reduction in seed vigour. Loss of storage tissue, injury to meristematic tissues, damage to vascular connections, presence of abnormal embryos and transverse breaks in cotyledons, use of seed treatment chemicals etc., lead to loss of seed vigour. A high vigour seed lot has good storage potential hence retain good germination after storage. Contrary to this, low vigour seed lots exhibit poor storage potential, rapid decline in vigour, especially under suboptimal storage conditions.

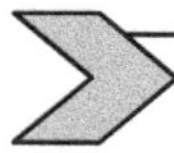 ## 5.6 Exercise

I. Long answer questions

1. Explain in detail how seed ageing causes loss to seed vigour.

2. Explain in detail why seed germination is a complex phenomenon.

3. How low vigorous seed lots establish only poor plant density in the field?

4. Explain in detail how the interaction of ageing with imbibitional damage cause loss to seed vigour.

5. Give a detailed account on the manifestations of differences in seed vigour.

6. Describe in detail various factors affecting seed vigour on mother plant.

II. Write short Notes on

1. Seed vigour and storage fungi,
2. Seed vigour and field fungi,
3. Transport and seed vigour,
4. Environmental factors and seed vigour
5. Differences in seed vigour
6. Genetic factors governing seed vigour
7. Physical damage,
8. Imbibitional damage,
9. Morphological abnormalities,
10. Morphological maturity indices

III. Short answer questions.

1. What are the morphological characters brought out by intrinsic variation for seed vigour?

2. Why intrinsic variation for seed vigour is considered important?

3. List out the extrinsic factors affecting seed vigour.

4. Why consistent relationship between seed size and vigour could not be established?

5. In pulse crops smaller sized seeds germinate faster than large sized seeds and appear more vigorous. Why?

6. Judicious application of nutrients forms the basis of seed crop management. Why?

7. List out the conditions governing the seed maturity.

8. How is it possible that seeds with varying degrees of maturity, storage potential and vigour are present in the same seed lot?

9. Briefly explain the genetic parameters at individual seed level that exert influence on seed vigour.

10. List out and explain the genetic parameters at population level that exert influence on seed vigour?

11. Briefly explain the physiological basis of deterioration of seed proposed by Delouche and Baskin, 1973.

12. How the cumulative effects of free radicles reduce seed vigour?

13. List out the changes brought out by peroxidation of lipid membranes.

14. Explain how pathological factors cause direct loss to seed vigour.

15. What is the role of field fungi in reduction of seed vigour?

16. What is the role of storage fungi in reduction of seed vigour?

17. List out the factors governing the field emergence of a seed lot.

18. List out the common morphological abnormalities observed during germination of low vigour seeds.

IV. Fill in the Blanks with suitable word or phrase

1. The causes of variation in seed vigour are -------- and --------.

2. Accumulation of maximum dry weight by seed during development coincides with -------- of the maternal plant.

3. During seed development, an arbitrary stage at which maximum dry weight is achieved by the seed is referred to as --------.

4. Presence of high moisture content in the seed at physiological maturity neither permits -------- nor --------.

5. The time gap between physiological maturity and harvest maturity is known as --------.

6. In the absence of dormancy and related factors, seeds attain highest -------- by the end of maturation period.

7. According to Nobbie, 1876, larger seeds always perform better in terms of production of vigorous --------.

8. Protein and amino acid composition is related to seed vigour was reported by --------

9. Increase in protein content by application of nitrogen to mother plant was found to increase both -------- and --------.

10. First formed seeds on a plant have a -------- advantage.

11. Immature and over mature seeds will have a -------- vigour.

12. Most seeds attain their maximum vigour at or around --------.

13. Recalcitrant seeds at low moisture contents loose -------- rapidly.

14. Orthodox seeds maintain their vigour for a long time because they can be stored either at -------- level or in a -------- storage condition.

15. Small and immature seeds are inferior to mature seeds with respect to -------- and --------.

16. Normal seed coat colour is indicative of -------- vigour.

17. Irrespective of the seed size, off coloured seeds were of -------- quality.

18. Dry conditions prevailing during seed crop harvesting enhances the proportion of --------.

19. Breeding for nutritional quality resulted in -------- problems associated with small, shriveled and shrunken and ultimately -------- seeds.

20. The degree of pre-harvest deterioration depends mostly on -------- and --------.

21. During storage, in addition to ageing, seed is likely to get   exposed to interacted with -------- and -------- damage.

22. The process of -------- and -------- leading to seed formation directly affects maximum attainable seed vigour.

23. --------stress during post fertilization stage results in poor seed development.

24. Plant growth inhibiting factors occurring during seed filling are likely to substantially reduce --------.

25. High incidence of hallow heart and low seed vigour in peas was a consequence of exposing seeds to high temperature coupled with --------.

26. The connecting link between developing ovule and maternal tissues is referred to as --------.

27. In case of high value seed it is always preferable to opt for -------- and -------- transport.

28. The seeds are not exposed to extremes of -------- and -------- in long distance transport.

29. The more common form of physical damage to seed is --------.

30. Reduction in vigour is expressed in the form of differences in -------- of germination.

31. Rapid imbibition is a common feature of soils with -------- moisture and -------- temperature.

32. Production of cracks on dry cotyledons leads to an increase in proportion of --------.

Answers

1. Several and diverse,
2. Senescence,
3. Physiological maturity,
4. Mechanical harvesting nor safe storage,
5. Maturation,
6. Potential vigour,
7. Seedlings,
8. Lowe *et al,* 1972
9. Seed vigour and seed yield.
10. Competitive,
11. Less,
12. Physiological maturity,
13. Vigour,
14. Low moisture, cool,
15. Viability and vigour.
16. High,
17. Poor,
18. Hard seeds,
19. Seed quality, poor vigorous,
20. Moisture and temperature.
21. Pathological and physiological,
22. cell division and differentiation
23. Moisture,
24. Seed vigour,
25. High moisture.
26. Funiculus.
27. Cooling and refrigerated.
28. Moisture and temperature.
29. Seed coat damage,
30. Speed.
31. High, low.
32. Abnormal seedlings.

 ## 5.6 References

Abdul-Baki, A., and Baker, J.E., (1973). Are changes in cellular organelles or membranes related to vigour loss in seed? Seed. Science and technology. 1: 89-125.

Abdullahi, A. and Vanderlip R.L. (1972). Relationships of vigour tests and seed source and size to sorghum seedling establishment. Agron. Journal. 64: 143-144.

Adam, N.M., M.B. McDonald Jr & P.R. Henderlong, (1989). The influence of seed position, planting and harvesting dates on soybean seed quality. *Seed Science and Technology* 17: 143-152.

Adebisi, M.A., Okelola, F.S., Alake, C.O., Ayo- Vaughan, M.A., and Ajala, M.O., (2010). Interrelationship between seed vigour traits and field performance in new rice for Africa (Nerika) genotypes. Journal of Agricultural Science and Environment. Vol. 10(2): PP: 15-2

Ali, S.S., Jafri, SJH., Khan, M.G., and Butt, MA., (1992). Relationship between seed weight and seedling vigour of rice. MARDI. Res. J. 20(2): 191-194.

Ali, S.S., Jafri, SJH., Ali, S., and Butt, MA., (1991). Genetic variability for seedling traits in *Oryza sativa*.L. Journal of Agricultural Plant Sciences. Vol.1. PP: 217-219.

Alison, A., Powell, W.D., Abdullah, N., Legesse, M., Deaolivaria and Matthews, S., (1990). Selection of grain legume genotypes for improved field performance . Paper presented at International Conference on Seed Science and Technology at IARI, New Delhi, India, February, 22-25.

Alscher RG, Erturk N, Heath LS. (2002). Role of superoxid dismutase (SODs) in controlling oxidative stress in plants. J. Exp. Bot. 53: 1331-1341.

Alberts, H.W. (1927). Effects of pericarp injury on moisture absorption, fungus attack and vitality in corn. Jour. Of American Society of Agronomy.19 : 1021-1030.

AOSA (1983). Seed Vigour testing Hand Book. Contribution No. 32 to the Hand Book of Seed testing. Ed. Clark, B.E.., McDonald, M.B., and Joo, P.K., pp 88., Lincoln. NE.   AOSA.

Arndt, C.H. (1945). Temperature - growth relationship of the rots and hypocotyls of cotton seedlings. Plant Physiology. Vol.20. pp. 200-220.

Aquino, A.L and Fernandez, P.G. (2001). Comparative productivity and seed quality of Mung bean grow under organic and Conventional production systems. Philippines Journal of Crop Science. 26(3): 45-51.

Won the performance of their progeny. Nature. 205: 819-820.

Bailly C, Bogatek-Leszczynska R, Come D, Corbineau F. (2002). Changes in activities of antioxidant enzymes and lipoxygenase during growth of sunflower seedlings from seeds of different vigor. Seed Sci. Res.12: 47-55

Baskin CC, Baskin JM. 2001. Seeds: ecology, biogeography, and evolution of dormancy and germination. San Diego, CA: Academic Press. Botanico de Madrid, 44: 97–107 Bartee , S.N. and Kreig, D.R., (1974). Cottonseed density associated physical and chemical properties of cultivars. Agron. J. 66: 435-437.

Barriga, C., (1961). Effects of mechanical abuse of Navy bean seed at various moisture levels. Agronomy Journal. 53: 250-251.

Borthwick, H.A., (1931). Carrot seed Germination. Proc. American Soc. Hort. Sci. 28: 310.

Burg, W.J. 1986. Aspects of seed quality control. In: *Seed Production Technology,* (Eds): J.P. Srivastava and L.T. Simarski, p: 82-887., ICARDA, Aleppo, Syria.

Burris, J. S. (1973). Effect of seed size on seedling performance of soybean. II. Seedling growth and Photosynthesis and field performance. Crop Science. 13: 207-210.

Campbell, C.A., Mc Bean, D.S., Green, D.G., (1969). Influence of moisture stress, relative   humidity and oxygen diffusion rate on seed set and yield of wheat. Canadian Jour. of Plant Sci. 49: 29.

Chester, K.S.   (1938). Gravity grading method, a Method of seed borne disease in cotton. Phyto - pathology. 28: 745.

Ching, T.M., (1973 a). Adenosine triphosphate content and seed vigour. Plant Physiology. 51: 400-402.

Ching, T.M., (1973 b). Biochemical aspects of seed vigour. Seed Science and Technology. 1: 73-78.

Cheplick G.P., Sung, L.Y., (1998). Effects of maternal nutrient environment and maturation position on seed heteromorphism, germination, and seedling growth in *Triplasis purpurea* (Poaceae). Intl. J. Plant Sci., 159:338–350.

Christensen, C.M., and Kaufman, H.H.., (1969). Gram storage, the role of fungi in quality loss. Minneapolis, University of Minnesota press

Collins, F., Cartter, I.L., (1956).Variability in chemical composition of seed from different portions of the soybean plant. Agron. J., 48, 216-219.

Copeland, L.O., and McDonald, M.B., (2001). Seed Science and Technology. 4th (ed.) Chapman and Hall, New York. Pp 409.

Copeland, L.O. (1976) Seed and seedling vigour. In Principles of Seed Science and Technology. Ed. L.O.Copeland. 149-184, Burgers Publishers. Ltd. Minnesota, USA

Datta SC, Evenari M, Gutterman Y (1970). The heteroblasty of *Aegilops ovataL*. Israel J. Bot., 19: 463-483.

Delouche, J. C., and Baskin, C.C., (1973). Accelerated ageing techniques for predicting storability of seed lots. Seed Science and Technology. 1: 427-452.

Delouche, J.C. and Caldwell, W.P. (1960). Seed Vigour and Vigour Tests. Proceedings. of AOSA, 50: 124-129.

Delouche, J.C., (1980). Environmental effects on seed development and seed quality. Horti. Science. 15 : 775-780.

Deshpande, V. K., G. N. Kulkarni and M.B. Kurdikeri. (1991). Storability of maize as influenced by time of harvesting. *Curr. Res.* 20: 205-207.

Dhillon, G.S. Kler, D.S. and Walia .A.S., (1978) Effect of seed size on growth, yield and quality of mung bean. Seed Research 5pp 37-43.

Dharmalingam,CV and Krishnen, V., (1978). Seed quality in relation to seed size and seed coat colour variations in blackgram. Seed Research. 6: 101-109.

Deshmukh,P.S., Shukla, D.S., Panwar, J.D.S., and Sairam, R.K., (1990). Seedling vigour and amylase activity in relation to seed size and temperature in wheat. Paper presented at International Conf. on seed science and Technology, at IARI, New Delhi, India. February 22-25

Dornbos, D.L., Jr. and Mc Donad,,M.B. Jr. (1986). Mass and composition of developing soyaben seeds at five growth stages. Crop Science. 26: 624-630.

Dornbos, D.L. Jr., and Mullen, R.E., (1991). Influence of stress during soybean seed fill on seed weight, germination, and seedling growth rate. Canadian Journal of Plant Science. 71: 373-383.

Duke SH, Kakefuda G (1981). Role of the testa in preventing cellular rupture during imbibitions of legume seeds. Plant Physiol. 67: 449-456

Edwards, M., (1976). Metabolism as a function of water potential in air dry seeds of charlock. Plant Physiology. 58:237-239.

Escalante, EE., and Wilcox, JR., Crop Sci, 1993, 33, 1166-1168

Ferguson, D., and Turner, JH., (1971). Influence of unfilled cotton seed upon emergence and vigour. Crop Sci. 11:713-715.

Finch savage, W.E., Come,D., Linn, J.R., Corbineau, F., (2005). Sensitivity of Brassica oleracia seed germination to hypoxia: a QTL analysis. Plant Sci. 169: 753-759.

Foolad, M.R, Arulsekar, S., Becerra, V and Bliss, F.A. (1995). A genetic map of Prunnus based on an interspecific cross between Peach and Almond. Theory of Applied Genetics. 91:262-269.

Fox R.L.and W.A.Albrecht.(1957). Soil fertility and the quality of Seeds. Mo. Agro. Expt. Sta.Res.Bull.619.

Gajbe, M.V., Musande,V.G.,and Verde , S.B. (1977). Interaction between some physical and chemical characteristics of seed in 22 cotton cultivars. Seed Science and Technology, 5: 539.

Gill, N.S., (1983). Effect of some simulated transportation environments of the quzlity of seed. Seed Research. 11 (2) 143 -148.

Geneve, R.L. and Kester, S.T. (2001). Evaluation of seedling size following germination usingcomputer-aided analysis of digital images from a flat-bed scanner. Hort Sci., 36, 1117-1120.

Green, DE., Pinnell, EL., Cavana, LV., and Williams LF., (1965). Effect of planting date and maturity date on soybean seed quality.Agron. J. 57: 165-168.

Green, D.E., and Pinnel,E.L., (1968). Inheritance of soybean seed quality.II. Heritability and visual ratings of soybean seed quality. Crop Sci. 8: 11- 15.

Gutterman, Y. (2000). Maternal effects on seeds during development. In. Fenner M. (ed.) Seeds: The Ecology of regeneration in plant communities, 2nd (edn.) Wallingford: CABI, pp. 59-84.

Guy, PA., and Black, M., (1998). Germination related proteins in wheat revealed by differences in seed vigour.Seed Science and research. 8: 99-1111.

Hampton, JG., Kahre,L., Van Gastel, A.J.G ., Boyce, KG., Leist, N., W.u Wen-Shi, W Loubser, W.J Van Der Burg, (1996). Seed Science and Technology, 24: 393-407.

Haskins, FA and Gorz, HJ., (1975) Influence of seed size, planting depth and companion crop on emergence and vigour of seedlings in sweet clover. Agron. Jour. 67: 652-654.

Hawthron, L.R., Toole, E.H., and Toole, V.K., (1962). Yield and viability of carrot seeds as affected by position of umbel and time of harvest. Proc. Amer. Soc. Hort. Sci. 30: 401.

Hong, C. K., Han, S.K., Ree, D.W., and Kim, K.S., (1982). Effect of seed size on growth and yield ofhybrid maize. Suweon. 24: 188-192.

Hunter, R. B., and Kannenberg, L.W., (1972). Effect of seed size on emergence and grain yield and plant height in corn. Canadian Journal of Plant Sciences. 53: 252-256.

Hussaini, SH., Sarada, P., and Reddy,BM., (1984). Effect of seed size on germination and vigour in maize. Seed Res. 12(2): 98-101.

Imlay JA. Pathways of oxidative damage. Annu Rev Microbiol. (2003). 57(1): 395-418. doi: 10.1146/annurev.micro.57.030502.090938.

ISTA (2011). International Rules for Seed Testing. ISTA, Bassersdorf.

Jafri, SJH., Ali, SS., Ijaz, M., and Butt, MA.,(1992). Relationship between germination percentage and some physical characeristics of rice genotypes. JPS 2(34): 122-123

Jenssen, E.H., Frelich, J.R and Gifford, R.O. (1972). Emergence force of Forage Seedlings. Agronomy Journal. Vol. 64(5). 635-639.

Kausal, R.T., Chanpada, S.P and Patil, V.N. (1992). Evaluation of threshing methods to assessing seed quality in Soybean. Seed Research. 20(1): 44-46.

Katiyar, R.P., Verma. J.R., Vaish, C.P. , Poonam Singh., (1990). Effect of seed weight on germination and seed vigour in maize and soybean. Paper presented at International Conference on seed science and technology, at IARI, New Delhi.Feb. 22-25.

Ketring, D.L.,(1984). Temperature effects on vegetative and reproducible development of Peanut.Crop Science: 24: 877-882.

Khatun, A., Kabir,G., and Bhuiyan, M.A., (2009). Effect harvesting stages on the seed quality of lentil during storage. Bangladesh J. Agrc. Res. 34(4): 565-576.

Khazaei, J., Shahbazi, F., Massah, J. (2007). Evaluation and modeling of physical and physiological damage to wheat seeds under successive impact loading mathematical and neutral network modelling. Jour of Crop Sci. 48(4): 1532-1544.

Kanneneberg, L.W and Allard, R.W., 1964. An association between pigment and lignin formation I the seed coat of Lima bean. Crop Science.4: 621-622.

Khatuni, A., Kabir,G., and Bhuiyan, M.A., (2009). Effect Harvesting stages on the seed quality of lentil during storage. Bangladesh J. Agrc. Res. 34(4): 565-576.

Knowles,TC., Dooerge,TA., and Ottaman,MJ., (1991). Improved nitrogen management in irrigated durum wheat using stem nitrate analysis.IIInterception of nitrate –N contents. Agron. J. 83: 353-356.

Kreig, DR., and Bartee. SN., (1975). Cottonseed density associated germination and seeding emergence properties. Agron. J. 67: 343-347.

Kreig, DR., and Carrol, JD., (1978). Cottonseed Metabolism as influenced by germination temperature, cultivar and seed physical properties. Agron J. 70: 21-25.

Kulkarni, G.N., (2002). Principles of Seed Technology. Kalyani Publishers, New Delhi, India.

Krishnaswamy V, Seshu DV (1989) Seed germination rate and associated characters in rice.Crop Sci.29: 902-907.

Levi, I., Anderson, J.A., (1950). Variations in protein contents of plants, heads, spikelets, and individual kernels of wheat. Can. J. Res. Sect. F., 28:71-81.

Legesse, N., and Powell,A.A., (1996). Association between the development of seed coat Pigmentation during maturation of grain legumes and reduced rates of imbibition. Seed Science and Technology. 24 : 23-32.

Lowe, L.B., Ayers, GS., and Ries,SK., (1972) Relationship of seed proteinand amino acid composition to seedling vigour and yield of wheat. Agron. J. 64: 608-611.

Lopez,A., and Grabe , D.F., (1973). Effect of protein content on seed performance of wheat.   Proceedings of AOSA. 3: 1606-116.

Lush, W. M., and Wien, H.C., (1980). The importance of seed size in early growth of wild and domesticated Cowpea. Journal of Agricultural Sciences. Cambridge. 94: 177-182.

Maun, M.A., Payne, A.M., (1989). Fruit and seed polymorphism and its relation to seedling growth in the genus *Cakile.* Canadian J. Bot., 67: 2743-2750.

Moravcova, L., Perglova, I., Pysek, P., Jitĕch, V., and Pergl, J., (2005). Effects of fruit position on fruit mass and seed germination in the alien species Heracleum mantegazzianum and the implications for its invasion. Acta Oecol., 28:1-10.

Mahesha, C.R., A.S. Channaveeraswami, M.B. Kurdikeri, M. Shekhargouda and M.N. Merwade. (2001a). Seed maturation studies in sunflower genotypes. *Seed Res.* 29 (1): 95-97.

Mahesha, C. R., A. S. Channaveeraswami, M. B. Kurdikeri, M. Shekhargouda and M.N. Merwade. (2001 b). Storability of sunflower seeds harvested at different maturity dates. *Seed Res.* 29(1): 98-102.

Matthews, S. and Khajeh Hosseini, S. (2007). Length of the lag period of germination and metabolic repair explain vigour differences in seed lots of maize (Zea mays). Seed Science and Technology, 35, 200-212.

Matthews, S., Beltrami, E., El-Khadem, R., Khajeh Hosseini, M., Nasehzadeh, M. and Urso,G. (2011a). Evidence that time for repair during early germination leads to vigour differences in maize. Seed Science and Technology, 39, 501-509.

Matthews, S., Wagner, M.-H., Ratzenboeck, A., Khajeh Hosseini, M., Casarini, E., El-Khadem, R., Yakhlifi, M. and Powell, A.A. (2011b). Early counts of radicle emergence during germination as a repeatable and reproducible vigour test for maize. Seed Testing International, 141, 39-45

Matthews, S., and Bradnock., WT., (1968). The relationship between seed exudation and field emergence in peas and French beans. Horti. Research.8:89-93.

Mac Key, D. B. (1970).   Relationship between laboratory germination and field emergence in some vegetable crops. J. Bot. Inst. Agr. Bot.12: 40.

Mehta, CJ., Kuhad, M.S., Sheoran, IS., Nandwal, A.S. (1993). Studies on seed development andgermination in Chick Pea cutivars. Seed Res. 21(2): 89-91.

Mc Daniel, R.G., (1973). Genetic factors influencing seed vigour Biochemistry of heterosis. Seed Sci. and Technology. 1(1): 25-50.

Donald, M.B., (1976). A review and evaluation of seed vigour tests. Proc. of AOSA, 65: 109-139

Mc Donald, M. B. (1985). Physical seed quality of sobean. Seed science and Tech. 13: 601-628.

Mishra, M.K., (1982). Soybean seed quality during conditioning. Proc. Mississippi. State seeds men's Short Course. 49-53.

Mohsen, Moussavi Nik, Mahdi Babaeian and Abolfazl, Tavassoli (2012). Effects of seed position on the parental plant on seed weight and nutrient content of wheat. Annals of Botanical Research. 3(1): 534-542.

Mayers, A., (1948). Hallow Heart, an abnormal condition of cotyledons of *Pisum sativum.* Proc. ISTA. 14: 35-37.

MeviSchutz, T., Goverde, J.M and Erhardt, A. (2003). Effects of fertilization and elevated Co2 on larval food and butterfly nectar, amino acid preference in *Coenonympha pamphilus* L. Behavioral ecology and Sociobiology. 54: 36-43.

Nautiyal PC, Misra JB and Zala PV. (2010). Influence of seed maturity stages on germinability and seedling vigor in groundnut. Journal of SAT Agricultural Research. 8.

Nautiyal PC, Ravindra V and Joshi YC. (1990). Varietal and seasonal variations in seed viability in Spanish ground nut. Indian Journal of Agricultural Science 60: 143-145

Nobbe, Fredric., (1876). Handbuch der Samen Kunde, Wiegandt - Hempel, Parey, Berlin.

Oliveira MDA, Matthews S, Powell AA (1984). The role of split seed coats in determining seed vigour in commercial seed lots of soybean as measured by the electrical conductivitytest. Seed Sci.Technol.24: 659-668

Osechas D, Torres A and Becerra L. (2002). Effect of nitrogen fertilization on the production and quality of seed of signal grass (Urochloadecumbens, Stapf). Zootecnica Tropical, 20(1): 135-143.

Paulsen, M.R and Nave, W.R. (1980). Corn damages from Conventional and rotary combiners. Transactions of ASAE. 23(5): 1110-1116.

Perry D.A., (1973). Interacting effects of seed vigour and the environment. In. W. Heydecler(ed.) Seed Ecology. Butterworths. London.

Perry, D. A., (1970). The relation of seed vigour to field establishment of garden pea cultivars. Journal of Agricultural Science Cambridge, 74: 343-348.

Perez MA and Arguello JA. (1995). Deterioration of peanut (*Arachis hypogaea* L) seed under natural and accelerated aging. Seed Science and Technology 23: 439-445.

Pollock, B. M. and Roos, E. E., (1972). Seed and seedling vigour, In T.T. Kozlowski (ed). SeedBiology Vol.I. Academic Press, NYpp 314-388.

Powell, AA., (1985). Impaired membrane integrity. A fundamental cause of seed quality differences in peas. In. PB. Hebblethwaite,MC., Heathand TCK Dawkins(eds). Butterworths. London. Pp. 383-395.

Powell, AA., (1986). Cell membranes and seed leachate conductivity in relation to the quality of seeds for sowing. our. Seed.Sci. and tech. 10: 81-100.

Powell, AA.,(1988). Seed vigour and field establishment. Advances in Research and Technology of seeds.11: 29-61.

Powell, A.A, Thronton, J.M., Mitchell, J.A., (1991). Vigour differences in brassica seeds and their significance to emergence and seedling variability. Journal of Agricultural Sciences, Cambridge. 116: 369-373.

Powell AA, Mattews S (1981) Association of phospholipids changes in the early stages of   seed   ageing. Ann. Bot. 47: 709-712.

Powell AA, Matthews S, Oliveira M de A (1984) Seed quality in grain legumes. Adv. App.Biol.10: 217-285.

Powell, A.A., and Matthews, S. (1979). The influence of test a condition on the inhibition and vigour of pea seeds. Journal of Experimental Botany. 30: 193-197.

Powell, A.A, and Matthews, S. (1978). The damaging effect of water on dry pea embryos during imbibition, Journal of Experimental Botany. 29: 1215-1229

Powell, A.A., (1988). Seed vigour and field establishment, Advances in research and technology of seeds. 11: 29-61.

Potts, H.C., Duangpatra, J.D., Hairston, W.G., Delouche, DC., (1978). Some influence of Hardseededness on soybean quality. Crop Science 18: 221-224.

Prochazkova D, Sairam RK, Srivastava GC, Singh DV. Oxidative stress and antioxidant activity as the basis of senescence in maize leaves. Plant Sci. (2001). 161(4): 765-771. doi: 10.1016/S0168-9452(01)00462-9.

Ries, S.K. (1971). The relationship of size and protein content of bean seed with growth and yield. Jour. Of American Society of Hort. Sci. 96: 557-560.

Sako, Y., McDonald, M.B., Fujimura, K., Evans, A.F. and Bennett, M.A. (2001). A system for automated seed vigour assessment. Seed Science and Technology, 29, 625-636.

Sahoo, A.K., Kulkarni, G.N. and Vyakaranahal, B.S., (1988). Effect of seed size on yield and quality in bunch groundnut. Seed Research. 16 (2): 136-142.

Senthil Kumar V., Selvaraj, J. A., Balamurugan, P., Selvaraj, U., and Masilmani, P. (1994). Effect of seed treatment and kind of sed on the storability of cotton Cv.LRA 5166 of TCB 313 Hybrid. Seed Technology News. 24 (4): 28.

Shieh, WJ., and McDonald, MB.,(1982). The influence of seed size, shape and treatmnent on inbred seed quality. Seed Sci and Technol. 10: 307-313.

Simon, E.W and Raja Harun, R.M. (1972). Leakage during seed imbibition. Journal of Experimental Botany. Vol, 23: PP-1076-1085.

Singh, A.R. and A. Lachanna. (1995). Effect of dates of harvesting, drying and storage on seed quality of sorghum parental lines. *Seed Res.* 13: 180-185.

Siddique, MA., and Goldwin., PB.,(1980). Seed vigour in bean cv. Apollo as influenced by Temperature and water regein during the development and maturation. Jour. Exptl. Botany. 31: 313-323.

Singh, J.N., Tripathi, S.K., Negi, P.S., (1972). Note on the effect of seed size on germination, growth and yield of soybeans. Indian Journal of Agricultural Sciences. 42: 83-86.

Smiciklas, K. D., Mullen, R.E., Carlson, R. E., and Knapp, A.D., (1992). Soybean seed quality response to drought stress and pod position. Agron. Journal. 84: 166-170.

Smiciklas, K D, Mullen,R.E. , Carlson R.E., and A.D., Knapp. (1989). Drought induced stress effect on soybean seed calcium and quality. Crop Sci. 29: 1519 – 1523.

Sosnowski, S and Kujinar, P. (1999). Effect of dynamic loading in the quality of Soybean. Dept of Agril Product. International Agrophysics. 13: 125-132.

Tashiro T, Takeushi T, Ishikawa M (1988). Studies on the temperature response of dwarf rice development and maturation of grain. Report of the Tokai Branch of the Crop Sci. Soc. Japan, 105: 7-8.

Vranova E, Inze D, van Brensegem F. Signal transduction during oxidative stress. J Exp Bot. 2002. 53(372):1227–1236. doi: 10.1093/jexbot/53.372.1227

Takayanai, K., and Murakami, K (1968). Rapid method for testing seed viability using iron-sugar analysis paper. Japan Agril Research. Quart. 4: 39-45. ,

Takeda,K. (1972). Relationship between the seed weight and seedling growth in the rice plant. Tech. Bullt. FAC. Agric. Hirosaki, Univ. 19: 10-21.

Tekrony,, D.M.; Egli, D.B.; Wickham, D.A. Corn seed vigour effect on no-tillage field performance: II. Plant growth and grain yield. **Crop Science**, n.29, p.1528-1531, (1988).

Taylor, D.L. (1942). Influence of oxygen tension on respiration, fermentation and growth of wheat and rice American J. Bot. 29: 721 – 738.

Thomson, J. R,, (1979). An Introduction of Seed Technology, Leonard Hill. Glasow. U.K.

Thronton, J.M and Powell, A.A., (1992) Short term aerated hydration for the improvement of seed quality in Brassica Oleracea. Seed science Research 2, 41-49.

Town Send C E, (1972). Influence of seed size and depth of planting on seedling emergence of two milk vetch species. Agron. J. 64: 627.

Tupper, G.R., (1970). Measurement of certain Physical characteristics of cotton seed related to rapid germination and seedling vigour in cotton seed. Proc. AOSA. 60: 138-148.

Tupper, GR., Kunz, O R., and Wilkes, LH., (1971). Physical characteristics of cotton seed related to seedling vigour and design parameters for seed selection. Trans. Am Soc. Agric. Eng. 1: 890-893.

Tung, L.D and Fernanadez, P.G. (2007). Yield and seed quality of modern and traditional soybean under organic biodynamics and chemical production practices in the Mekong delta & Vietnam. Omonrice: 15. Pp: 75-85.

Vieira, R.D., TeKrony, D.M., and Egli, D.B., (1992). Effect of drought and defoliation stressin the field on soybean germination and vigour. Crop Science. 32: 471-475

Warraich, EA., Basra SMA., Ahamed, N., Ahamed, R., and Aftab, M., (2002). Effect of nitrogen on grain quality and vigour in wheat. International jour. Of Agricluture and Biology. 04-4pp 517-520.

Wester, R.E. (1970). Chlorophyll, a genetic marker for quality seed in Lima bean. Seed World.101.24

Williams, W.A., (1956). Evaluation of emergence forces exerted by seedlings of small seeded legumes using probit analysis. Agron. Journal. 48: 273.

Willey,R.W., and Hath ,S.B., (1969). The quantitative relationship between plant population and crop yield. Advances in Agronomy. 21: 281-321.

Wilhelm, W.W., (1998). Dry matter partitioning and leaf area of winter wheat grown in a long term fallow tillage comparisions in US central Great Plains. Soil and Tillage. Res. 49: 49-56.

Wanjura, DF., Hudspech, EB., and Bilbro, JD., (1969). Emergence time, seed quality and planting depth effects on yield and survival of cotton. Agron.J. 61: 63-65.

Walley, D.B. Mc Kell, C.M., and Green L.R., (1966). Seedling vigour and the early non-photosynthetic stage of seedling growth in grass. Crop Sci. 6: 147-150.

Zakaria,MS., Ashraf, HF., Serag, EY., (2009). Direct and residual effects of nitrogen fertilization, foliar application of potassium and plant growth retardant on Egyptial cotton growth, seed yield and seed viability and seedling vigour. Acta. Ecologia. 29: 116-123.

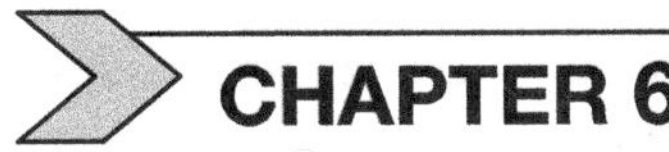

CHAPTER 6

Relationship of Seed Vigour with Yield

(The hands that help are holier than the lips that pray)

6.1 Introduction

Quality of seed is rather a broad term that encompasses around several factors. As a consequence, the seed industry is facing a chronic problem of production of seeds with low vigour since its inception, TeKrony, Egli and Balles (1980). Though, this phenomenon of production of seeds with low vigour could be attributed to the environmental conditions prevailing at the time of the commercial crop establishment, growth and development on one hand, and those inherent factors of such as the seed size, vigour, and germination ability of the planting material on the other hand, together exert strong influence on the vegetative / biological growth and reproductive / economic yield of the succeeding crop both directly and indirectly.

The environmental stresses prevalent such as those extremes in temperature and moisture, either in the field or even during storage, are bound to decrease the viability in general and vigour in specific and hence causes loss to the yield of succeeding crop in different ways. The relationship between laboratory germination, seed and seedling vigour, field emergence and yield has been the subject matter of numerous studies of the past one-century. Despite a lot of efforts, some researchers have reported close correlation between standard germination and seedling field emergence, while some other studies have clearly proved the over estimations of seedling's field emergence, TeKrony and Egli, (1991).

Seshu and Dadlani (1993) reported that, in order to realize the full genetic potential of any variety, first of all, the three chief aspects of seed quality, viz., seed size, germination and vigour should be high at the time of planting. Despite larger size, seeds lose their germination over a period of storage. Therefore, in order to maintain the vigour and ultimately the yield level of seeds, Dornbos (1995) has identified and suggested the adoption of the following four strategies.

1. Adoption of sound agronomic management practices in the seed production fields so as to minimize the effects of stress as and when it occurs.

2. Wider distribution of the seed production areas so as to minimize the risk of local unfavourable growing conditions.

3. Confining the seed production programmes to irrigated areas only as far as possible minimizes the effects of drought and unfavourable conditions, and

4. Adopt uninterrupted development of new and stable genotypes with potential yield that can withstand in variable production environments.

Despite best planning and execution, one should always keep in mind that it is not possible to predict mid and long term weather conditions, stress situations. Therefore, the development of new varieties should always be oriented at minimizing the negative effects of environment on both yield and quality.

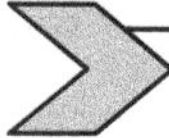

6.2 Factors Responsible for the Effect of Poor Seed Vigour on Crop Yield

The two distinct factors whereby poor seed vigour exerts affect on crop yield are as follows.

1. First of all, it could reduce the field emergence potential such that, even if the subsequent performance of the individual plants were unaffected, yield could be reduced through the establishment of a suboptimal plant-population density

2. The poor seed vigour might affect yield because individual plants that subsequently emerge perform less well than those from a better-quality seed lot.

Generally, low vigour seed will not be acceptable for cultivation because of their failure to maintain and sustain growth of seedlings sufficient enough for giving adequate seed yield. Seeds with low vigour produce crops, which may yield less when compared to the crops grown from high vigour seeds. Low vigour seeds could decrease the speed of seedling emergence and percentage of emerged seedlings, initial crop growth, leaf area and dry matter accumulation, Schuh *et al.* (2000).

6.2.1 Initial Crop Growth

Hofs *et al.,* (2004), and Mielerzski *et al.,* (2008) in rice; Dias *et al.,* (2010) and Mondo *et al.,* (2012) in maize; Schuh *et al.* (2009) and Dias *et al.,* (2011) in soybean reported that differences in initial crop growth is related to seed vigour.

6.2.2 Seed Size

Pal *et al.,* (1991) reported that high vigour large jute seeds had high germination percentage and fully uniform establishment of plants compared to low germination and low establishment from small seeds probably due to several abnormalities in seedlings and estimated that the loss in plant stand from small seeds reduced fiber yield to around 40 percent when compared to the crop produced from the large seeds.

6.2.3 Environment

6.2.3.1 Environmental Stress during Vegetative Growth

In addition to the above, environment too exerts great influence on seed size. A multitude of factors interact and affect the trait, seed size such as, high temperatures, short days, red light, drought and high nitrogen levels (Fenner 1991). Environmental stress occurring during vegetative growth affects yield less than stress occurring in later stages of development. Therefore, the loss of yield depends on crop type and the effect on plant population. In case of soybean, drought stress, before seed filling did not affect yield. Whereas in case of maize, drought in early vegetative phase reduced the yield significantly because this phase, determines both number of rows and number of kernels per row of a maize cob, and thereby appears to be very much sensitive to stress. No consistent relationship between seed mass and viability or vigour has been found. An indirect relationship between seed mass and vigour existing is by reducing emergence ability of large seeds. Generally large seeds are more prone to mechanical damage during harvesting.

6.2.3.2 Environment Stress during Seed Filling

The environmental stress during seed filling stage is also associated with reduced seed quality and thereby gives raise to small, light, flattened, shriveled and shrunken seeds, which exhibit only poor vigour, Delouche (1980).

6.2.3.3 Environment Stress during Seed Development

Similarly, the high temperature during seed development is bound to reduce visual seed quality, germination ability and ultimately the seed vigour. High night temperatures, on the other hand, are also associated with low seed corn germination ability and vigour. After attaining physiological maturity, the exposure of soybean seed to varying degrees of environmental conditions/ stresses is expected to cause deterioration and loss of seed quality in three different ways

1. Seeds exposed to alternate drying and wetting in the field has reduced the seed yield as a consequence of rapid and differential absorption of water by different tissues,

2. Exposure of seed to hot and dry weather after attaining physiological maturity too reduced the seed vigour significantly and

3. Exposure of seeds to warm and humid weather prolonged the desiccation period as moisture has to be reduced from 50 to 15 per cent and hence reduces vigour.

 ## 6.3 Direct and Indirect Effects of Seed Vigour on Crop Yield

Seed vigour is the cumulative impact of complex genetic and physiological factors on the growth and development of endosperm and embryo, Sun *et al.,* (2007). Quicker and uniform germination facilitates superior crop yields, whereas slow and non-uniform germination due to low seed vigour decreases crop yields, Basra *et al.,* (2002). However, obtaining optimum and uniform plat stands in the real field conditions require rapid seedling emergence. But in reality, several environmental stresses such as low and high temperatures, mechanical stress, and high salinity too, at times, exert impact during stand establishment. In case, in the open fields, when seeds are sown in deeper (7-10 cm) in the direct planting system, no doubt these seeds utilize moisture retained beneath the soil for germination successfully. But, at the same time, such seeds are likely to encounter mechanical resistance till they emerge out of the soil surface. Moreover, the saline seedbed conditions arising out of aridity or utilization of saline ground water for germination, may either delay germination or increase the spread of germination in the field, Foolad and Lin, (1997). In general, seed and seedling vigour exerts influence on the yield of field crops, Siddique and Wright (2004). Seed vigour exert its influence both directly and indirectly on the economic / biological yield.

6.3.1   Direct effects of Seed Vigour on Crop Yield

Seed vigour directly influences the plant growth processes involved in the production of yield, i.e., by exerting influence on the amount of solar radiation intercepted by the crop canopy, its conversion into dry matter and economic yield, photosynthetic efficiency *etc.,* in a given dynamic environment over a period of time, Charles and Edwards, (1982). The direct effects on subsequent plant performance are more difficult to discern, Ellis, (1991).

6.3.2   Indirect effects of Seed Vigour on Crop Yield

Vigour of a seed lot indirectly exerts influence on the crop performance in the field. The indirect effects include those on percentage emergence and time from sowing to emergence. These exert influence on yield either by altering plant population density, or spatial arrangement and even the duration of the crop. Enormous published literature clearly showed that seed lots with low vigour contributed directly to reduced forage and grain yield or had no effect (TeKrony and Egli, 1991) because it is highly unlikely that low seed vigour directly limits the ability of a plant population to express their genetic potential for realization of maximum yields. Therefore the yield loss, whenever occurs, could be due to the indirect relationship of seed vigour with either through the reduction in final field emergence or rate of emergence or even by causing uneven emergence depending on the plant type and the form of crop being harvested.

6.4   Effect of Seed Size on Seed Vigour and Crop Yield

Seed vigour is a multi-faceted quality. Therefore, the applied definition will depend to a large extent on the desirable features of that particular crop and the environmental conditions which the seed in turn is expected to tolerate. The success of germination, plant stand establishment, and growth and ultimately the final yield of any crop, no doubt, depend to a large extent on the quality of the seeds used to raise that crop. It has long been estimated that, use of good quality seed alone is capable of increasing crop yield by 15-20%. Seed size being an important indicator of seed quality, exerts affect on vegetative growth of the plant initially and hence is frequently related to yield. Usually genetic variation is the chief cause for variation in size of seeds between varieties. Seeds can be classified on the basis of size as very large, large, medium, small and very small and this variation could be attributed to the flow of nutrients into the seed from the mother plant. A wide array of different effects of seed size have been reported for seed germination, emergence and related agronomic aspects in many crop species. Seed size is a widely accepted measure of seed quality and in general large sized seed has better field performance than small seed. Jerlin and Vadivelu, (2004) have reported that, large sized seeds have exhibited higher seedling survival, growth and establishment. A wide range of diverse effects of seed size in several crop species have been published in respect of seed germination, seed emergence and other related aspects, Kaydan and Yagmur, (2008).

On the other hand, the seeds of a seed lot immediately after harvesting and threshing differ significantly in respect of their size, weight and density due to the impact of the production environment and cultivation practices adopted. Seed size, is one of the most important characteristics of seeds that can directly affect the seed development and duration. Adebisi *et al.,* (2011) also reported that seed size is an important component of seed quality and hence affects the performance of the crop. The existence of positive correlation between seed size and vigour was reported by Ries and Everson, (1973), Cookson *et al.,* (2001), where larger seeds produced more vigourous seedlings. Similarly, Roy *et al.,* (1996) also reported an increase in rate of germination and seedling vigour index values with increasing seed size in rice and suggested to opt for the selection of larger sized seeds for better plant stand establishment. Whereas Ambika *et al.,* (2014) concluded that vegetative growth is frequently related to yield, market grade factors and harvest efficiency.

6.4.1 Effect of Large Size of Seeds on Seedling Vigour

The size of seeds is expected to exert direct influence in respect of germination, ground cover and ultimately on the performance of plants and the same has been confirmed by immense studies on seed size in various species. The size of seeds is dependent on the age of mother plant, nutrients supplied by the mother plant, nutrient filling duration and several other factors including environmental conditions prevailing at that particular point of time. The accumulation of food reserves in corn seeds was explained through regressive equations as a function of time, Hanway (1963). But, according to Rench and Shaw (1971), the point of maximum physiological quality of corn seeds with moisture content and dry weight varied with genotype and seeding date. In sunflower, Nagaraju (2001) reported higher germination percentage, seedling vigour index, dry weight and field emergence in large size seeds compared to small seeds. Lowest germination percentage, emergence and seedling growth in small sized muskmelon seeds has demonstrated that no association exists between physical parameters of seed quality, Nerson, (2002). Similarly, Balamurugan *et al.,* (2004) in sunflower; Willenborg *et al.,* (2005) in oat; Varma *et al.,* (2005) in gram, reported larger seeds expressed high seedling vigour index than small sized seeds. Sulochanamma and Reddy, (2007) reported seedling vigour of shrivelled and small sized seeds of groundnut was less than bold seeds. In safflower, Sadeghi *et al.,* (2011), reported larger seeds produced highest germination per cent, fresh coleoptile weight, radicle fresh weight and test weight when compared to small sized seeds. Nik *et al.,* (2011), also reported that large wheat seeds gave plants with more vigour and dry matter. Hojjat, (2011) reported early and better germination from large lentil seeds than small seeds. Gunaga *et al.,* (2011) reported that higher and quicker germination in bigger sized seeds could be due to the presence of higher amount of carbohydrates and other nutrients than in medium and small sized seeds. Contrary to these studies, Ahirwar, (2012) reported lowest germination percentage in small seeds.

Singh and Kailasanathan (1976) reported that larger seeds of spring wheat produced higher yields under late sown conditions. Contrary to this, Kavitha and Choudhury, (1984) reported higher yield from large sized wheat seeds under optimum management conditions. Whereas, Baalbaki and Copeland (1997) reported that seed size exerted influence on emergence, establishment, yield components and ultimately grain yield of wheat. Similarly, Simmone *et al.,* (2000) also reported that seed size has a strong effect on germination as well as growth and biomass increment of a plant. Whereas, Rukavina *et al.,* (2002) in Croatian spring malting barley reported that increasing seed spike size and density had increased number of tillers, increased main stem length, increased thousand kernel mass, seed vigour and increased yield in addition to early decrease in moisture content at harvest that has obviously facilitated advancement in maturity duration. However, no consistent relationship between seed mass and viability or vigour was found. The relationship between seed mass and vigour is always indirect, i.e., by reducing the emerging ability of large seeds. Moreover, large seeds are highly prone to mechanical damage during harvesting and even processing because of the higher amount of prevailing moisture content.

6.4.2   Effect of Small Seeds on Seedling Vigour

Dar *et al.,* (2002) reported that small and medium sized seeds produced better germination and seedling vigour than those of bigger seeds. Similarly, Pekson *et al.,* (2004) also demonstrated that cultivars with low-test weight had registered higher germination percentage than larger pea seeds. In mustard, Kumar et al., (2005) reported that even the ungraded seed category was better than small sized seeds for sowing purpose. Adebisi *et al.,* (2013) in tropical soybean seed lots reported that Small seed size had registered 97% germination and 90% emergence, whereas those with large seed size produced highest seed/plant, pods/plant and seed yield /plant. Contrary to this, Zarejan *et al.,* (2013) reported that smallest seed size had lowest emergence and hence one can assume that plants grown from small seed had less fertile tillers than those grown from large seed and accordingly there is a decline in biological and grain yield in smallest seed size.

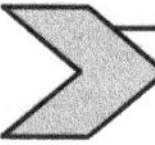 **6.5 Relationship between Seed Vigour and Crop Yield**

The relationship between seed vigour and the ultimate yield of the plants have not been clearly indicated and several published reports concluded that differences in yield were primarily attributed to the differences in the plant stand. The establishment of a predetermined level of population of plants particularly in vegetable crops is critical especially under direct drill sowing. The vegetable crops may not be in a position to compensate the yield as readily as crops with tillering habit. Differences in percentage emergence is expected to influence plant population density and the spatial arrangement of plants. One should always keep in mind that the uneven emergence could also be due to (a) limited moisture present in the soil medium upon sowing, (b) placement of seeds at irregular depths, (c) compaction of soil after seeding, (d) excessive amounts of plant residues in the soil surface and ultimately (e) the low seed vigour. If seed quality has affected only percentage emergence, then the growers could theoretically overcome such difficulty by adjusting the seed sowing rates. Though it is possible to quantify the field emergence of seed lots on different seedbeds, in practice it is very difficult to make adjustments to seed sowing rates because of the difficulties encountered in forecasting the environment of a particular seedbed. Positive associations between seed quality, vigour and yield in barley (Perry and Harrison, 1977), Cotton (Bishnoi, 1971), Groundnut (Basrin, 1970) and Rice (Sitisrung, 1970) have been reported. In several crops, the effects of seed vigour on total emergence, rate of emergence and uniformity of emergence and plant stand establishment have been well documented by Heydecker (1972).

Since, AOSA has defined seed vigour as "all those seed properties that determine potential for fast and uniform emergence, and development of seedlings under a wide range of field conditions". Similarly, 'quality of seeds' too have been defined as an attribute of viability and vigour that enables the emergence and establishment of normal seedlings under a wide range of environments. Therefore, Hampton, (2002), has aptly reported that seed vigour exerts high influence on the establishment of initial plant population as well as their adequate development which will affect the crop yield consequently. This invariably means, low germination speed, high sensibility to stress of seeds and seedlings during the process of germination and ultimately results in plants with slow, low and irregular growth or with less or poor root development and hence have been considered as typical characters of seed with low physiological potential, Markos-Filho, (2005). Whereas Geneva (2008) reported that seed vigour is a complex feature, which is dependent on plant genetics, environmental position of the plant, crop harvesting and processing. Therefore, seed vigour can be considered one of the most important parameters of seed quality.

A seed attains its maximum seed vigour mostly at the end of the seed filling phase on the mother plant and is widely known as mass maturity or physiological maturity (TeKrony and Hunter, 1995; TeKrony and Egli, 1997) but according to authors like Demir and Ellis, (1992); Sanhewe and Ellis, (1996); Ghassemi-GoleZani and Mazllomi-Oskooyi, 2008) this could be slightly after physiological maturity. These contradictory published results so far could be either due to the use of a different species or probably relates to differences in the harvesting time, TeKrony and Egli, (1997). After attaining this highest point of vigour, seeds will remain with this quality only for some time and thereafter begin to deteriorate even while still intact on the mother plant itself or during storage and continue to deteriorate till planted in the soil, Ghassemi - - Golezani and Hosseinzadeh - Mahootchy, (2009). The rate of deterioration due to ageing is positively related to the ambient temperature, relative humidity and seed moisture content, Ellis and Roberts (1981). When ageing is advanced, germination rate, uniformity and tolerance to environmental stress and consequently seedling emergence and post emergence seedling growth are decreased, Hampton and TeKrony, (1995); Khan *et al.,* (2003).

Seed yield is considered to be the characteristic of an individual plant (photosynthetic efficiency), the plant community (leaf area index, plant density, leaf angle *etc.,*), the environment and interaction of growth processes over a period of time in a changing environment. Nagaraju, (2001) in sunflower; Teleghani *et al.,* (2002) in sugar beet; Roorzrokh *et al.,* (2005) in chickpea; reported higher seed yield and yield traits from large size seeds. In addition to the environment, modern agronomic practices like choice of suitable tillage, desired plant population and required seed rate, planting time, maintenance of higher soil fertility status, coupled with good insect and weed management strategies aimed at minimizing the effects of stress on crop growth should be considered. One should always remember that optimum plant population of a seed production field is a function of genotype, soil tilth, and water holding capacity of the soil and irrigation capability. Despite all this, seed vigour has been reported to have no relationship with yield by TeKrony and Egli, (1991) in soybean, Rodo and Markos - Filho, (2003) in onion, Larsen *et al.,* (1998) in oilseed rape and pea, Pedersen and Joy, (2001) in spring and winter wheat; Pedersen and Toy, (2001) in winter barley and Kanouni *et al.,* (2009) in chickpea.

 ## 6.6 Relationship between Seed Vigour and Plant Growth

6.6.1   Effect of Seed Vigour on Early Plant Growth

Seed vigour exerts its affect on initial stages of plant growth through characters such as lower germination, higher susceptibility of seeds and seedlings to stresses coupled with plants exhibiting slow, low and irregular growth and less root system development, which clearly indicates the low physiological potentiality.

6.6.1.1   Initial Seed Growth

TeKrony *et al.,* (1989) declared that seed vigour affects initial crop growth but the differences should decrease during subsequent developmental stages and disappearing totally by the end of the crop life cycle. Crop varieties and hybrids differ substantially in respect of their response to weed competition. Therefore, those that grow faster are considered to be more competitive.

6.6.1.2   Efficient Resource Capture

On the other hand, crop development is directly related to efficient resources capture and utilization with greater competitive ability, therefore, necessitates initial rapid and uniform growth of plant populations. However, in the stressful situations such as initial crop - weed competition, the degree of difference in respect of seeds originating from high and low vigour seeds will always tend to be high. No doubt, this particular situation could be avoided by using high quality seeds. Most vigorous seeds of soybean produced seedlings with longer length of primary and full root. Vanzolini and Carvalho (2002) and Kochinski *et al.,* (2005) reported differences in initial plant growth and grain yield from plants originating from seed lots with varying levels of seed vigour. However, Khan and Hassan (2007) reported that, varieties differ in their competitive ability and correlated with a capacity with rapid resource uptake by the crop when supported by characters like higher leaf area index and more extensive root tissues, Callaway, (1992).

Plenty of published literature is available in respect of differences in initial crop growth related to seed vigour such as Dias *et al.,* (2010) in corn; Schuch *et al.,* (2009) in Soybean; Mielerzski *et al.,* (2008) in black oats; Hofs *et al.,* (2004), Melo *et al.,* (2006) in rice, are only a few to mention. Therefore, Marcos *et al.,* (2011) concluded that, where crop yield is correlated with efficient resource uptake, including water, nutrients and solar radiation, faster initial growth will probably result in better competitive ability and ultimately higher crop grain yield.

6.6.1.3   High Productivity

Rapid and synchronized seed germination and seedling emergence are crucial for achieving an optimal crop stand and ultimately high productivity. Optimum plant population of a seed production field is a function of genotype, soil tilth and water holding capacity, and ultimately the irrigation capability.

6.6.1.4   Canopy Development

Despite all these provisions, heterogeneity during seedling emergence is also expected to affect the canopy development, Pommel *et al.,* (2002). No doubt, establishment of good seedlings obviously gives a guarantee for uniformly strong plants, Harris *et al.,* (2009). Therefore, good seedling establishment and seedling vigour are primarily essential for sustainable and profitable crop production and hence is considered the most critical stage of a developing crop. This appears to be truer in case of crops like maize, Ghiyasi *et al,* (2008). However, Egli and Rucker, (2012) have confirmed that high-vigour seeds always have more uniform emergence than low vigour seeds.

6.6.2   Effect of Seed Vigour on Late Plant Growth

The affect of seed vigour on late vegetative growth is frequently related to yield in crops that are either harvested mostly at the vegetative stage or even during early reproductive stage. However, there usually no such relationship exists, in crops harvested at full reproductive maturity, because seed yields at full reproductive maturity are usually not closely associated with vegetative growth. Therefore, the use of high-vigour seeds can be justified for all crops. Rapid emergence of seedlings from high vigour seed lots were reported in Kenaf by Mentis and Smith, (2003), in Sorghum by Damavandi *et al.,* (2007), in Safflower by Khavari *et al.,* (2009), in Common bean by Kolasinka *et al.,* (2000) and in Maize by Tekrony and Egli, (1989). Contrary to this, Peksen *et al.,* (2002) in Pea; Kumar *et al.,* (2005) in Mustard; Rastegar and Kandi, (2011) in Soybean reported that smaller seeds performed fast and gave better and uniform germination than larger seeds.

Low seed vigour greatly influences both the number of emerging seedlings, and the timing and uniformity of seedling emergence. Plants with delay in emergence, at any given stage will have a lower development than plants with early emergence. Under such situations, the differences pertaining to competitive ability for light and other resources like water, nutrients of soil will become pronounced and leads ultimately to uneven plant growth. This has a major impact upon many aspects of crop production that determine cost effectiveness and the inputs required, and also has direct influence on the yield and marketing quality of a crop (Finch-Savage, 1995) and (Bleasdale, 1967). Any type of subsequent efforts or amount of inputs used during later stages of crop development will not compensate for this outcome. Most of the plant tissues involved in the production of dry matter and ultimately the yield are produced only after the emergence of seedlings. Hence it seems unlikely that the seed vigour would influence their ability to carry out physiological processes and accumulate dry matter.

Though low vigour seeds yield less than high vigour seeds at same planting densities, this cannot be demonstrated convincingly as at the time of emergence, in the near absence of competition for growth resources, plant growth is nearly exponential. Hence, even a few days delay in emergence will lead to substantial difference in growth in subsequent stages of plant growth especially when the seedlings have high relative growth rate or when the crop is sown at high densities (between plant competition), or when the crop is harvested not long after seedling emergence. Egli *et al.,* (1990) reported that soybean seedlings with necrotic lesions or physiological injury on cotyledons have registered greatly reduced growth rates. Since, a greater proportion of this type of seeds is the result from low vigour seed lots, the average of size of seedlings may be smaller after emergence at any time. Therefore, obviously it takes longer time to attain maximum growth rates.

6.7 Effect of Seed Vigour on Crop Yield

Glenn and Daynard (1974) had successfully demonstrated that irregularity among plants development in a population resulted in drastic decrease of grain yield and hence suggested uniform population development to maximize final yield. Khah *et al.,*(1989) also reported that faster emergence of wheat plants is associated with higher development of seedlings originated from highly vigour seeds and thereby resulted in positive advantages that reflect on grain yield. High quality seeds may improve crop yield via high and rapid emergence of seedlings. Therefore, cultivation of high quality seeds is essential for satisfactory yield production. Several published reports have clearly shown that poor stand establishment caused by low quality seed consequently resulted in low yield in corn, Cruz- Garcia *et al.,* (1995); Copeland and McDonald, (2001) in barley; Iqbal *et al.,* (2002) in cotton seed; Samrah and Al-Kofahi (2008) in soybean; and Ghassemi- Golzani *et al.,* (2010) in oilseed rape. Further, it depends on the form of crop harvested also. Forage crops are harvested at vegetative and /or early reproductive stage of growth gives positive relationship between seed vigour and yield, TeKrony and Egli, (1991). In case of meadow broom grass, the total emergence was correlated with standard germination and both total emergence and forage yield were correlated significantly with accelerated ageing percentage, Hall and Wiesner, (1990).

Markos-Filho, (2005) reported that in a general way seed vigour would affect initial growth, however, those differences should decrease in subsequent development stages disappearing until the end of the crop cycle.

Reason: Slower rate of emergence is frequently associated with low vigour seed. It means, the plants thus produced are smaller in size and therefore have only reduced vegetative yield when compared to the yield of high vigour seeds, Ellis, (1989).

In case of crops harvested for grain / seed at harvest maturity under normal conditions hardly shows any relationship between seed vigour and yield, TeKrony and Egli, (1991).

Reason: Probably the photosynthetic capacity of a single leaf is not affected by seed vigour but by its genetic composition. Assume when a plant population attains optimum crop canopy, leaf area index of six, irrespective of seed vigour, photosynthetic capability of canopy is sufficient to maximize the yield, but in reality seeds with low vigour produce only smaller plants, therefore, it is difficult to achieve a leaf area index of more than six by the time the reproductive development is initiated and hence results in yield loss.

6.7.1   Seeds with Low Vigour

Low vigour seeds, on the other hand, cause the development of uneven seedlings in a plant stand and thereby reduce final grain yields. Ford and Hicks, (1992), reported that grain yield to the tune of less than 11 per cent was observed in maize fields due to the failure to emerge uniformly and this could be attributed to the fact that the initial differences in plant size caused by cold stress in the early stages continued into the later stages of plant development (Landi and Crosbie, (1982). Similarly, in rice, Siddique (1986) classified seeds of five cultivars into high, medium and low vigour based on conductance and reported 7.89 and 14.14 percent reduced grain yield respectively from medium and low vigour seeds when compared to high vigour seeds.

6.7.2   Seeds with High Vigour

High vigour seed does not necessarily mean higher yields. Because, depending on the original vigour level of the seed, crop/ variety planted, plant stand achieved etc., influence yield during growing season. High vigour seeds gave increased grain yield only when plant densities were suboptimal for maximum yields either under late sown conditions or emergence under highly stressful environments. Mondo *et al.,* (2012) on maize reported effects of seed vigour on intra specific competition between plants originated from high and low vigour seeds resulted in unequal growth and less yield.

Competition between crop plants and weeds is a challenging task in crop production, and in case of maize, barring environmental variables; it is the chief cause of yield loss, Rajcan and Swanton, (2001). Selection of genotypes with morpho-physiological characters such as fast initial growth and height could increase the ability of plants to compete with weeds, Balbinot-Junior and Fleck, (2005). Use of high vigour seeds in integrated weed management could be one of the possibilities that need to be explored to reduce the amounts spent on herbicides, Monto *et al.,* (2012).

Plant establishment, growth and final yield are affected by many factors such as plant densities and environmental influences. Therefore, the effect of vigour on crop performance is complex; Pedersen and Joy, (2001) and it may be difficult to find a general relationship between results of a vigour test and actual crop performance, Larsen *et al.,* (1998). Since seed quality is the most important physiological indicator of germination, organizing successful seed production will be the main prerequisite for achieving the economic benefits.

If seed quality did not have some potential to influence crop yield, then there would be no need for seed certification too. Therefore, it is necessary to put great effort not only in seed production technology but also in the seed quality testing technology. Again it is possible through better understanding the biology, physiology, biochemistry and molecular biology in understanding the complex processes undergoing in the seed in addition to the characterization of important agronomic traits. Studies to analyze the impact of plant to plant variability in maize by Liu *et al.,* (2004), Andrade and Abbate (2005) and Tillenaar *et al.,* (2006) resulted in the conclusion that non- uniform special distribution of plants along the planting row as well as uneven seedling emergence results in variability in biomass among adult plants and reduction in grain yield at canopy level. Egli and Rucker (2012) also reported slow emergence from the low vigour seeds resulted in unevenness in plant stand sufficient enough to reduce crop yield.

 ## 6.8 Summary

The quality of seed used determines the success of germination, development and ultimately of the yield of the succeeding crop. Seed vigour exerts potential influence on yield through seedling emergence. For sustainable and profitable crop production, good seedling establishment capacity or vigour is highly essential. Number and speed of seedling emergence determines cost effectiveness, inputs required and marketing of a quality crop.

Seed size directly influences germination, development and duration and performance of plants. But, no consistent relationship between seed mass and viability or vigour was found. An indirect relationship exists via reduction in enzyme ability in large seeds. In addition to environment, a number of agronomic practices also influence plant growth and development and ultimately yield.

Genotype, soil tilth, water holding capacity of soil and irrigation capability etc., determines the optimum plant population establishment capacity of field. Low vigour seeds are not acceptable for cultivation, because they emerge slowly, produce smaller sized plants, reduced vegetative yield.

Seed vigour exerts both direct and indirect influence on yield. Positive relationship between vigour and yield was reported in forage crops harvested at early - reproductive stage, whereas no relationship exists between vigour and yield in crops harvested for grain at physiological maturity.

High vigour seed does not necessarily mean high yields, but this could be achieved under certain conditions. A clear understanding of agronomic traits together with various complex processes undergoing in the seed on one hand and focused efforts in seed production and quality testing technology on the other hand renders economic benefits easily. Use of high vigour seeds in integrated weed management needs to be explored.

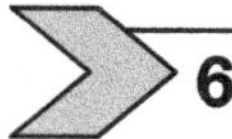 **6.9 Exercise**

I. Answer the following questions briefly

1. List out the conditions on which seed vigour depends.

2. List out the factors on which seed size depends.

3. Why use of high vigour seeds justified for all crops.

4. List out the factors exerting influence in minimizing plant growth.

5. What are the strategies recommended by Dormbos to maintain yield of vigourous seeds.

6. Why development of new varieties should always be oriented at tackling the negative effects of environment on yield and quality.

7. How environmental stress during seed filling stage reduces seed quality.

8. List out the factors that contributed to loss of seed quality after physiological   maturity in soybean.

9. Why low vigour seed is not acceptable for cultivation.

10. Explain how seed vigour exerts influence on yield directly and indirectly.

11. List out the causes for uneven emergence.

12. Explain how seed vigour affects vegetative growth.

13. Drought in early vegetative phase causes reduction in yield of maize. How?

14. List out the two distinct factors for poor seed vigour that affect crop yield.

15. Seed production should be located in regions with favourable conditions specific to species. Explain why.

16. Why slower rate of emergence is frequently associated with low vigour seed?

17. Why crops harvested at full maturity hardly show any relationship between seed vigour and yield?

18. Relationship between weed competition and early seedling vigour.

19. Direct and indirect effects of seed vigour on yield.

20. Impact of poor vigour seeds on crop yield.

II. Fill in the blanks with suitable words

1. Seedling vigour exerts direct influence on crop performance through ---------.

2. --------- and --------- should be high at planting to realize full genetic potential.

3. Large seeds are more prone to --------- damage during harvesting.

4. Loss of yield always depends on --------- and effect of vigour on ---------.

5. High temperature during seed development addition to visual quality reduces --------- and ---------.

6. The minimum requirements of a seed production field for getting optimum plant population are water holding capacity, soil tilth, irrigation capability and ---------.

7. Warm and humid weather prolongs --------- period as the moisture has to be reduced from 50 to 15 per cent in maize.

8. Low seed vigour exerts influence on ---------, --------- and --------- in seedling emergence.

9. Positive relationship exists between seed vigour and yield in --------- crops harvested at reproductive stage.

Answers

1. Seedling establishment
2. Viability, Vigour
3. Mechanical
4. Crop type, plant population
5. Germination and vigour
6. Genotype
7. Desiccation
8. Number, timing and uniformity
9. Forage

6.10　References

Adebisi, M. A., Kehinde, T.O., Salau, A.W., Okesola, L.A., Porbeni, J.B.O., Esuruoso, A.O., and Oyekale, K.O. (2013). Influence of different seed size fractions on seed germination, seedling emergence and seed yield characters in tropical soybean. Int. J. Agric. Res. 8: 26-33.

Adebisi, M.A., T.O. Kehinde, M.O. Ajala, E.F. Olowu and S. Rasaki, (2011). Assessment of seed quality and potential longevity in elite tropical soybean (*Glycine Max* L.) Merrill grown in Southwestern Nigeria. Niger. Agric. J., 42: 94-103

Ahirwar, J.R., (2012). Effect of seed size and weight on seed germination of *Alangium lamarckii*, Akola, India. Res. J. Recent Sci., 1: 320-322.

Ambika, S., Manonmani, V., and Somasundaram, G., (2014). Review on Effect of Seed Size on Seedling Vigour and Seed Yield. *Research Journal of Seed Science,* 7: 31-38.

Andrade, F.H., and 1Abbate, P.E., (2005). Response of maize and soybean to variability in sand uniformity. Agronomy Journal. 97(4): 1263-1269.

Awan, S.I., Niaz, S., Malik, MFA., and Ali, S., (2007). Analysis of variability and relationship among seedling traits and plant height in semi-dwarf wheat. J. Agril. Soc. Sci. 3(2): 59-62.

Balamurugan, P., P. Srimathi and K. Sundaralingam, (2004). Influence of seed size on vigour and productivity of safflower. Sesame Safflower Newslett., Vol. 19.

Balbinot-Junior,A.A. and Fleck, N.G., (2005). Weed management in the corn crop through plant spatial arrangement and characteristics of genotypes. Ciencia Rural. 34(6): 245-252.

Basra, SMA., Zia, MN., Mehmood, T., Afzal, I., and Ahmed, N., (2002). Comparison of different in vigouration techniques in wheat seeds. Pakistan J Arid Agric. 5: 11-16.

Basrin, C.C. (1970). Relation of certain physiological properties of peanut seed to field performance and storability. Ph.D Thesis. Mississippi State University. USA.

Bishnoi, U.R., (1971). Deterioration of cotton seeds under warm mist storage conditions and itsconsequences in terms of seed and plant responses. Ph. D Thesis, Mississippi State Univ. USA.

Bleasdale, J. (1967). The relationship between the weight of a plant part and total weight as affected by plant density. J Hortic Sci 42: 51-58.

Callaway, MBA., (1992). A compendium of crop varietal tolerance to weeds. American Journal of Alternative Agriculture. 7(4): 169-180.

Charles- Edwards, D.W., (1982). Physiological determinants of crop growth. Acad. PressLondon. Pp 161. Copeland, L. O., and Mc Donald, M.B. (2001). Principles of Seed Science and Technology 4[th]edition. Academic publishers, Boston.

Cookson, W.R., J.S. Rowarth and J.R. Sedcole, (2001). Seed vigour in perennial ryegrass (*Lolium perenne* L.): Effect and cause. Seed Sci. Technol., 29: 255-270.

Curz- Garcia, F., Gonzalez-Hernandez, V.A., Molina- Moreno, J., Vazquez-Ramos, J.M., (1995). Seed deterioration and respiration as related to DNA metabolism in germinating maize. Seed Sci. Tech. 23: 477-486.

Dar, FA., Gera, M., Gera, N., (2002). Effect of seed grading on germination pattern of some multipurpose tree species of Jammu Region. Indian Forest. 128: 509-512.

Damavandi, A., Latifi, N., and Dashtban, A.R., (2007). Analysis of seed vigour tests and its performance in Sorghum, J. Agric. Sci. and Natur. Resou., 14(5): 16-23.

Delouche, J. C., (1980). Environmental effects on seed development and seed quality. Horti. Science. 15: 775-780.

Dias, M.A.N., Mon do, V.H.V., Cicero, S.M., (2010). Maize seed vigour and weed competition. Brazilian seed Jour. 32 (2): 93-101.

Dias, M.A.N., Pinto, T.L.F., Mondo, V.H.V., Cicero, S.M., Pedrini, L.G., (2011). Direct effects of soybean seed vigour on weed competition. Brazilian seed Jour. 33 (2): 346-351.

Dornbos, D. L., (1995). Seed Vigour. In AS Basra (ed) Seed Quality basic mechanisms and agricultural implications. Pp 45 - 80. Food Products Press. New York.

Efeito do vigor de sementes de soja sobre o sen desempenho cm campo. Reviota Brasileira de sementes Brasilia.

Egli, D.B., and Rucker, M., (2012). Seed vigour and uniformity of emergence of corn seedlings. Crop Sci. 52(6): 2774-2782.

Egli, D.B., TeKrony, D.M., Wiralagi, R.A., (1990). Effect of soybean seed vigour and size on seedling growth. J. of Seed Tech. 14(1): 1-12.

Ellis, R. H. (1989). The effects of differences in seed quality resulting from priming or deterioration on the relative growth of onion seedlings Acta. Horti. 253: 203-211.

Ellis, R.H., and Roberts, E.H., (1981). The quantification of ageing and survival in orthodox seeds. Seed Science and Technology. 15: 1-17.

Fenner, M., (1991). The effects of the parent environment on seed germinability. Seed Sci. Res 1: 75-84. Finch-Savage W (1995). Influence of seed quality on crop establishment, growth and yield. Seed Quality: Basic Mechanisms and Agricultural Implications: 361-384.

Ford, J.H., and Hicks, D.R., (1992). Corn growth and yieldin uneven emerging stand. J. Prod. Agric. Madison. 5: 185-189.

Geneva, R.L.,(2008).Vigour testing in small seeded horticultural crops.Acta Horticulture. 782 :77- 82. Ghassemi-Golezani, K., J. Bakhshy, Y. Raey and A. Hossenizadeh-Mahotchi, (2010). Seed vigour and field performance of winter oilseed rape (Brassica *napus* L.) cultivars, Not. Bot. Hort. Agrobot. Cluj., 38(3): 146-150.

Ghassemi-Golezani, K. and A. Hossenizadeh- Mahotchi, 2009. Change in seed vigour of fababean (Vicia *faba* L.) cultivars during development and maturity, Seed Sci.Tech., 37: 713-720.

Ghassemi- Golezani K, Mazloomi-Oskooyi R (2008). Effect of water supply on seed quality development in common bean (Phaseolus *vulgaris* var). J Plant Production 2:117-124.

Ghiyasi, M., Miyandoab, M.P., Tajbakhsh, M., Salehzade,H., Meshkat, M.V., (2008). Influence of different osmo- priming treatments on emergence and yield of maize (*Zea mays* L). Res. J. Biol. Sci. 3: 1452-1455.

Glenn, F.B., Dynard, T.B., (1974). Effects of genotype, planting pattern and planting density on plant to plant variability and grain yield of corn. Canadian journal of Plant Science. 54(2): 323-330.

Hall,R. D., and Wiesner, L.E., (1990). Relationship between seed vigour tests and field performance of 'Regar: meadow broomegrass. Crop Science. 30: 967-970.

Hampton, J.G., (2002). What is seed quality? Seed Sci. and Tech. 30: 1-10.

Hampton, J.G., (1995). Methods of viability and vigour testing- A critical appraisal. In A.S. Basra (ed.) Seed Quality 81-118.Food Products Press, Bringhamton.

Harries, D., Joshi,A., Khan,P.A., Gothakar, P., Sodhi, P.S., (1999). On farm seed priming in semi arid agriculture: development and evaluation in corn, rice and chickpea in India using participatory methods, Expt.Agric. 35: 15-29.

Heydecker, W. (1972). In E.H. Roberts (ed.) Vigour and viability of seeds. pp 209-252. Chapman & Hall, London.

Hofs, A., Schuch, L.O.B., Peske, S.T., Barros, A.C.S.A., (2004). Emergence and initial growth of rice seedlings according to seed vigour. Brazilian seed Jour. 26(1): 92-97.

Hojjat, S.S., (2011). Effect of seed size on germination and seedling growth of some lentil genotypes. Int. J. Agric. Crop Sci., Vol. 3.

Iqbal, N., Basra, S.H.M.A., Rehman, K., (2002). Evaluation of vigour and oil quality in cotton seed during accelerated ageing. Int. J. Agric. Biol. 4: 318-322.

Jerlin, R. and K.K. Vadivelu, (2004). Effect of fertilizer application in nursery for elite seedling production of Pungam (*Pongamia pinnata* L. Picrre). J. Trop. Agric. Res. Ext. 7: 69-71.

Kanouni, H., M. Khalily and R.S. Malhotra, 2009. Assesment of cold tolerance of chickpea at rainfed highlands of Iran, American-Eurasian J. Agric. and Environ. Sci., 5(2): 250-254.

Khan, I.A., Hassan, G., (2007). Competitive ability of various wheat cultivars with wild oats. In : African Crop Congress Proceedings. 8. African Crop Sci. Soc. 1901-1904.

Khan, M.M., Iqbal, M.J., Abbas, M., and Usman, M., (2003). Effect of accelerated ageing on viability, vigour and chromosomal damage in pea (Pisum sativum L). Seeds. Pakistan J. Agric. Sci. 40: 50-54.

Khawari,F., Ghaderi Far F, and Soltani , E., (2009). Laboratory tests for predicting seedling emergence of safflower (Carthamus tinctorius, L.) cultivars. Seed Tech. 31: 189-193.

Khah, E.M., Roberts, E.H., and Ellis , R.H., (1989). Effects of seed ageingon growth and yield of spring wheat at different plant population densities. Field Crop Research. 20(3): 175-190.

Kolchinski, E.M., Schuch, L.O.B., and Peske, S.T., (2005). Seeds vigour and intra-specific competition in soybean. Ciencia Rural. 35(6): 1248-1256.

Kolasinska, K., J. Szyrmer and S. Dul, 2000. Relation between seed quality tests and field emergence of common bean seed. Field Crop Res., 40: 470-475.

Kumar, A., R.P.S. Tomer, R. Kumar and R.S. Chaudhary, (2005). Seed size studies in relation to yield attributing parameters in Indian mustard [*Brassica juncea* (L) Czern and Coss]. Seed Res., 33: 54-56.

Kaydan, D. and M. Yagmur, (2008). Germination, seedling growth and relative water content of shoot in different seed sizes of triticale under osmotic stress of water and NaCl. Afr. J. Biotechnol., 7: 2862-2868.

Landie, P., and Crossbie, T.M., (1982). Response of maize to cold stress during vegetative growth. Agron Journal. 74: 765-768.

Larsen, S. U., Poulsen, F.V., Eriksen, E.N., and Pedersen, H. (1998). The influence of seed vigour on field performance and evaluation of the applicability of controlled deterioration vigour test in oilseed rape (*Brassica napus)* and Pea (*Pisum sativum*). Seed Science and Technology. 26: 627- 641.

Liu, W., Tollenaar, M., Stewart, G., Deen, W., (2004a). Within plant spacing variability does not affect corn yield. Agron. J. 96(1): 275-280.

Liu, W., Tollenaar, M., Stewart, G., Deen, W., (2004b). Response of corn grain yield to spatial and temporal variability in emergence. Crop Sci. 44(3): 847-854.

Melo, P.T.B.S., Schuch, L.O.B., Assis, F.N., Concenco, G., (2006). Individual behaviour of rice plants from different physiological quality seeds in rice populations. Brazilian seed J. 28(2): 89-94.

Marcos- Filho, J., (2005). Fissiologia de sementes de plantas cultivadas. Piracicaba. Fealq. Pp. 495.

Mentis, P.D. and C. A. Smith, 2003. Kenaf seed storage duration on germination, emergence and yield Industrial Crops and Products, 17: 9-14.

Mielezrski, F., Schuch, L.O.B., Peske, S.T., Panozzo, L.E., Peske, F.B., Carvalho, R.R., (2008). Field performance of isolated plants of hybrid rice in function of the seed physiological quality. Brazilian Seed Journal. 30(3): 86-94.

Mondo, V.H.V., Cicero, S.M., Dourado- Neto, D., Pupium, T.L., Dias, M.A.N., (2012). Maize seed vigour and plant performance. Brazilian Seed Jour. 34(1): 143-155.

Nagaraju, S., 2001. Influence of seed size and treatments on seed yield and seed quality of sunflower cv. Morden. M.Sc. Thesis, University of Agricultural Sciences, Dharwad, Karnataka, India.

Nerson, H., (2002). Relationship between plant density and fruit and seed production in muskmelon. J. Am. Soc. Hort. Sci., 127: 855-859.

Nik, M.M., M. Babaeian and A. Tavassoli, (2011). Effect of seed size and genotype on germination characteristic and seed nutrient content of wheat. Sci. Res. Essays, 6: 2019-2025.

Peksen, E., A. Peksen, H. Bozoglu and A. Gulumser, (2004). Some seed traits and their relationships toseed germination and field emergence in pea (*Pisum sativum* L.). J. Agron., 3: 243-246.

Pedersen, J.F., and J.J. Toy., (2001). Germination, Emergence and yield of 20 plant colour, seed-color, near-isogenic lines of grain sorghum, Crop Sci. 41: 107-110.

Perry, D. A., and Harrison, J.G., (1977). Effects of seed deterioration and bed environment on emergence and yield of spring- sown barley. Ann. Appl, Bio. 86(2): 291-300.

Pommel, B., Mouraux, D., Cappenllen, O., Ledent, J.F., (2002). Influence of delayed emergence and canopy skips on the growth and development of maize plants; A plant scale approach with CERES- Maize European Jour. Of Agronomy. 16(4): 263-277.

Rajcan, I., and Swanton, C.J., (2001). Understanding maize- weed competition: resource competition, light quality and the whole plant. Field Crops Res. 71(2): 139-150.

Rench, W.E., and Shaw, R.H., (1971). Black layer development in maize. Agronomy journal, madison, 63(2): 303-305.

Rastegar, Z. and M.A.S. Kandi, (2011). The effect of salinity and seed size on seed reserve utilization and seedling growth of soybean (*Glycin max*). Int. J. Agron. Plant Prod., 2: 1- 4.

Ries, S.K. and E.H. Everson, 1973. Protein content and seed size relationships with seedling vigour of Wheat cultivars. Agronomy Journal.65:884-886.

Rodo, A.B. and L. Marcos-Filho, 2003. Onion seed vigour in relation to plant growth and yield, Horticulture Brasileria, Brasilia, 21(2): 220-226.

Roozrokh, M., Shams, K and Vghar., (2005). Effects of seed size and seedling depth on seed vigour of chickpea. Proceedings of the 1st International Congress on Legumes, Nov. Mashhad, Iran.

Roy, S.K.S., A. Hamid, M.G. Miah and A. Hashem, (1996). Seed size variation and its effects on germination and seedling vigour in Rice. J. Agron. Crop Sci., 176: 79-82.

Sadeghi, H., F. Khazaei, S. Sheidaei and L. Yari, (2011). Effect of seed size on seed germination behavior of safflower (*Carthamus tinctorius* L). J. Agric. Biol. Sci., 6: 5-8.

Siddique, A.B., and Wight, D., (2004). Effects of time of harvest at different moisture contents on seed fresh weight, dry weight, quality (viability and vigour) and food resources of peas(*Pisum sativum* L). Asian J. Plant Sci. 13: 983-992.

Sitisrung, P., (1970). Deterioration of rice seed in storage and its influence on field performance. Ph D., Thesis Mississippi State University, USA.

Samrah, N.H., and Al-Kofahi, S., (2008). Relationship of seed quality tests to field emergence of artificial aged barley seeds in the semiarid Mediterranean region. J. Agric. Sci. 4: 27-230.

Schuch,L.O.B., Kolchinski, E.M., Finatto, J.A., (2009). Seed physiological quality and individual plants performance in soybean. Brazilian Seed Journal. 31(1): 144-149.

Seshu, D.V., and Dadlani, M., (1993). Seed Research. 2: 700-706.

Sulochanamma, B.N. and Y.T. Reddy, (2007). Effect of seed size on growth and yield of rainfed groundnut. Legume Res., 30: 33-36.

Sun, Q., Wang, J., and Sun, B., (2007). Advances in seed vigour physiological and genetic mechanisms. Agril. Sci. China, 6(9): 1060-1066.

TeKrony, D. M., and Egli, D.B. (1997). Accumulation of seed vigour during development and maturation in Ellis, R.H, Black, M., Murdoch,A. J. and Hong, T.D. (eds.) Basic and applied aspects of seed biology. Academic Publishers. 369-384.

TeKrony, D. M., and Egli, D. B., (1991). Relationship of seed vigour to crop yield, A Review. Crop Science. 31: 816-822.

TeKrony, D.M. and D.B. Egli, (1989). Corn seed vigour on No-village field performance, Plant growth and grain yield. Crop Sci., 29: 1528-1531.

TeKrony, D.M., and Hunter, J.L., (1995). Effect of seed maturation and genotype on seed vigour inmaize. Crop Sci. 35: 862-868.

Taleghani, F., M. Dehghanshooar, A. Ghasesmi, A. Yousefabadi, M. Chegini and F. Mohamadi., (2002). Determination of optimum seed size and quantity of coating materials for monogerm sugar beet seed. Iran. J. Sugarbeet, 18: 10-12.

Tollenaar, M., Deen, W., Echart, L., Liu, W., (2006). Effect of crowding stress on dry matter accumulation and harvest index. Agronomy Jour. 98 (4): 930-937.

Vanzolini, S., and Carvalho,NM., (2002). Effects of Soybean seed vigour on field plant performance. Journal of Seed Science. 24(1): 33-41.

Verma, S.K., Bajpai,G.C., Tewari, S.K., and Jogendra, S., (2005). Seedling index and yield as influenced   by seed size in pigeonpea. Legume Res., 28: 143-145.

Willenborg, C.J., Wildeman,J.C.,. Miller, A.K., Rossnagel, B.G., and Shirtliffe, S.J., (2005). Oat germination characteristics differ among genotypes, seed sizes and osmotic potential. Crop Sci., 5: 2023-2029.

Zareian, A., Hamedi, A., Sadeghi,H., and Jazaeri, M.R., (2013). Effect of seed size on some germination characteristics, seedling emergence percentage and yield of three wheat in laboratory and field. Middle East Journal of Science and Research. 13: 1126-1131.

CHAPTER 7

Seed Deterioration and Mechanism of Seed Vigour

(The impossible is often untried)

7.1 Introduction

Despite having great economic importance, several multipurpose tree species of agro-forestry origin, including those fruit yielding trees of tropical and temperate countries, yet suffer from reduced shelf life of their seeds and thereby rendering the reforestation programmes a difficult task. Same is true with the tree and crop species of arid and semi arid areas, where desertification is a major problem and issues pertaining to reforestation remained an absolute failure due to the inability to supply quality seed, leave alone the preservation and conservation of such genetic resources. This is so because, in some species, it is not possible to store the seeds till sowing in the next season, while some other species lose their viability either during collection or even before their transportation. Such seeds are often sensitive to low temperatures and get damaged by desiccation and are called as recalcitrant seeds. Since the long-term genetic conservation or storage of such seeds poses a still bigger challenge, short and medium term strategies are the need of the hour for optimum use of these species in reforestation programmes.

A mature seed, upon being fully loaded with molecular reserves of food and energy, is basically designed to initiate its function only under favourable conditions. Until such conditions are made available, seeds will remain in a physiologically inactive state, barring the maintenance of respiration that too at the lowest permissible rate. In between harvesting and sowing in the ensuing season, the seeds deteriorate in quality over a period of time as a consequence of attack and spoilage by microorganisms and insects during production as well as in storage, processing and transportation. Cracked and broken seeds deteriorate more rapidly than undamaged seeds. Despite the absence of physical damage, the seeds might become susceptible to rapid deterioration due to physiological impairment. In addition to this,

LEARNING OBJECTIVES

- Seed deterioration, Viability, and Classification of seeds
- Storage requirements of seeds
- Mechanism of seed vigour
- Physical and chemical reactions involved in seed deterioration
- Changes in ultra-structural properties of organelles
- Changes at genetic, physiological, biochemical and molecular Levels.
- Theories proposed on seed deterioration.

extremes of temperature, moisture, deficiency of minerals and a number of environmental stresses during seed development and maturation can also reduce the longevity of seeds. When compared to mature and large seeds, it is difficult to store immature and small seeds present in a seed lot, Patil and Andrews (1985) reported that hard-seededness extends the longevity of seeds. The loss due to deterioration of seeds has become a major problem for agriculture in both developing and under developed countries.

 ## 7.2   Seed Deterioration

Deterioration is a natural phenomenon of all living organisms and seed is no exception to this rule. No doubt, deterioration and death are the two inevitable consequences of life. Yet, it is possible to retard the rate of deterioration, that too by providing the optimum storage conditions. This makes seed deterioration an inexorable process. Therefore, Delouche (1973) has aptly characterized seed deterioration as an inexorable and irreversible process of seed populations. Seed deterioration, an inexorable process that varies significantly depending on the species and storage conditions, basically refers to all those processes which contribute to the loss of vigour and viability of seed. Deterioration in seed can occur in the field after they attain physiological maturity, particularly when the harvesting is delayed in wet weather and is commonly referred to as weathering, a condition which is as bad for seed quality as seed storage conditions. But before this stage is attained, the individual seeds lose their vigor at different rates. This is what we generally observe as increased sensitivity to adverse storage conditions resulting in changes at both physiological, biochemical and molecular levels. Roberts (1972) differentiated these changes occurring during seed deterioration on the basis of genetic, cytological and metabolic activities. The relationship between physiological and biochemical changes and loss of germination ability was reported by Priestley, (1986) and Smith and Berjak, (1995). Keeping in view the overlapping occurrence and interrelated nature, the changes occurring during seed deterioration were classified as physiological and performance indicators by Copeland and Mc Donald (2001).

At Physiological level the changes are represented by

    (i)   Slower rate of seedling growth and development

    (ii)   Reduced tolerance to biotic and abiotic stresses and early seedling growth.

    (iii)   Lack of hypocotyls elongation and stunting of radical.

And ultimately there is an unexpected increase in the number of abnormal seedlings. At Biochemical level the earliest possible changes occur chiefly at sub-cellular level leading to

    (i)   Reduced membrane integrity resulting in enhanced solute leakage from imbibed seed.

    (ii)   Deterioration of membranes affect several metabolic activities including respiration rate in respect of gaseous exchange, enhanced RQ, reduced P/O ratio, and enzyme activity in respiratory pathways, etc.

    (iii)   Decrease in the ability to retain inorganic electrolytes and organic solutes such as amino acids, organic acids and water soluble sugars.

    (iv)   Reduced enzyme activity particularly in dehydrogenases, carboxylases amylases, lipases esterases was recorded during seed germination. Basavarajappa *et al*, 1991 reported increase in the activity of degradative enzymes like proteases and phosphatases with seed ageing.

However, the rates of deterioration of seeds vary greatly not only depending on species but also on several other management factors. Some varieties exhibit less deterioration than other varieties in a given species. Even within a variety also, storage potential of individual seed lots differ significantly. It is not an exaggeration to state that even within a seed lot also the individual seeds will exhibit varying levels of storage potentials.

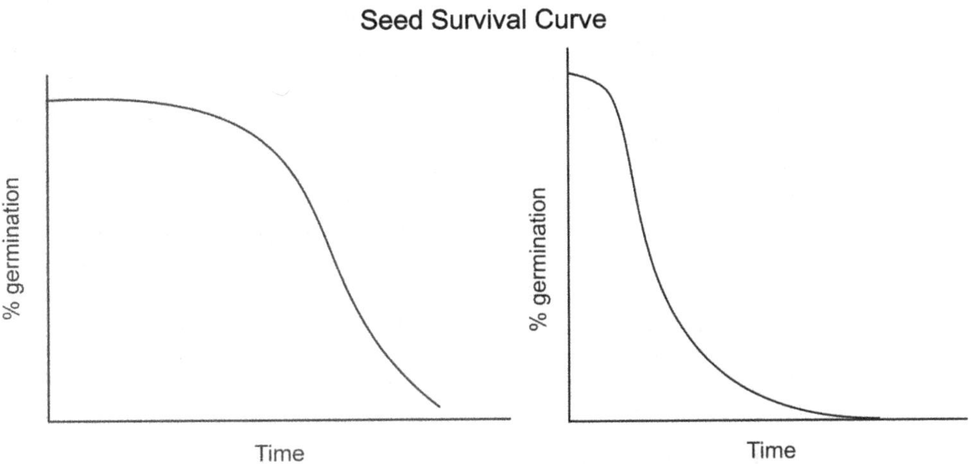

Fig. 7.1 Before natural ageing   **Fig. 7.2** After natural ageing

The term seed deterioration refers to all those processes, which directly or indirectly contribute to the loss of viability and vigour aspects of seed quality. Loss in vigour and viability could be either due to ageing or also due to effect of prevailing adverse environmental conditions and ultimately ending up in the death of seed. Therefore, Seed deterioration is nothing but a permanent reduction in the level of expression of a character or vitality from a higher to lower level over a period of time (Fig. 7.1). It means, after attaining physiological maturity, the seed moves from one stage of deterioration to another in a set pattern with specific symptoms at each point. Even after ageing, seeds maintain germination for some more time. In other words, seed deterioration being a catabolic process, once initiated, it cannot be reversed, a low quality seed lot can never be made in to a high quality seed lot (Fig. 7.2). As a consequence, one can easily find evidence of deterioration at physiological level in the form of (i) discolouration of testa, (ii) poor germination, (iii) poor growth of seedlings, (iv) production of more number of abnormal seedlings, and (v) ultimately resulting the poor yield. Natural ageing is the chief cause of deterioration. Temperature and moisture content are the two important factors affecting seed ageing. Every 1% increase in moisture content and 1^0C increase in temperature doubles the rate of ageing. Therefore, once initiated, the process of seed deterioration becomes irreversible and hence cannot be prevented in any way. However, the process can only be regulated or slowed down to some extent that too under a specific set of conditions only. Seed deterioration though initiates immediately after attaining physiological maturity, it exerts influence on viability only when the seeds have reached an advance stage beyond which natural repair is not possible by the seed itself.

7.2.1 Seed Viability

A good majority of seeds have been uniquely equipped to survive till a right place and time are available for beginning the subsequent generation. This does not necessarily and automatically mean that the seeds will retain their viability indefinitely. The physical condition together with the physiological status, determines the lifespan of seeds. In the expression of seed longevity, the expression of maternal genotype appears to be more pronounced. No doubt, the genetic constitution of the seeds themselves does have some direct impact. The genetic change involved in the expression of undesirable character is caused by either recombination or mutation or even the alterations are caused by pathogenic fungal infections. But still, in the post fertilization era, since the development of seed is associated with physiological activity of loading the ovule with desired nutrients as a function of the sink, wherein, the translocated carbohydrates condense in the form of either starch, or amylose, or amylopectin's and hemicelluloses and accumulate in various storage tissues. Similarly, proteins and fatty acids are also stored in the form of complex polypeptides and triglycerides respectively. In addition to these, a number of auxins, inhibitors, vitamins, different kinds of minerals, enzymes etc., exert direct influence on seed deterioration. On one hand, the auxins promote germination, while on the other hand, the inhibitors inhibit germination and thereby play an important role in the maintenance of seed viability. Potassium nitrate, kinetin, thiourea, and gibberellins too play an important role in germination promotion, where as ferulic acid, para-scorbic acid, Coumarin, ammonia and abscisic acid as found in some seeds, are expected to prevent the germination. One should always remember that the maintenance of longevity of seed depends on the destruction of some of the essential components required for triggering the biochemical steps needed for initiation of germination like Gibberellic acid, Cytokinin, Ethylene etc.,

Just like all living organisms, the seeds too will deteriorate and die eventually. Under a given set of conditions, the time required by a population of seeds, when free from pests and diseases, to become non-germinable has been referred to as longevity. Considerable amount of variation existed in respect of species and cultivars for longevity, Stedman *et al.,* (1996) and ranging from few days in *Acer saccharum* (Jones, 1920) to few hundred years in Indian lotus (Ohga, 1926). Seeds belonging to the species such as *Canna, Lotus* and *Lupines* are viable even after 500 years because they are better equipped both genetically and chemically. Shen-Miller *et al.,* (2002) reported the oldest germinated seed of *Nelumbo nucifera* collected from northern China is 1300 years old.

The long-term natural longevity experiments of Beal, W.J., (1879) at Michigan and J. W. T. Duvel, (1902) at Rosylin, on one hand and artificial longevity studies of seed ageing process by accelerated ageing of Delouche and Baskin (1973) and relative humidity control experiments with saturated salt solutions by Copeland and McDonald (2001) have clearly revealed that longevity of seeds is under the control of seed moisture, temperature and genetic factors.

7.2.1.1   Classification of Seeds

Seeds were classified into different groups depending on the life span, storage duration and tolerance to desiccation as detailed below.

7.2.1.1.1   Ewarts's Classification (1908)

More than a century ago, Ewart (1908) classified seeds into three categories. (i) Microbiotic, (ii) Mesobiotic and (iii) Macrobiotic with a life span of below 3 years, 3-15 years and 15-100 years respectively. This classification carries very little meaning in the absence of information pertaining to availability of its storage environment.

7.2.1.1.2   Robert's Classification (1973)

Seeds of different species in general and even within a species when harvested at different stages of maturity in particular, vary considerably in respect of their extent of tolerance to desiccation, i.e the ability to germinate upon rehydration following desiccation. Roberts (1973) based on desiccation tolerance and storage behaviour of seeds proposed the terms Orthodox and Recalcitrant seeds as detailed below.

7.2.1.1.2.1   Recalcitrant Seeds

Recalcitrant seeds are usually found in warm humid climates, some in temperate and very few in dry areas. These seeds are intolerant to desiccation and low temperatures. Seeds are remarkably short lived and tolerance level is species specific. Dormancy is either absent or weak in recalcitrant seeds. The potential storage period too varies from few weeks in extremely recalcitrant species to several months for more tolerant species. The seeds are usually medium to large in size. Larger size of the seeds could be attributed partly to the prevailing high moisture content. These seeds, being desiccation sensitive, moisture content is always high, 30-70 per cent at maturity, with large variation between individual seeds and cannot be dried to moisture contents below 30% without any injury. Therefore, these seeds cannot conform to the viability equation of Roberts, (1973). Since, the seeds must retain relatively high moisture, are concurrently active metabolically. Recalcitrant seeds accumulate dry mater till the time of dispersal hence they have no or very little maturation drying and are therefore, unable to tolerate freezing. Examples included in this group comprises of large seeded species like *Quercus, Salix, Juglans, Corylus;* Plantation species like *Hevia, Coffee, Theobroma* and aquatic species like Zizania; horticultural species such as Mango, Litchi, citrus. etc.

By storing the seeds at reduced moisture content and temperature the rate of metabolism can be minimized. When fresh recalcitrant seeds begin to dry, viability is first slightly reduced, as moisture is lost, but then begins to be reduced considerably at certain critical moisture content- often termed as lowest safe moisture content. Viability is reduced to zero if further drying continues beyond the lowest safe moisture content and hence survive a limited amount of desiccation. The critical moisture contents vary greatly among species, among cultivars, among seed lots, King and Roberts, (1979); Chin (1988) and method of drying, Farrant *et al.,* (1985); Pritchard and Prendergast, (1986); Pritchard (1991). The lowest safe moisture content vary between extremes of 23.0 per cent for cocoa (*Theobroma cacao*) and 61.5 per cent for *Avicennia marina,* Farrant *et al.,* (1986).

7.2.1.1.2.2   Orthodox seeds

Usually, species with Orthodox seeds dominate arid and semiarid environments; forerunners in humid climates and prevalent as temperate and tropical high altitude species. The seeds of these species can sustain intense desiccation by the end of their maturation period on the mother plant as well as during processing, and hence retain their germination potential over long periods of dry storage. Seeds are long lived because metabolic activity needs water. Optimum storage temperature is 0-5°C at 5-7 per cent moisture content for conventional storage. Most species can be stored for several years and some species can be stored even for few decades. Seeds are usually small to medium in size and are often with hard seed coat. Seeds can be successfully dried to moisture content as low as 5 % without injury or loss to viability, because this low moisture content is practically immobile and does not enter into chemical reactions as it is bound in the macromolecules. As such the seeds remain viable even in extremely dehydrated state and hence are capable of producing normal seedlings upon hydration. Accumulation of dry weight completes before maturation. At maturity moisture content can be typically brought down to 6-10 per cent. Variation between individual seeds is very much limited. Tolerates freezing temperatures. Cryopreservation of these seeds is possible at 2-4 per cent moisture content and −15 to −20 °C temperature. By virtue of having a unique property of survival during dry storage, facilitates the use of these seeds in agriculture. Mostly annual temperate species adopted to open fields, Rape seed, Tobacco, onion, *Vigna sesquipedalis, Hibuscus esculentus,* etc Immature seeds of orthodox species are desiccation intolerant, whereas mature seeds not only survive desiccation to low moisture contents, but in addition to their longevity in air dry storage increases in a predictable way with reduction in seed storage moisture content and temperature, (Roberts, 1973). The orthodox seeds, even if uninfected, despite sheltering inoculum of a spectrum of fungal pathogens, will deteriorate eventually during storage as a result of inherent processes and the rate of degeneration is directly correlated with relative humidity and storage temperature, Roberts (1983). As the seeds deteriorate naturally, the fungal pathogen proliferates its activity, seed moisture content increases and seeds will become completely decayed. The storage fungal species like *Aspergillus* and *Penicillium* are capable of maintaining their metabolic activity at 12-18 per cent moisture content. The orthodox seeds of higher plants are nothing but the true representatives of highly sophisticated biological systems capable of interrupting their development on the mother plant itself through intense desiccation and restart their cellular activity upon imbibition.

7.2.1.1.3   Justice and Bass Classification (1978)

Most of the cultivated species including vegetables are short lived. Therefore, based on the Relative Storability Index (RSI), Justice and Bass (1978) classified crops in to three classes, viz., RSI-1, RSI-2 and RSI-3 with the expectation that 50% of seeds will germinate even after 1-2 years (eg. Sorghum, bajra, soybean, sunflower, groundnut, cotton, onion etc.,), 3-5 years, (eg. wheat, barley, oats, cabbage, carrot, etc.,) and beyond 5 years (eg. tomato, sugarbeet, Alfalfa etc.,), respectively.

7.2.1.1.4   Classification of Seeds, Ellis, (1990)

From evolutionary point of view, seed desiccation tolerance appears to be associated with plants grown in drier environments whereas desiccation sensitivity is mostly confined to moist environments. Therefore, Ellis *et al.,* (1990) added an intermediate type to the classification proposed by Roberts, (1978).

7.2.1.1.4.1   Intermediate Seeds

Teng and Hor (1976) demonstrated that the seeds of star fruit (*Averrhoa carambola*) and papaya (*Carica papaya*) are capable of withstanding desiccation to around 10-12 percent moisture content and could be stored successfully in hermetic containers at these moisture contents, but lost viability much more rapidly in air dry storage at 0°C than at warmer temperatures of 12-21 °C. Hong and Ellis, (1995) reported that, the seeds of *Coffea arabic; Coffea canephora*; oil palm, papaya and several other Citrus species have exhibited a type of storage behaviour which is intermediate between the orthodox and recalcitrant categories of Roberts (1973). The essential feature of intermediate seed storage behaviour is that negative relation between seed longevity in air-dry storage and moisture content is reversed at values below those in equilibrium at 20°C with about 40-50 per cent relative humidity. The seeds though could be stored in certain air dry environments successfully but their behaviour did not satisfy Robert's definition of orthodox seed storage behaviour, are placed in intermediate seeds. Desiccation tolerance and storage behaviour of intermediate seeds will be between the orthodox and recalcitrant seeds. Often but not always, intermediate seeds exhibit damage when tested for viability immediately after desiccation to moisture contents below those in equilibrium at 20°C. The typical feature of intermediate seeds of tropical origin is that the longevity of dry seeds at 7-10 per cent moisture content is reduced with reduction in storage temperature below 10°C, Hong and Ellis (1992). Chilling sensitive at more than 10 % moisture content, Sensitive to imbibitional stress below 10 % moisture, e.g., Neem, Papaya. At high water contents, the seeds are expected to be metabolically active and able to trigger protection mechanisms like phytoalexin production in response to associated microorganisms.

 ## 7.3 Storage Requirements of Seeds

Depending on the crop species, every seed lot produced has to undergo a specified period of storage before being planted in the ensuing season. In order to produce an optimum plant population per unit area after storage, the seed populations must be in a position to give a reasonably high germination upon sowing. In order to achieve this, care should be taken to minimize or control the decline in germination to a desirable level till sown in the ensuing season by providing appropriate storage conditions. Usually around 70-80 per cent of seed produced is utilized in the next season. Only around 20 per cent seed is stored for more than two seasons and such seed generally include either breeder seed or foundation seed of parental lines of hybrids. Usually this could be due to the excess seed produced either on account of unpredictable market demand or fluctuations in cropping shifts and may also be due to unlifted seed by the indenters. Therefore, it becomes mandatory for the seed companies to predict deterioration in seed quality from time to time, so as to replenish the stocks in a systematic manner. Type of seed, initial germinability, moisture content, temperature, relative humidity of storage environment etc., determine the rate of deterioration. Seeds with low moisture and high quality stored under cool dry conditions are better than high moisture low quality seeds under hot humid conditions.

Kind of seed, value of seed, volume of seed, class of seed, climatic conditions of the area determine the type and duration of storage required. Storage is of two types, viz., ambient (uncontrolled) storage and controlled storage. The former is useful for bulk storage for short period packed in impervious (5-8% seed moisture) and pervious (12-13% seed moisture) containers. In the later, both temperature and humidity are controlled and hence suitable for both medium and long-term storage of high value and low in volume seeds. Breeding materials and germplasm lines need to be stored on long-term basis because of their volume and value. Depending on the length of storage required, seeds are dried to less than 5% moisture, packed in hermitically sealed moisture proof containers and stored at 10°C to - 20°C. Germination can be fairly well maintained by storing seeds at 30 % RH and 15 °C at which most species of seeds attain equilibrium moisture content of 5 %.

Seed moisture directly exerts great pressure on the seed longevity than temperature. At physiological maturity, most of the species have around 40 -50 per cent seed moisture content. At 40%-50% moisture content, most of the biochemical and cytological deterioration will take place. The seeds, though vigourous and germinable, deteriorate very fast due to physiological, mechanical and pathological reasons, if not dried properly. Usually, a moisture level of 10%-

13% for open storage and 4%-8% for sealed storage are considered safe. Bhaskaran *et al.,* (2004) suggested the following moisture levels to be safe for open storage of cereal seed. Seed storage moisture of 8-10, 9-11, 10-12 and 11-13 respectively gives a storage life of 4.0, 2.0, 1.0 and 0.5 years. Further, the seeds stored at high moisture content cannot tolerate low oxygen pressure, as the same is highly necessary for respiration.

Seed storage temperature also plays an important role in deciding the longevity of seed lots. Higher temperature increases the rate of metabolic processes, particularly in the presence of high moisture. This in turn facilitates invasion of pathogens and insects, increases the rate of deterioration, and ultimately exerts influences on the seed longevity. Most of the biological processes get slowed down at low temperatures, the lower the temperature, the slower the process of deterioration. Copeland and McDonald (2001) also reported that at low moisture content the effect of temperature is very little, and seeds with high moisture need to be stored only at temperatures below $10^{o}C$. Even to inactivate most of the seed borne fungi and insects, the temperature should be less than $8^{o}C$. In nature, ionizing radiation is also expected to exert some influence on the process of ageing of seeds and ultimately leading to their deterioration.

The rules of thumb of James (1961), that relative humidity and temperature in ^{o}F should not exceed 100 in long-term storage, and Harrington (1972) for every 1 per cent increase in seed moisture and for every $5^{o}C$ increase in temperature, life span of seed is halved. No doubt, these thumb rules are over simplified estimates of a process under the influence of several factors. Hence, Agarwal (1980) has aptly pointed out that different plant species exhibit characteristic longevity patterns. In addition to seed moisture and temperature, there are several factors contributing significantly towards variation in seed viability, and are presented in Table 7.1.

Table 7.1 Factors and their effects responsible for variation in seed viability.

Factors	Effects
1. Pre-harvest factors	
1.1 Nutritional deficiencies,	Ca, B, Fe deficiency affects germination;
1.2 High rain fall and wind	Poor fruit and seed setting, washout of pollen
1.3 High temperature and RH	Decrease in seed mass
1.4. Varying stages of maturity	Varying degrees of maturity and ultimately storage potential.
2. Post-harvest factors	
Factors	**Effects**
2.1. Genetic factors	
2.1.2. Reduction in Size and shape of seed	Poor germination percentage;
2.1.3.Temperature and moisture stress during flowering, seed filling, maturation.	Reduction in source sink relations, reduction in germination and vigour, loss of seed physical quality, shriveled seed
2.1.4. Chemical composition of seed and changes in macromolecules	Delayed and uneven emergence
2.2.Physiological status of seed	Hard Seededness
2.3. Pathological factors like fungi, RH	Quick loss of seed viability
2.4. Mechanical status of seed during harvesting and post harvest handling operations	Quick loss of seed vigour
2.5. Storage environment factors like temperature and Relative humidity	Deterioration and ageing

7.3.1 Relationship between Longevity-Temperature and Moisture Content of Seeds (The Viability Equation)

A basic viability equation (1) developed by Roberts and Abdala (1968) was modified with equation (II) proposed by Ellis and Roberts (1981) is as follows.

Equation I: $\log p = Kv - C_1 m - C_2 t$

where, p = mean viability period; m = seed moisture content; and t = temperature in $^{\circ}C$,

Equation II: $\{K_E - (C_w\} \log m - (C_H) t - (C_g) t2\} V = K_i - P/100$

where,   V = probit final viability after P days of storage,

        P = storage time days,

        t = temp $^{\circ}C$,

        m = % moisture fresh weight basis,

        K_i = initial probit viability at 0 days,

C_H, C_g and K_E C_w are species-specific temperature and moisture constants respectively.

The relationship between longevity-temperature and moisture content for orthodox seeds in air- dry storage can be explained with the help of the following viability equation.

$$V = K_i - p/10^{KE - C_w \log 10 m - C_H t - C_Q t2}$$

where, V is the probit percentage viability after p days in storage at moisture content (% w.b.) and temperature t ($^{\circ}C$), K_i is a constant specific to the seed lot, and K_E, C_W, C_H and C_Q are species constants, (Ellis and Roberts, 1980) and this equation is subject to both high and low moisture content limits (Roberts and Ellis, 1989).

The gradient of negative logarithmic relation between seed moisture content (m) and longevity provides the value of C_W (sensitivity of seed longevity to seed moisture content). Considerable variation exists for the values among different species. It is high in cereals around 6, much lower in oil seeds around 3.5, onion and soybean it is 4. In tree species it is 2.15 in *Terminalia brassii*. Its mean value for forest tree species is 2.8 whereas for herbaceous species is 4.72, Tompestt (1994). This means, in order to obtain the same relative increase in longevity, it becomes mandatory to dry the orthodox tree and oily seeds more than starchy seeds from a given moisture content and hence could be attributed to the differences in seed chemical composition, thereby equilibrium relative humidity and seed water potential at same moisture content, Roberts and Ellis (1989). Longevity doubles with each 8.7 percent reduction in equilibrium relative humidity or in other words seed longevity increases by a factor of 2.22 for each 10 percent reduction in equilibrium relative humidity, Ellis *et al.,* (1990) and Zanakis, *et al.,* (1993). Seeds of orthodox species imbibe and begin to germinate not only is this desiccation tolerance gradually lost but seed longevity in subsequent air-dry storage also reduces, (Hong and Ellis, 1992a).

Seed deterioration does not occur in any set pattern in different varieties. Deterioration in maize was observed in scutellum and axis. In wheat it begins with root and moves through embryo. In general, deterioration is initiated in meristematic tissues of seeds. The sequence of events occurring during ageing was initially proposed by Delouche (1969) and was modified by Heydecker (1972) subsequently. Roberts (1972) considered these events only symptoms rather than the causes of seed deterioration. From the above discussion, Delouche (1973) concluded that (i) all seeds loose viability hence seed deterioration is an unavoidable process, (ii) seed viability is an irreversible process and (iii) seed population contains mixture of seeds at varying levels of ageing. Roberts and Ellis (1981) model is an invaluable tool for prediction of seed storage. The published reports of Priestly (1986), Smith and Berjak (1995) helped us in understanding seed ageing process only partially. Copeland and McDonald (2001) classified these symptoms of seed deterioration into performance and physiological manifestation as detailed below.

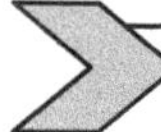

7.4 Manifestation of Seed Vigour

Manifestation of seed vigour can be broadly discussed at physiological and biochemical level as follows.

7.4.1   Physiological Manifestations of Seed Vigour

After the attainment of physiological maturity, deterioration symptoms appear in a set pattern. As seeds age, germinability is maintained for some time and is followed by a drastic reduction in which most seeds fail to germinate completely while only a few germinate and grow normally. The major symptoms observed during physiological manifestation of seed deterioration are listed below. (i) Delayed emergence, (ii) Slow rate of growth and development, (iii) Decreased germination percentage, (iv) Increased number of abnormal seedlings, (v) Lack of elongation of hypocotyl and stunting of radicle, (vi) Lowered tolerance to environmental stresses during emergence and early seedling stage, (vii). Decreased yield potential and (viii) Complete loss of germinability and death of seed, (ix) Reduced tolerance to suboptimal environmental conditions, (x) Decreased tolerance to adverse storage conditions and (xi) Changes in seed colour.

7.4.2   Biochemical Manifestations of Seed Vigour

In the life cycle of any seed, during quiescent stage, often misrepresented as dormant stage, loss of vigour occurs initially at a rate determined by both the genetic makeup as well as internal and external environment. Subsequently, while in storage also, all the cytoplasmic constituents of cells begin to degrade and the extent to which each affects seed viability depends only on the importance of that particular constituent to the cell function. Since, the germination vigour is driven by the ability of a plant embryo embedded within the seed. The success of germination quality is determined by the quality of messenger RNA stored in the embryo during maturation by the mother plant in order to resume its metabolic activity in a coordinated and sequential manner. In addition to this, proteostasis and integrity of DNA also play a major role in the germination phenotype because of its pivotal role in cell metabolism, relationship with hormonal signaling pathways regulating seed germination. Further, the sulfur amino acid metabolism pathway also determines the commitment of seed to initiate development toward germination and hence germination vigour depends on multiple biochemical and molecular variables, Loic Rajjou *et al.,* (2012). These changes responsible for the loss of seed vigour are basically associated with the ultra-structural degeneration of the cell together with biochemical changes and include (i) changes in the quantity of chemical constituents, (ii) accumulation of growth inhibitors (iii) genetic makeup including the aberrations in chromosome frequency etc. There are several physical and chemical reactions that contribute significantly to seed ageing by adversely affecting the molecular function.

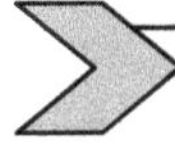 ## 7.5 Physical and Chemical Reactions involved in Seed Deterioration

7.5.1 Physical Reactions

Physical reactions are those reactions that are dependent on the availability of moisture and temperature and hence these reactions either alter the structure of macromolecules or bring out changes in the distribution of molecules within the cells. Basically these reactions are described by using phase diagrams when they are in equilibrium. The physical reactions are reversible when the seeds are removed from storage. Phase changes in the membranes of polar lipids are the best-known example of physical change in the cells of seeds in storage. Availability of water and temperature exert direct affect on hydrophilic and hydrophobic interactions, which in turn affect the partitioning of amphiphilic molecules. Similarly, during dry storage several enzymes lose their activity and proteins become less extractable because the protein denaturation is controlled both by high temperature and ionic concentration. However, it is very difficult to demonstrate the direct predominant link between protein denaturation and loss of enzyme logistically. Slow crystallization of sugars such as sucrose has been implicated in ageing whereas raffinose sugars, that retard crystallization may promote seed longevity. Under typical storage conditions, the rate of deterioration increases with the increase in moisture level of seed and this process continues till a level is reached when additional ageing almost ceases. Villers and Edgecombe (1975) observed that the storage of imbibed seeds prevented the accumulation of apparent nuclear damage in dry stored seeds and also reported reversing of ageing damage already occurred. Hydration was found to prevent and correct the damage caused by gamma irradiation and hence can be concluded that the loss of viability in dry storage seeds is the result of inability of repair system to operate in dry seed tissues. This only suggests that the imbibing seeds undergo extensive repair.

7.5.2 Chemical Reactions

Depending on the availability of high, intermediate and low relative humidities to the seed in storage, the chemical reactions taking place in the stored seed can be respectively classified as hydrolytic, oxidative and peroxidative and changes with the change in substrate concentration within the cell as detailed below.

7.5.2.1 Hydrolytic Reactions

These reactions occur in seeds stored at higher relative humidity and lead to accumulation of free fatty acids in the presence of water molecules. De-esterification of lipids into free fatty acids and glycerol, de-polymerization of starches into simple sugars are the best examples.

7.5.2.2 Oxidative Reactions

Oxidation and reduction reactions either remove electrons from the carbon or directly insert oxygen and cause degradation. Oxidative degradation is caused by non-enzymatic mechanism catalyzed by oxygen, light, heat and metals in the presence of oxidizing agents like NAD, metal ions and sulph-hydril groups. The purpose of these reactions is to retrieve energy from glycolysis through β-oxidation of fatty acids. When cells are water stressed when the water potential becomes less than -2MPa, it results in the accumulation of partially reduced oxygen or high energy intermediates either due to incomplete or imbalanced metabolism, or changes in enzyme activity, or shifts in substrate concentration and respiration without ATP production. The by-products of oxidative reactions are referred to as reactive oxygen species or free radicals. The dehydration and rehydration processes taking place respectively during seed development and germination are associated with high levels of oxidative stress resulting in DNA damage, Dandoi *et al.,* (1987).

7.5.2.3 Peroxidation Reactions

The oxygen molecule, being extremely electrophilic, forms hydro-peroxides by directly combining with electron dense moieties like unsaturated fatty acids. Since the peroxide bond is single, it is weak and can be easily broken by reducing agents to form the strongest known oxidant, hydroxyl radical. Depending on the catalyst and fatty acid, a number of by-products are formed in lipid peroxidation, which may include, pentane, ethane, ethylene, aldehydes and ketones. Malandialdehyde and hydroxyl ions of non-metals are known to cause damage to DNA. On the other hand, oxygenation of fatty acids and lack of double bonds also cause decreased fluidity and increased leakiness of membranes.

After imbibition, respiration increases rapidly and provide ATP required for protein synthesis from the pre-existing substrates and biochemical systems in response to instructions either pre-existing or received from embryo and embryonic axis and in turn supplies enzymatic, structural and soluble protein required for a number of processes of growth. Based on the increased activity of protein synthesis, mRNA, tRNA, ribosomes, polysomes, glyoxysomes, enzymes, coenzymes and cofactors etc., result in catabolically produced new cells and tissues. If the seed is vigourous, all three phases perform well, subject to the condition that the structural and functional integrity of cell organelles, in discharging various roles of cellular maintenance, repair, growth and continuity. If the initial reaction speed is faster and the more functional the conserved system, the greater will be the force of germination. Changes in the ultra structural properties of one or more organelles, resemble the late maturation phase of a seed, may be associated with loss of vigour, Abdul Baki and James Baker (1973).

7.5.2.4 Millard Reaction

This reaction represents the complex interactions occurring between glycated Amadori products to form brown coloured polymeric compounds. These polymeric compounds are likely to cause damage to DNA due to inactivation of proteins and thereby cause the loss of viability of seeds. Hence, the term 'Browning reaction is expected to significantly increase the formation of reducing sugars like fructose, glucose and glalctose etc. Increased presence of reduced sugars are the key driving force behind the Millard reactions for initiating non-enzymatic protein degradation during seed ageing. Damage caused to genome during storage lead to loss of seed vigour and viability, Cheah and Osborne, (1978). The cytotoxic volatile aldehydes are produced during degradation of Millard products, Danehy (1995). This theory of seed ageing appears to be conceptually attractive because the reaction occurs when the moisture content is between 6-15 per cent, where the seeds deteriorate very fast. Contrary to this, Backer and Bradford (1994) reported lack of correlation between seed viability and accumulation of Millard products through fluorescence assay.

7.5.2.5 Amadori Reaction

It is a simple non-enzymatic reaction, wherein the reducing sugars attack amino groups to form fructosyl derivatives or glycated proteins. The Amadori products are complex sugar- protein compounds present in seeds that display deterioration changes. They consists of an isomer of a glycosyl-amine (R-C (=0) CH2-NH-R' where, R is a part of reducing sugar and R' is usually a protein formed from dehydration and rearrangement of a Schieff's base. A Schieff's base is formed from the reaction of an amine group (NH2) with an aldehyde (HC=0). The products of Amadori reaction are intermediates of Millard's reaction, which are degraded into a variety molecules depending on moisture, temperature and pH of the reaction medium.

7.6 Changes in the Ultra-structural Properties of Organelles

7.6.1   Nucleus

In maize, the nuclei of aged seeds developed into a peculiar lobed condition and became more prominent with the increasing period of imbibition, Berjak and Villers (1972). Imbibed seeds not only prevented the apparent accumulation of nuclear damage in dry stored seeds but also reversed the damage that had already occurred. Therefore, Villiers and Edegcumbe, (1972) have concluded that the loss of viability has resulted from the inability of repair system to operate in dry tissues,

7.6.2   Mitochondria

The electron micrograph indicated dilated cristae and dense coagulated matrix of mitochondria isolated from aged seeds of soybean, Abu Chakra and Ching (1967). The mitochondria isolated from fresh and aged seeds have shown some respiratory uncoupling in aged seeds. The mitochondria isolated from the root cap of maize indicated irregular profiles during early stages of degeneration with frequent appearance of peculiar internal membrane bound space and occasionally elongated swollen cristae are associated with decrease in matrix density Berjak and Vilers (1972). The reduced electron density in the matrix of mitochondria even after hydration, only suggests the occurrence of irreversible changes leading to degradation processes during the drying and storage of seed (Abdul baki and Baker (1975). Functional changes in mitochondrial degradation, decrease in number with advancement of age, appear to play a major role in seed deterioration, because the mitochondria lose their swelling-contracting ability, breakdown of ATP into ADP, reduced levels of ATP in aged seeds in a number of crops Banerjee *et al.,* (1981).

7.6.3   Golgicomplex

Reduction in protein and carbohydrate synthesis brings changes in polysomes and golgi bodies (Abdul Baki, 1969)

7.6.4   Plastids

Berjak and Vilers (1972) reported swollen, less dense and pronounced inner membrane of plastids, an indication of irreversible change.

7.6.5   Ribosomes and Lysosomes

In aged seeds, Berjak and Vilers (1972) also observed ruptured membranes in root cap cells of lysosomes and found only monosomes or reduced ribosomal RNA. Whereas   missing polysomes were reported by Abdul Baki and James Baker (1973).

7.6.6 Endoplasmic Reticulum

Presence of short distended profiles in severely aged seeds and occurrence of hypertrophy during germination is a symptom of mechanism to compensate the mitochondrial damage and suggests a connection between respiratory metabolism and proliferation of endoplasmic reticulum. Seeds loose viability due to early occurrence of mitochondrial damage leading to reduction in respiration rate in seeds set to germinate, Berjak and Vilers (1972).

7.6.7 Cell Membranes

Loss of membrane integrity appears to be the root cause of degradation of most of the cell organelles of the protoplast including the membrane covering the protoplast. Basically, the cell wall comprises of middle lamella, primary wall and secondary wall and the loss of integrity of cell membrane, though not observed directly so far, and only implied for quite a long time from now, allows leakage of substances from aged seeds. Ching and Schoolcraft (1966), while working with rye and red clover seeds reported lack of perfect correlation between loss of viability and electrical conductance of leachates, and the same was questioned by Abdul Baki and Anderson (1970) while working with aged barley seeds. The inability of non-viable seeds to resist penetration of heavy metal salts like barium chloride is nothing but an expression of loss of membrane integrity during the process of ageing. Further, it is not possible to attribute vigour loss to the malfunctioning of any one specific organelle, because a number of biochemical changes might be taking place simultaneously in seeds while they lose vigour and may be some of them might be associated with the changes in the properties of membranes and therefore, the sequential biochemical events which culminate in loss of vigour are still an enigma. Ghosh *et al.,* (1981) proposed that decreased incorporation of ^{3}H leucine was due to a decrease in membrane integrity. Contrary to this, Aquilar *et al.,* (1992) reported specific damage to transcription rather than general damage to protein synthesis.

7.7 Basis of Seed Deterioration

7.7.1 Changes at Genetic Level

The information pertaining to seed longevity and differences in longevity is scanty because this trait is polygenic in nature and is also under the influence of several internal and external environmental factors. Though environmental factors affecting longevity was researched thoroughly, genetic aspects of longevity could not draw much attention earlier. Past research on a number of seeds by a number of researchers has clearly demonstrated that any combination of time, temperature and moisture content that leads to loss of viability during storage also leads to genetic damage in the survivors due to the fragmentation of DNA into low molecular weight components due to the activation of DNAse (s). Fusion and fragmentation aberrations in aged barley seed was reported by Gustafson, (1937) with x-ray analysis.

Buildup of chemical mutagens during storage results in damage to nucleic acids. D' Amato (1952) reported that mutations in aged seeds are the consequence of mutagens accumulated during the ageing process and probably are the decomposition products resulting from metabolism. The two observations made by James (1960) that (i) properties of chromosomes change with age and (ii) mutations in seeds are highly correlated with seed age and decline in viability of ageing seeds are highly significant. Roberts (1973) proposed the existence of close correlation between chromosomal aberration and point mutations. Further, a sharp increase in chromosomal aberrations in the form of deletions and chromosome bridges with increase in moisture content of seeds in dry storage was demonstrated by Villers and Edgecombe (1975) successfully. Considerable indirect evidence is available to show that seed deterioration is associated with genetic mutation that impairs cellular functions of vital seed tissues and chromosomal damage was provided by Orlova and Soldatova (1980) and claimed that no mutation occurs in lag phase during storage. Priestley *et al.,* (1980) reported that accelerated ageing of soybean seeds were little affected with tocopherol content, a lipophilic antioxidant synthesized in abundance in seeds in specific. The little loss of viability and little visible chromosome aberration are considered to be the consequence of normal distribution of deaths, Ellis and Roberts (1981). Subsequently, the frequency of chromosomal aberrations leading to increasing damage of chromosomes that in turn is associated with the loss of viability has been reported, Lanteri and Belleti (1990). Identification of mutants that either improve or reduce longevity or identification of transcription factors with specific and multiple effects on genes involved in longevity needs to be identified. Larson (1997) reported that free radicals are also suspected to assault on chromosomal DNA.

1. Purine and Pyrimidine bases together with de-oxyribose sugar moieties are the potential targets of oxidative damage in DNA chain.

2. Modification of sugar residues results in strand shrinkage, the chief reason for propensity of genetic mutations as seeds age.

3. Increase in chromosome damage increases with the period of storage.

4. Extensive chromosomal damage results in DNA template impairment and reduced protein and RNA synthesis.

5. Nuclear DNA being highly conserved is free from radical attack as it is enclosed by a nuclear membrane and surrounded by histone proteins. Contrary to this mitochondrial DNA appears to be most susceptible to free radical attack in the absence of protective layer as well as histone proteins.

6. Damage to segments of DNA containing genes for the repair enzymes could be more deleterious than damage to less essential DNA fragments.

7. The integrity of DNA has little effect on transcription of earliest events of germination.

At molecular level studies on this typical trait was initiated with QTL mapping by (Betty *et al.,* 2000) in cabbage; Miura *et al.,* (2002) in rice; Clerkx *et al.,* (2004) in Arabidopsis. The poor longevity of Arabidopsis mutants is evidenced by rapid loss of viability upon storage. Various factors exerting influence on genetic aspects of longevity are briefly presented here.

Genetic studies on vitamin E loci mutation resulted in tocopherol deficiency in all tissues in Arabidopsis, Stadler, *et al.,* (2004). When compared to wild types, mutations either increased or decreased the longevity and indicated the role of tocopherol in maintaining the viability during quiescence. In Arabidopsis, Clerkx *et al.,* (2004) reported that gene redundancy is responsible and offers protection to Reactive Oxygen Species (ROS) through several pathways. Sun *et al.,* (2002) reported that mutants with reduced storage activity exert influence on small heat shock protein genes in embryo. Kotak *et al.,* (2007) reported potent activators and promoters of heat shock protein genes. Accumulation of anthocyanins, and pro-anthocyanidins in the inner integuments of seed coat offered a protective barrier for the embryo and there by increased seed coat imposed dormancy and hence seed longevity, Baudry *et al.,* (2004). Bentsink *et al.,*

(2000) reported that seed storability is not affected by either sucrose content or raffinose sugars. Therefore, Hoekstra (2005), perceived that the rate of ageing in seeds is determined by deteriorative processes in membranes rather than the glossy state of cytoplasm. Devaiah *et al.*, (2007) reported that gene for membrane lipid hydrolyzing phospholipase in Arabidopsis enhanced seed germination and oil stability after storage exposure of seeds to adverse conditions.

The accumulation of unknown products during storage may be detrimental to seed during germination or accelerated ageing and since these products are not related to structural damage and since the hydration-dehydration reaction occurs too rapidly for significant repair to occur. The intrinsic factors-1, responsible for denaturation of macromolecules (Fig.7.1) and Intrinsic factors-2 responsible for production and accumulation of essential toxic metabolites (Fig. 7.2) leading to denaturation of macromolecules and events of deterioration leading to seed ageing and the extrinsic factors responsible for loss of seed viability are presented in Fig. 7.3. So, logically, it could be inferred that seed deterioration may be mainly caused because of the accumulation of toxic metabolites, deteriorated macromolecules, damage to genome and lipid peroxidation. The accumulated toxic chemicals during germination or accelerated ageing may get enzymatically activated and results in seed death.

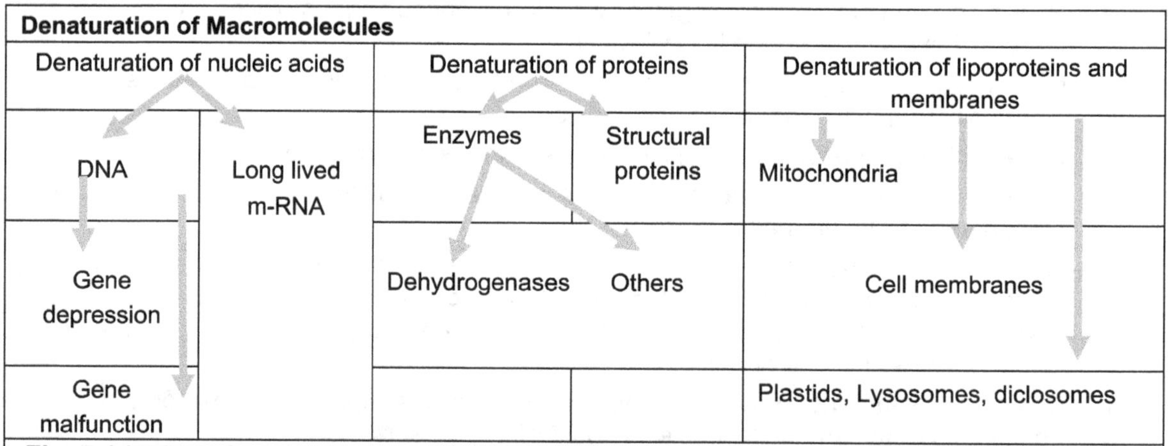

Denaturation of Macromolecules					
Denaturation of nucleic acids		Denaturation of proteins		Denaturation of lipoproteins and membranes	
DNA	Long lived m-RNA	Enzymes	Structural proteins	Mitochondria	
Gene depression		Dehydrogenases	Others	Cell membranes	
Gene malfunction				Plastids, Lysosomes, diclosomes	

Fig. 7.1 Intrinsic factors-1 denaturation of macromolecules and events of deterioration leading to seed ageing

2. Production of essential metabolites		**3. Accumulation of Toxic Metabolites**			
Respiratory losses of food reserves	Harmones, Vitamins, losses	Mutagens		Growth inhibitors	
Proteins | Lipids		DNA		1.Fermentation products	2.Phenolics
Carbohydrates		Gene depression	Gene malformation		
				3. IAA	4. ABA

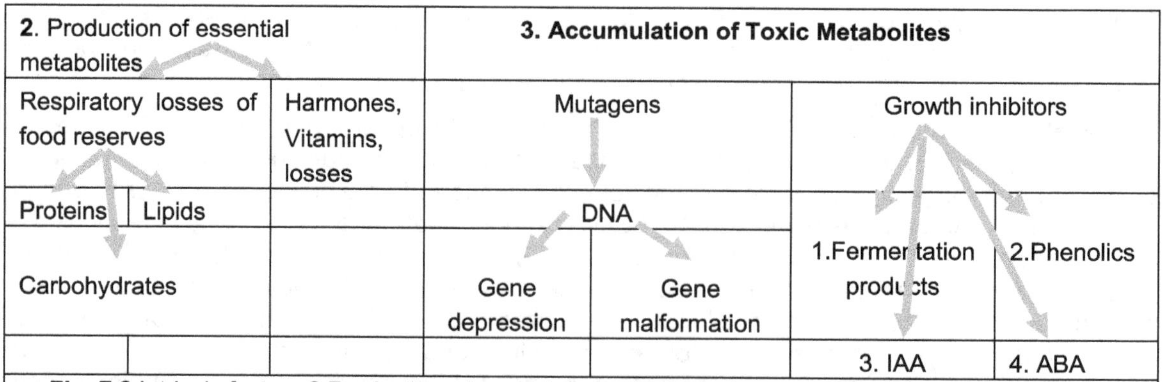

Fig. 7.2 Intrinsic factors-2 Production of essential metabolites & 3 accumulation of toxic metabolites leading to denaturation of macromolecules and events of deterioration leading to seed ageing

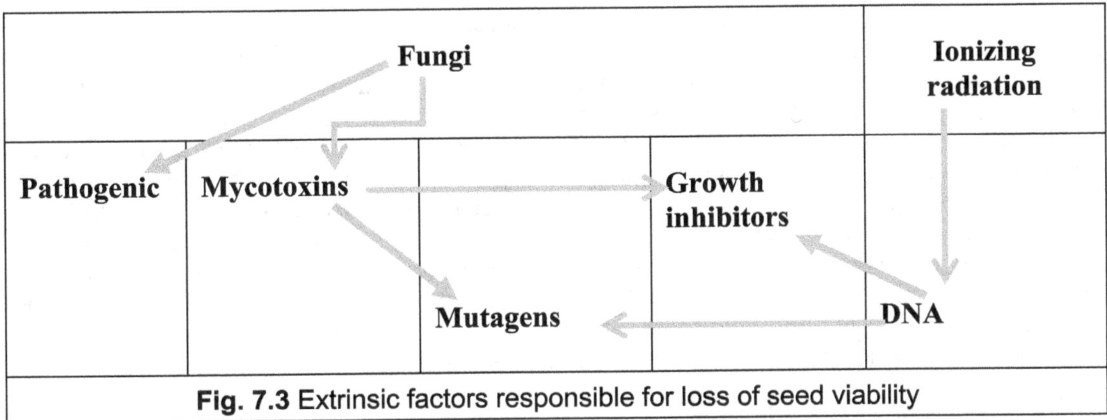

Fig. 7.3 Extrinsic factors responsible for loss of seed viability

7.7.2   Basis of Seed Deterioration - Changes at Physiological Level

Deterioration of seed during various post harvest treatments including storage under adverse conditions is an unavoidable physiological process leading to loss of viability and vigour. It is in no way different from either degeneration or senescence and includes all changes that take place during the death of a seed. The process of seed deterioration, at physiological level, is associated with changes such as gradual decrease in germination capacity, higher mean of germination time, enhanced number of abnormal seedlings and reduced tolerance to adverse storage conditions etc. This could be attributed to several biophysical and biochemical changes simultaneously occurring in various components of seeds. These changes at physiological level can be broadly divided into four stages as detailed below.

7.7.2.1   Changes in Enzyme Activity

The metabolism of seeds is greatly influenced during storage, as a consequence of diverse alterations taking place in various enzyme activities simultaneously brought about by configurational as well as compositional changes in their structure. Partial folding or unfolding, degradation into subunits and condensation to form polymers constitute the three chief configurational changes.

Activation of hydrolytic enzymes especially during storage is highly seed moisture dependent and hence results in normal germination at higher moisture contents and vice versa. Therefore, deterioration of seeds is a consequence of accumulation of breakdown products and energy expenditure, Basavarajappa *et al.,* (1991). Various enzymes associated with respiration like Catalase, Peroxidase, Cytochrome oxidase and Glutamic acid oxidase decrease in activity while Phytase and, Proteases tend to increase with the loss of viability. Contrary to this, a reduction in the activity of dehydrogenases, responsible for the generation of ATP was also observed during ageing of seeds. The enzymes play an important role in mobilizing and utilization of food reserves during germination. Published literature is available pertaining to the changes in the activity of amylase, oxidase and decarboxylase in deteriorating seed of different crops.

Bailly *et al.,* (1996) reported a reduction in the activity of free radical and peroxide scavenging enzymes like catalase, peroxidase, superoxide dismutase, ascorbate peroxidase and glutathione reductase during accelerated ageing. The increase in activity of hydrolytic enzymes, which are preponderant, is associated with loss of viability and enhances the possibility of rupture of lysosomes. Ageing inhibits the activities of peroxide scavenging enzymes like superoxide dismutase, ascorbate peroxidase etc., (Bernal-lugo *et al.,* (2000). A reduction in the activities of these enzymes play a key role in free radicals, which are significant to seed deterioration in general, McDonald (1999) and seed ageing in specific. Changes in enzyme activity though observed in several seeds but good correlation with loss of viability could not be observed so as to use it as a test of viability barring dehydrogenase activity test with tetrazolium salts. The enzymes play an important role in mobilizing and utilizing food reserves during germination. Reduced enzyme activities with natural and accelerated aged seeds in several species have been reported by many workers. Vijay and Dadlani, (2003). The reactive oxygen species are derived from mitochondrial oxidative metabolism during germination. Despite having its own limitations still suggests that enzyme activity is the specific cause of loss of viability and ultimately vigour. Changes in the levels of specific enzymes may not always provide an accurate indication of seed deterioration.

7.7.2.2   Changes in Respiration Rate

Respiration is important for all life processes to continue. Seeds retain their viability for a long time by maintaining low respiration rate. Respiration assumes significance especially when the seeds respire and liberate energy efficiently upon hydration necessary for germinability. Respiration rates in respect of exchange of gasses, intake of O_2 and release of CO_2 weakens progressively and makes the seeds deteriorate and ultimately leads to loss of germination because respiration is a complex expression of the activity of a group of enzymes that react together while breaking down the food reserves. The respiratory metabolism upon oxidation of food reserves yields a number of intermediates necessary for the synthesis of protoplasmic components like proteins, nucleic acids and lipids. Simultaneously, respiratory metabolism also releases the much needed biochemical energy in the form of ATP in addition to the specific amount of heat. Subnormal heat is an indication of low vigour, whereas excess heat is indication of infection with microorganisms. Changes in respiration under natural ageing Abdul –Baki, (1969); under accelerated ageing Anderson (1970); and Chilling injury, irradiation conditions (Woodstock and Feeley 1965) have been reported. RQ of more than 1.5 have been reported in aged seeds Woodstock *et al.,* (1984). Reduction in the rate of respiration is closely related to an associated with the earliest symptoms of seed ageing and ultimately seed deterioration and decrease in seed vigour and could be attributed to the breakdown of mitochondrial membrane, cristae, which would reduce total respiration. Lowered rates of oxygen consumption, higher rate of RQ values during imbibition only reflect decline in viability and possibly an index of deterioration of seeds.

7.7.2.3   Changes in the Quality of Chemical Constituents

Even after loss of viability, seeds still contain large amount of food reserves, Barton, (1961). Seeds with more food reserves deteriorate more quickly than seeds with low food reserves, Janes, (1967). The increase in temperature, moisture content and oxygen pressure tends to increase the rate of respiration. Similarly, accumulation of fatty acids in seeds, reducing sugars and decrease in protein content and total sugars in cereal seeds impairs the functioning of mitochondria, Earnshaw, Trulove and Butler, (1970); accumulation of supra optimal quantities of indoleacetic acid derivatives, Sircar and Biswas, (1967); Phenolics including ferulic acid and Caumarin, Dey *et al., (*1967); auto toxic substances and auto mutagenic substances accumulate in seeds as they age and loose viability, Roberts, (1973).

7.7.2.4   Impairment of Membrane Integrity

Breakdown in bio membranes integrity results in loss of metabolic activities, and brings in both structural and functional changes in organelles. Enhanced solute leakage is the direct manifestation of reduced membrane integrity and earliest indication of seed deterioration. This linear relationship is the basis for development of several tests, ISTA (2004).

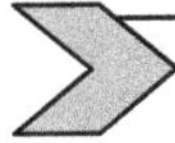 ## 7.8 Basis of Seed Deterioration - Changes at Biochemical and Molecular Level

Though the precise cause for the loss of seed viability is not yet fully understood, lipid peroxidation and associated free radical oxidative stress could be considered as the chief contributors to seed deterioration at biochemical and molecular levels. No doubt, prevalence of different mechanisms of seed deterioration under diverse storage conditions cannot be ruled out. But free radical damage appears to be the chief event of deterioration at comparatively low temperatures whereas higher temperatures are associated with the thermal inactivation of proteins.

7.8.1   Release of Volatile Compounds

For the first time, Fielding and Goldsworth, (1982) reported the release of volatile compounds as a sensitive indication of physiological age and potential vigour of wheat seeds. Subsequently, Pal and Basu, (1989) also reported that during early stages of rice germination, the production of volatile aldehydes was significantly lower in high vigour rice seeds than medium and low vigour seeds. Hailstones and Smith (1989) reported that, volatile aldehydes are the breakdown products of lipid peroxidation in soybean seeds that have significant correlation with germination, seed vigour and levels of hydro-peroxide and concluded that a definite relationship exists between lipid peroxidation and seed vigour. Zhang *et al.,* (1993, 1994 and 1995) reported that dry seeds release as many as 59 volatile compounds including n-type aldehyde like hexanal, responsible for non-enzymatic oxidation of fatty acids, isotope aldehydes like 2 methyl propanal and 3 methyl butanol. Similarly, acetaldehyde, the most toxic volatile compound present in dry seeds in abundant and is having the capacity to modify amino base in proteins and DNA. In dry seeds, the alcohol dehydrogenase can transform acetaldehyde into a non-poisonous ethanol, which is probably a detoxification path in dry seeds. Further, the amount of ethanol thus accumulated can also be converted back to acetaldehyde in the presence of alcohol dehydrogenase. In dry seeds some of the most common volatiles like methanol, acetaldehyde, acetone and ethanol etc., cause loss of seed germinability during storage. Generally it was observed that, the increase in temperature and relative humidity during storage increases the deleterious effects of these volatile compounds. Acetaldehyde had the strongest deleterious effect followed by ethanol and acetone while methanol was almost in many seeds. Ageing of seeds is a consequence of non-enzymatic reaction between the volatile aldehyde groups of glucose with amino group of proteins. The volatile compounds thus evolved in the seed, would have low molecular weight and hence have high mobility and thereby interact with several substances without any restriction even within the dehydrated seeds and hence the chief cause of deterioration even under lower relative humidity and temperature during the storage period.

7.8.2   Free Radicals

A free radical is an atom or group of atoms, capable of independent existence, which contains one or more than one unpaired electrons. In a biological system, these free radicals, which are highly reactive and short lived, can be initiated by any one of the following three methods viz., ionizing radiations or metabolic processes or lipoxigenases. Once a free radicle is produced, it is attacked by oxygen and initiates a chain reaction only to provide further additional cycles of reaction and ultimately free radicals. O, (mono oxygen atomic, a transient form); O_2 known as Dioxygen, is available in abundance, acts as a strong oxidizer, readily accepts electrons and contributes structurally to several organic and inorganic compounds; O_3 (Trioxigen, also known as Ozone and is confined to high altitudes). In the ground state, oxygen has two unpaired electrons with parallel spin and hence necessitates acceptance of additional electrons one at a time in accordance with the laws of spin restriction. Further, the transition elements having the capacity to share single electron to oxygen to promote production of free radicals by way of overcoming spin restrictions. O_2^+, O_2^-. In addition to oxygen, there are around 20 or so elements with biological significance as free radicals, which include, H, HO, C, S, N, P, Fe, Cu and Mn. The formation of hydroxyl radical is through the reduction of H_2O_2 by Fe^{+2} in Fenton reaction ($H_2O_2 + Fe^{+2} = HO + HO + Fe^{+3}$) or $O_2 + H_2O_2 = O_2 + HO + HO$. Hydroxil radicals are by far the most reactive oxygen radicals and cause extensive macromolecular damages. Lipoxygenase enzymes also generate free radicals when the seed moisture content is more than 14 percent.

7.8.3 Auto–oxidation

The unsaturated and polyunsaturated fatty acids like oleic and linoleic acids, which are abundantly found in phospholipid bilayer in seed bio-membranes, Gille and Joenje, (1992), Dussert *et al.,* (2006) when subjected to continuous oxidation produce highly reactive free radicals, often hydrogen, H^+ from α– Methylene group of fatty acid which is adjacent to the double bond, free radical intermediates called hydro-peroxides and a wide variety of products from its decomposition. The lipid auto oxidation is accelerated at higher temperatures and increases the concentration of oxygen, which according to Harrington (1973) is the cause of seed deterioration at less than 6 percent moisture content. The plants are capable of synthesizing a complete component of molecular defenses against oxidative attack. Chloroplast is considered to be the richest source of antioxidants like vitamin-E, vitamin-C and Pro-vitamin-A. The superoxide radical is produced through the autoxidation of hydro-quinones', thiols or enzymatically by flavor-protein dehydrogenase such as mitochondrial NADH dehydrogenase. Whereas, Hydrogen peroxide, is produced either spontaneously or dis-mutation of superoxide radicals.

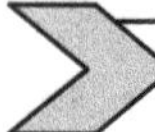

7.9 Basis of Seed Deterioration and Aageing-Changes at Molecular Level

7.9.1 Changes at Protein Level

Seed proteins are also subjected to ageing injury. Priestley (1986) reported reduction in protein extractability, whereas Kalpana and Rao reported total decrease in protein content. Nautial *et al.,* (1985) also reported a lower level of protein in non-viable seeds than in viable seeds as a result of hydrolysis. The same is evident from the increase in total free amino acids consequent to ageing treatment. Further, leaching of amino acids into the imbibing medium as reported by several workers finds its support in the form of high protease activity in seeds subjected to accelerated ageing conditions, Basavarajappa *et al .,* (1991).

7.9.1.1 Protein Synthesis

Anderson (1977), based on radioactive studies of amino acids in several species, reported that seed deterioration is associated with reduced capacity for protein synthesis. Whereas, Bray and

Dasgupta, (1976) reported that the inability to synthesize proteins is accompanied by an excessive loss of capacity to synthesize RNA and this could be the consequence of damage caused to nuclear DNA. The impairment of protein synthesis at the level of translation results from either structural modifications of ribosomes or loss of one or more rRNA or ribosomal proteins, Roberts and Osborne (1977). Contrary to this, Bray and Chow (1976ab) reported that dissociation failure of polyribosomes results in retardation of protein synthesis. Loss of protein synthesis is not associated with changes in tRNA or aminoacyl-tRNA synthetase activity. Further, reduction in ATP synthesis of non-viable seeds is likely to exert its effect on protein and RNA synthesis. Dell' Aquilla and Spada, (1994) reported reduced and delayed de novo protein synthesis due to the loss of viability and vigour. Eshahi et al.,(1997) reported that aldehydes resulting from lipid peroxidation during seed ageing interacts with amino acids and produce stable aldehyde protein abducts, which in turn leads to decrease in solubility of proteins and their aggregation. In seeds of several species, alteration in polypeptide profiles cause deterioration upon ageing, Vijay (2005). This could be attributed to the high protease activity either from cross-linking of protein or glycation of functional proteins and their breakdown. Decreased amount of dehydrins and other desiccation associated proteins and increase in glycolytic enzymes were reported by Rajjou et al., (2008) in the seeds deteriorated naturally as well as in controlled deterioration. From this information on protein synthesis and turnover that determine age related decline in repair, maintenance and ultimately the very survival of the seeds and hence is used extensively by the seed industry and researchers.

7.9.1.2   Protein Inactivation

Addition or deletion of certain functional groups results in inactivation of proteins, either by oxidation of sulfhydryl groups or conversion of amino acids within the protein structure. Inactivation of proteins during seed storage is likely to depress the metabolic capacity to repair the damages incurred during storage. Protein L-isoaspartyl – O methyl-transferase (PIMT) initiates the repair of isomerized aspartyl and asparaginyl residues that accumulate spontaneously with age in proteins of a variety of organisms and have been shown to alter the structure and function of proteins. Changes in the protein structure could also be attributed to the attack of free radicals. Oge et al., (2008) reported that decreased accumulation of methyl-transferase in *Arabidopsis* decreased seed longevity and germination and vice versa. It means, the PIMT repair pathway permits active elimination of deleterious protein products during germination and seedling establishment.

7.9.1.3   Protein Modification

The non-enzymatic glycation through Amadori and Millard reactions appears to be the most probable cause for inactivation of proteins during seed deterioration in the absence of active enzymatic metabolism in dry seeds. Larson (1990) reported different classes of oxidants attacking soluble proteins than membrane proteins. The reactive amino acids susceptible to oxidative damage appear to be in the order of cysteine, tryptophan, methionine and phenylalanine. Bernal-lugo and Leopold (1992) reported a marked increase in reducing sugars like glucose and galactose in deteriorating maize seeds that are responsible for enhancing Amadori and Millard reactions through the presence of reactive semi aldehyde groups.

7.9.1.4   Changes at DNA Level

Ageing causes cleavage in embryonic tissues, Osborne (2000). Accumulation of breaks in DNA was observed in aged seeds, which will be repaired during the phase –I of imbibition and facilitates seed germination. As deterioration advances, accumulation of cellular lesions and

DNA breakage reaches a threshold level beyond which germination and normal development is not possible. This could be attributed to the impairment of polymerase and ligase enzymes associated with DNA, Sen and Osborne, (1977), Edler *et al.,* (1987). The process of deterioration occurring under natural ageing is different from the one occurring under accelerated ageing conditions, Boubriak *et al.,* (2000) and was confirmed by SSR and AFLP analysis, Vijay *et al.,* (2009). In *Arabidopsis*, inactivation of the plant specific DNA Ligase VI entails a delay in seed germination, Waterworth *et al.,* (2010) and also reported that *atlig6* mutants too display hypersensitivity to seed ageing and therefore, implicates AtLIG6 as a major determinant of seed quality and longevity. The observation pertaining to the involvement of enzymes in the repair of oxidative damaged DNA in *Medicago truncatula* seed germination offers further proof of the necessity of DNA repair during early germination phase, Macovei *et al.,* (2011). Hunt *et al.,* (2007) reported that nicotinamide, is an inhibitor associated with DNA repair, must be degraded during germination. NIC 2, an Arabidopsis gene, encoding for nicotinamidase enzyme is expressed in higher levels in mature seeds and relieves inhibition of PARP activity and allows DNA repair to occur.

7.10 Theories proposed on Seed Deterioration

Carbohydrates, proteins and lipids besides nucleic acids are the chief constituents of any seed, albeit their proportion varies between different species. Cereals are rich in carbohydrates while pulses and oilseeds are rich in proteins and fatty acids respectively and all these compounds are liable for damage during deterioration. Therefore, the involvement of numerous ultra-structural and biochemical compounds in one hand and incomplete understanding of numerous reactions related to seed deterioration on the other hand have rendered the identification of prime cause more difficult and hence could be attributed to the fact that much of the age related metabolic changes are only symptoms rather than the cause. In the absence of strong evidence for a single determinant of any one of these symptoms of seed deterioration, several hypotheses have been forwarded to explain the mechanism of seed ageing by researchers working on different plant systems. Though lipid peroxidation model has displayed great interest among the scientific community around the world, it could better explain seed deterioration in oil seeds. May be, the exact mechanism of seed deterioration could be the culmination of all these models. Some of the most important models are as follows.

7.10.1 Depletion of Food Reserves

This is one of the oldest theories proposed on deterioration. Most of the seeds contain food reserves that last thousands of years. Ewart (1908) proposed that longevity of seed not depends on the available food reserves because the protein molecules will disintegrate upon drying seeds to very low moisture levels. Crocker (1938) proposed protein coagulation theory, which is not tenable because of the presence of a number of proteins in the embryo. Even the non-viable seeds contain enough food reserves for seedling growth and development (Barton (1961). The biochemical degradation processes in seeds are too small to account for depleting the food reserves within the life span of most seeds hence this theory has not been accepted. Therefore, Harrington, (1967) proposed that depletion of oxidizable material in meristematic tissues might cause deterioration even when the adjacent tissues like cotyledons and endosperm still have plenty of food reserves. Probably this could be attributed to lack of mobilization facility in dry seeds might have led to the starvation of these meristematic cells. Lars Schmidt, (2000) proposed that depletion of essential metabolites, either through respiratory loss of food reserves such as carbohydrates, lipids and proteins or direct los of vitamins, hormones etc., is responsible for seed deterioration and ageing.

7.10.2 Reducing Sugars

Madhava Rao and Kalpana (1994) proposed the hypothesis based on the existence of correlation between the accumulation of reducing sugars and loss of seed viability while working with rapidly aged pea seeds. They have indicated the possibility of occurrence of Amadori Reaction and Millard Reaction in rapidly ageing seeds, which in turn cause protein inactivation and damage to nucleic acids and ultimately leads to seed deterioration. Bentsink et al., (2000) also reported that neither raffinose series sugars nor sucrose content had a specific affect on seed storability, contrary to this, Hoekstra (2005) reported that it is the deteriorative processes in membranes rather than the glassy state of cytoplasm that determines the rate of ageing of seeds. According to Rosnoble et al., (2007), the genetic modification of sucrose non-fermenting kinase (SnRK1) complex exhibited impaired accumulation of oligosaccharides of raffinose family and greater sucrose content when compared to control seeds. It clearly shows that the genetically modified seeds had a phenotype with low germination percentage, low seedling vigour and hence reduced seed longevity when compared to the wild type seeds. In the absence of differences in dry weight or reserve accumulation of starch, proteins and lipids, it was concluded that SnRK-1 complex functioning can regulate the seed longevity and also affects the non-reducing sugar content in subsequent stages of seed maturation. The genetic analysis of sucrose and raffinose content present in the seed of Arabidopsis clearly points out the role played by SnRK complex in regulating other pathways for control of seed longevity in addition to bringing changes in the composition of sugars.

7.10.3 Loss of Glossy State

An amorphous, non-equilibrium condition in which a liquid achieves high viscosity that resembles a solid is referred to as a glassy state. Williams and Leopold (189); Bruni and Leopold (1991); Sun and Leopold (1993); Maki et al., (1994), have reported glassy state of cytoplasm in dry seeds. The raffinose famili oligosaccharides such as sucrose, stachyose, verbascose that accumulate during seed maturation facilitates the formation of glassy state in cytoplasm, Leprince et al., (1993). When compared to reducing sugars, these sugars are preferred for the formation of glassy state because of their high molecular weight. As a consequence, the cytoplasm will have a high trasnition temperature at which the glassy state will be lost. This is one of the most important factors that determine the rate of physical and chemical changes associated with the seed deterioration. Because of the lesseer tendency for chrystallization, these sugars suppress the formation of glassy state, not only maintains the structural integrity of membranes and proteins during dry conditions but also prevents damage by the increased level of chaotrophic ions like Na+, K+, Cl- within the cytoplasm during water withdrawl. Williams and Leopold (1989) reported that in glassy state, cytoplasm is so viscous that both diffusional movements as well as deteriorational reactions are arrested. At the same time it prevents lipid peroxidation, Amadori and Millard reactions and hence decline in germinability and vigour during seed ageing due to its association with the loss of glassy state. Franks et al ., (1991) reported that glossy state may be lost by either increasing water content or by increasing the temperature.

7.10.4 Accumulation and Repair of Cellular Lesions Model

Early researchers like de Vries, (1901) and Nillson, (1931) considered the increase in seedling abnormality with aged seeds as a mechanism of survival. The earlier observation of de Vries (1901) has been discussed by Priestly (1985) as manifestation of trisomics and Polyploids peculiar to *Oenothera* and not due to seed deterioration. Cellular membranes lose their selective permeability while deterioration is progressing and permits the leaching of cytoplasmic metabolites into inter cellular spaces. The degradation of membranes is possible either through hydrolysis of phospholipids by phospholipase or autoxidation of phospholipids and hence can be considered only as a result of ageing rather than cause. Rosy *et al.,* (1964) reported that addition of unsaturated fatty acids, phospholipids and chelating agents together with binding compounds like albumin repairs partly or in whole functional changes in mitochondria. Similarly some growth regulators are also expected to help in maintaining the integrity selective permeability of mitochondrial and other cytoplasmic membranes.

Several point mutations are mostly recessive mutations and hence are not manifested phenotypically. Since several evidences have confirmed that gene mutations and chromosomal aberrations induced during seed storage led to seedling abnormality. Roberts (1972) also reported close association between accumulation of subcellular damages and loss of viability. Further, Roberts (1983) also reported a negative correlation between percent aberrant anaphase cells exhibiting chromosomal damage and viability of seeds. Whereas, Elders and Osborn (1993) demonstrated with rye seeds that repair of single strand breaks in the nuclear DNA of dry seed was one of the earliest events during the imbibition of rye seed. The repair gets progressively less efficient with seeds stored for longer duration and DNA fragmentation in embryos of dead seeds up on imbibition results in the accumulation of nucleosomal multimers (Boubriak *et al.,* (2000). Copeland and McDonald (2001) reported that cellular aberrations increase progressively with the advancement of the ageing process because Berjak and Villers (1972) have already reported a progressive increase in the cellular aberrations with the advancement of ageing process.

Smith and Berjak (1995) compiled the sub-cellular abnormalities reported by several researchers while working on the membrane ultra structure as follows. (i) Abnormalities in inner and outer membranes of mitochondria and plastids, (ii) Lobbing of nuclear envelope, (ii) loss of endoplasmic reticulum and golgi bodies, (iv) dissolution of membranes of vacuole and protein bodies, (vi) discontinuities in plasma membrane, (vii) withdrawal of plasma membrane from the cell wall and (ix) occasional appearance of floccular material in the extra protoplast space.

Osborne *et al.,* (2003) further reported that loss of highly conserved oligonucleotide sequences of the telomere region leads to loss of viability. Whether or not the cellular degenerative changes are the cause or consequences of seed ageing, it is widely accepted that these damages get repaired upon subsequent hydration and became the basis for the seed priming technology.

7.10.5 Loss of Antioxidant Molecules

All plants are gifted with antioxidant molecules, either lipid soluble (Tocopherols, vitamin E, Beta carotene) or water soluble (Vitamin C and glutathione). Puntarulo *et al.,*(1991) reported that free radicals are produced during metabolism naturally in chloroplasts and mitochondria. The Tocopherols being a chain blocking antioxidants are expected to block the lipid peroxidation. Short supply of these antioxidant molecules during oxidative stress leads to seed deterioration.

The water-soluble antioxidant, tripeptide glutathione, is highly necessary for the storage and transport of sulfur required for protein synthesis. It is highly susceptible to oxidation due to the presence of sulfhydryl groups, Smith *et al.,* (1990). Several reports are indicating that ageing involves oxidation of thiols like glutathione. Since it is harmful to cellular metabolism, it is likely to contribute to aging induced seed deterioration.

7.10.6 Acquisition of Desiccation Tolerance Model

7.10.6.1 Desiccation Damage

During seed drying, the two types of non-mutually exclusive damage envisaged are (a) damage caused to seed due to the direct removal of water and (b) metabolically derived damage. The former is associated with the mechanical strain and removal of the hydration shell of the molecules. This could be attributed to the progressive shrinkage of cells. As a consequence, the plasma membranes get exposed to mechanical strains and get damaged due to fluctuations in the surface to area ratios leading to lesions in membrane and ultimately results in the decreased volume of cell, molecular crowding of cellular components and condensation of chromatin which make the cytoplasm more viscous. This in turn decreases the diffusion of metabolites and oxygen progressively within the cytoplasm and organelles. The metabolically derived damage is the result of uncontrolled chemical reactions of metabolism due to the accumulation of by products at toxic levels. It could be due to the decreased availability of oxygen to mitochondria during drying below -6 mPa or accumulation of ethanol and acetaldehyde before the loss of membrane integrity or the inability to switch on the fermentation process in desiccation tolerant species, or oxidative stress. Sometimes, it is difficult to distinguish between the damage caused by both types not because but because each type will depend on a number of factors including eco-physiological origin of species, nature of storage reserves and composition of the lipid membranes, degree of development of tissues, and the rate at which water is lost (hours, days and months). Usually the membranes, proteins and nucleic acids are the most common targets of damage by both types of damage.

7.10.6.2   Desiccation Intolerance or Sensitivity

Recalcitrant seeds, comprising of commercially important tropical species like rubber, tea, cacao *etc.,* being desiccation sensitive or desiccation intolerant, cannot be stored under dry and cold conditions and hence exhibit a very poor pattern of longevity because they cannot with stand a partial or total loss of water during maturation. Recalcitrance is a quantitative feature dealing with the sensitivity to drying rather than the ability to survive during drying, and depends on several inherent characteristics of seed like morphology, composition, stage of development, drying conditions including rate of moisture loss and temperature, rehydration conditions *etc.* At times slow drying is more deleterious than fast drying because the tissues reach low water contents and Glassy state before damage occurs. Usually, critical water content is determined to quantify the extent of desiccation sensitivity, Pammenter and Berjak, (1999).

7.10.6.3   Desiccation Tolerance

Desiccation tolerance refers to all those mechanisms that enable plants to survive despite severe water loss. Good majority of plant species have seeds with a well defined and pre-programmed desiccation phase towards the end of the developmental stage set for acquiring desiccation tolerance and all such seeds are called as orthodox seeds. The maturation drying results in reduced storage reserve deposition, brings the seed embryo into glassy state, which in turn leads to metabolic quiescence, storability and resistance against detrimental environmental changes and hence this trait has adaptive importance with role in dispersal and survival of species. The ability to survive complete water loss at water potentials below -150 MPa requires several biochemical adaptations related to tolerance and hardiness.

7.10.6.4   Acquisition of Desiccation Tolerance

Majority of plant species begin to lose water at some point during the course of their development and dry down to water content of 5-10 percent on fresh weight basis. Since this loss of water does not kill them, they can be considered as desiccation tolerant. The acquisition of desiccation tolerance depends on the two things viz., the rate of drying and the amount of water content achieved after desiccation. It means if the rate of drying is faster, the seed may become able to tolerate water loss. For example, the seeds of French been harvested 26 days after pollination hardly survive when water is made to lose immediately. But the same seed if allowed to dry slowly over a period of one week will be able to germinate normally. This clearly indicates that further development took place in the seed during this one-week period, which is sufficient enough not only to promote germination but also for the synthesis of protective mechanisms. Usually, the seeds acquire the ability to survive the initial stress of desiccation obviously possess the ability to survive in the desiccated state. Therefore, most of the mechanisms proposed for desiccation tolerance are also implicated for seed longevity.

Orthodox seeds too show poor desiccation tolerance mechanism upon priming and ultimately have a shorter shelf life period. Several researchers have considered the maintenance of cellular integrity after drying as indication of desiccation tolerance and considered it as a developmental programme in germinated seeds. Further it was observed that it is lost progressively and sequentially in the radicle followed by hypocotyl, epicotyl, cotyledons and finally coleoptile. Hence, several recent researchers considered the relationship between acquisition of desiccation tolerance by the seed and its longevity potential as one of the probable criteria. Ascorbic acid and osmotic stress are the two factors that have been identified so far to trigger the signal transduction chain leading to acquisition of desiccation tolerance. Usually, the content of abscisic acid reaches its peak value half way through the development and decreases gradually upon attaining maturation drying, Black and Pritchard (2002). Desiccation tolerance can be considered as a multifaceted trait because seeds were able to survive only when all mechanisms and conditions necessary tolerance are present. This clearly shows that there is no obvious relationship existing between the level of desiccation tolerance and amount of moisture of any tolerance factor. Accumulation of late embryogenesis abundant protein (LEAs), ABA and oligosaccharides are among the several factors that determine the role in acquisition of desiccation tolerance, Obendorf (1997).

7.10.6.5   Desiccation Tolerance- Protection

Avoidance of accumulation of damage induced by loss of water in dry state and activation of repair mechanism are the two strategies adopted to explain the mechanism of desiccation tolerance. The former occurs in the presence of preexisting or newly synthesized protecting factors like non-reducing sugars, stress proteins like LEAs and HSPs, and free radicle processing systems etc., which act both during seed desiccation and seed drying state. Whereas the later, activation of repair of desiccation induced damage mechanisms occurs up on rehydration. It was observed in several plant species that compatible solutes accumulate in higher concentrations and offer protection to macromolecules after water is removed. It means the concentration of these compatible solutes determines the degree of tolerance to stress by counteracting and destabilizing the effects of molecules and ions like Na^{+2}. As a consequence of the loss of water concentration of amino acids like argenine increases in cytoplasm. Hokestra *et al.*, (2001) concluded that during drying, different mechanisms of protection appear to act at different stages of water loss. During early dehydration stage survival is through avoidance of protein unfolding and thereby restricting the membrane disturbance through preferential hydration because membrane lipids in fully hydrated cells will be in an undisturbed crystalline state. Hence loss of water not only results in the increased concentration of cytoplasmic amphiphilic compounds but also partition in to membranes owing to preferential interaction. Amphiphile partitioning into membranes is complete with the dissipation of bulk water. Subsequent removal of water in the intermediate water range (0,3gH2O/g dry weight) from the hydration shell, result in (a) sugar molecules replacing water at hydrogen bonding sites,(b) preserves the native protein structure (c) maintains spacing between phospholipids in tolerant cells, (d) keeps the membrane surface preferentially hydrated and (e) prevents membrane fusion.

The bi layer remains in the liquid-crystalline phase. Contrary to this absence of these solutes in sensitive cells result in (a) phase transition into the gel phase, (b) fusion of membranes due to packing of phospho-lipid molecules into transition phase, (c) phase separation and irreversible membrane damage, (d) glassy matrix resulting from immobilization of cytoplasmic components exhibit differences in properties between tolerant and sensitive cells below 0.1 *g* water/*g* of dry weight.

Simultaneous curtailing resulting in the production of ROS through down regulation of the metabolism and free-radical scavenging was considered as an important mechanism of survival. In the study of plant desiccation tolerance, it has often been found that one specific mechanism does not confer tolerance on its own, but that the interplay of several mechanisms simultaneously is essential. Therefore, a sensitive component is protected in different ways so as to guarantee optimal survival. Most biophysical investigations concerning anhydrobiosis in plants have been focused on phenomena in the dried state.

7.10.6.6   Desiccation Tolerance- Repair Mechanism

An hydro-bites, in addition to protecting the cellular structures, have also acquired another tolerance mechanism of repairing the desiccation-induced damage upon hydration due to the presence of a series of genes called rehydrins. During the germination of grass seeds, ubiquitin based repair mechanism gets activated and converts abnormal L-isoasprtyle residues in to normal L-isoaspertyle residues in the presence of an enzyme methyl- transferase. In order to ensure survival, the seeds upon hydration must get the proteins accumulated during drying or storage repaired. Similarly seeds also have acquired DNA repair mechanism during rehydration so as to maintain genomic integrity. Usually, the embryos fail to survive when certain drugs block DNA repair during rehydration. Therefore, the invigouration treatments like priming are basically aimed at reducing the chromosomal aberrations and thereby decrease the frequency of abnormal seedlings even after storage under detrimental conditions because these physiological treatments induce repair mechanisms before radicle emergence, (Osborne and Boubriak (1994) and Osborne *et al.*, (2002).

7.10.6.7　Desiccation Tolerance – Response of Metabolism

Coordinated regulation of metabolic pathways is nothing but the adoptive mechanism of tolerance. It is possible through the regulated use of carbon sources which in turn allows the use of accumulation protective enzymes as well as   production of energy necessary for the synthesis of protective proteins. The differential level of hydration levels, allow neither the activities of enzymes nor metabolic pathways to shut uniformly and simultaneously. For example, at 10 percent moisture level, only lipids hydrolyzing enzymes remain active whereas amylases and dehydrogenases involved in oxidation and reduction are active at intermediate levels of moisture. Contrary to this all enzymes are able to perform catabolic activities in the presence of bulk of water. The signals and biochemical mechanisms leading to reduction of metabolism and their importance in desiccation tolerance are unknown, (Leprince *et al.*, 2002).

7.10.6.8　Lipid Peroxidation Model

Wilson and McDonald (1986) proposed this model of Seed Deterioration. Kaloyereas (1958), for the first time proposed that oil rancidity is the cause of reduction in pine seed germinability and demonstrated this by treating the seeds before storage with starch phosphate, an antioxidant and for the first time indicated that lipid peroxidation could be a direct cause of seed deterioration during storage. The phenomenon of early death during germination of aged seeds suggests the presence of destructive elements, which become active only after imbibition of the seed. Hence, this model is based on the destructive element, the oxygenated fatty acids, produced either through autoxidation or enzymatically, Wilson and McDonald (1986). The lipid autoxidation occurs in all fully imbibed cells because water acts as a buffer between the macromolecules and reactive compounds. The polyunsaturated fatty acids, the chief constituent of phospholipid bilayer in bio membranes, Gille and Joenje, (1992), Dussert et al., (2006), with a methyl group between two double bonds, are highly susceptible and liberates hydrogen free radical very easily. The fatty acid hydrocarbon chains undergo spontaneous oxidization in the presence of oxygen and produce reactive free radical intermediates called as hydro-peroxides. This reaction is accelerated at high temperature and causes an increase in the concentration of oxygen. The rate of this reaction is greatly enhanced by Lipoxygenase group of enzymes present in the seeds. The process of lipid peroxidation is initiated by a free radical, whose identity is not yet clear. The creation of a short lived; highly reactive free radicals within the biological system can be initiated either by the ionizing radiations or metabolic processes or even lipoxigenases. Once a free radical is produced, it initiates a chain reaction and creates additional reaction cycles and more free radicals. The free radical may be the hydroxyl radical itself at times, which in the presence of Cu^+ and $Fe^{+2.}$ may have been formed in *vivo* by a preceding activated oxygen-transitional reaction, the initiation event with hydroxyl radical results in the formation of a carbon centered lipid radical, which undergoes random rearrangement and attack by oxygen molecules only to form lipid peroxy radicals, whose random rearrangement forms a fat soluble chain breaking antioxidant like tocopherol and terminates in lipid peroxidation. This reaction gets terminated when two radicals combine to produce a stable end product. The abstraction of hydrogen from a lipid membrane decreases the fluidity analogous to dehydrogenation, which at room temperature forms solidified margarine and results in the inactivation of membrane bound proteins including transport proteins and receptors and ultimately brings in changes in membrane permeability.

Depending on the moisture content of seeds, the mechanism of lipid peroxidation will also change, because the process of seed deterioration occurring during natural and accelerated ageing are different and the same is evidenced by the fact that different nucleases get activated at different levels of moisture contents, Boubriak *et al.,* (2000). At low seed moisture content

(<6%) autoxidation could be the primary cause of seed deterioration. Contrary to this, when moisture content exceeds more than 14%, lipoxigenase activity increases, causes multiple nuclease degradation of DNA. Vijay *et al., (2009)* reported differences in mechanism of seed deterioration with natural and accelerated ageing with the help of SSR and AFLP analysis. Fast fragmentation of DNA when compared with slow DNAase nicking in the random single strand cleavage of DNA held at 9 per cent moisture content. Between 6 to 14 percent moisture content, lipid peroxidation is most likely to be at its minimum because sufficient water is available to serve as a buffer against autoxidative free radical attack but not enough water is present to activate lipoxigenase mediated free radical production. As such, the mechanism of lipid peroxidation is different under natural (low RH) and accelerated (high RH) ageing.

The inherent capacity of hydroxyl radicals to combine well with most molecules present in the living system including lipids, proteins and nucleic acids (Halliwell and Cutteridge (1990) causing mutations or strand breaks in DNA by combining the bases and initiates lipid peroxidation directly or indirectly through the inactivation of enzymes involved in processing of oxygen in various forms. In the presence of transition metals, the lipid hydro peroxides decompose to form ethane and several other cytotoxic unsaturated aldehydes and this is what is referred to as deterioration of membranes. Heat enzymes, cytochrome, transition metal ions and hydrogen bonding trigger hydro peroxide breakdown resulting in smaller secondary products or more free radicals. The chemistry of lipid peroxidation was discussed in detail by Gutteridge and Halliwell (1990).

7.10.6.9   Biological Effects of Lipid Peroxidation

Herman (1956) for the first time hypothesized that ageing was a consequence of free radical attack on cellular constituents and connective tissues and in 1969 proposed that lipid peroxidation is the major source of free radicals in tissues. The mechanism of damage caused to organisms through lipid peroxidation can be broadly grouped in to three ways as detailed below. The three ways in which lipid peroxidation cause potential damage to seeds are as follows.

7.10.6.9.1   Destruction of membrane lipids

Bio membranes, conferred with inherently large surface area and usually have more unsaturated fatty acids than storage lipids will definitely become a key site of direct injury of lipid peroxidation. Simon, (1974) reported that lipid peroxidation of membranes may lead to decline in membrane integrity and increases membrane permeability. Ghosh *et al.,* (1981) reported loss in integrity of the plasma membrane in aged seeds by way of leakage of cytoplasmic components into the external medium. For example, the inner membranes of mitochondria are composed of higher proportion of unsaturated acyl chains than other membranes and renders membrane permeability more common, Moreau, Dupont and Lance, (1976). Linnane and Crowfoot (1975), reported that increased mitochondrial membrane permeability breaks down the proportion of gradient required for maintenance of respiratory coupling leading to decreased respiratory competence and hence the decline in membrane integrity making it the primary cause of lipid peroxidation.

7.10.6.9.2   Co-oxidative assault by free radicals

The crucial enzyme systems including the electron transport system are basically membrane bound and are intimately linked to a potential source of damaging free radicals. Only stable molecules are incorporated into the autoxidative chain reaction by the abstraction of hydrogen atoms. In addition to lipid peroxide, their secondary products can also react with the terminal groups of amino acids in proteins and enzymes and stimulates the formation of Schiff bases between peroxidized phospholipids and membrane proteins, Casthilo *et al.,* (1994) and thus lead to polymerization of proteins.   Esterbauer (1987) reported that free radicals, due to their short life span cause potential short distance damage whereas the long distance detrimental effects are caused by more stable chemicals formed during lipid peroxidation.

7.10.6.9.3   Formation of cytotoxic aldehydes

When compared to fresh seeds, deteriorated seeds will produce 20 times more volatile aldehydes during imbibition. A wide variety of cytotoxic effects were produced by 4- hydroxyl-alkenals due to the formation of aldehydes during hydro-peroxide breakdown. These aldehydes in turn react with sulfhydryl groups and inactivate proteins. The interaction studies between peroxidation and respiration provide the possible consequences of lipid peroxidation in seed as detailed below.

The inner membrane of mitochondria represents a logical site of attack by oxygen is specifically composed of a high proportion of unsaturated fatty acids and hence require a higher degree of fluidity, Moreau, Dupont and Lance (1974). The oxidation of membranes directly interferes with respiration in at least three possible ways viz., (i) by changing the polarity/viscosity of inner membranes, (ii) interference with mitochondrial assembly during imbibition and (iii) Co-ordination of critical enzymes and cofactors.

The bio membranes must exist in a liquid crystalline state so as to function properly but lipid peroxidation either interprets or alternates the liquid crystalline phase through the introduction of polar oxygenated moieties on the hydrophobic acyl chains of phospholipid molecules and thereby increases membrane permeability. Leopold and Musagava (1979) suggested that lowering of respiratory activity and shifting of respiratory pathway plays a major role in decline of germinability and vigour.

Lipoxigenase activity, which is endogenous to mitochondria, resulted in the damage of fatty acids in cauliflower due to the formation of volatile aldehydes and accumulation of malandialdehyde in the presence of phospholipase. Similarly, the cytochrome molecules present in the mitochondria capable of catalyzing both fatty acids and fatty acid hydro peroxides and initiates the random attack on polyunsaturated fatty acids located in the plasma membrane. As a consequence, a chain reaction free radicals result in polar bridges across the hydrophobic barrier of plasma membrane, which in turn leads to an increase in the permeability associated with ageing. The two approaches, (i) measuring the changes in phospholipid content and (ii) detection of malandialdehyde as a secondary product of lipid peroxidation can be used to demonstrate the association of lipid peroxidation with deterioration of ageing of seeds can be discussed under the following five heads viz., (i) monitoring of lipid loss, (ii) detection of free radicals, (iii) detection of hydroperoxides, (iv) detection of secondary products and (v) exogenous modulation of peroxidation.

7.10.7 Monitoring Lipid Loss

According to Harman and Mattick (1976), measurement and monitoring of polar and non-polar levels, phospholipids and total lipids and relative changes in different levels of saturation, the preferential loss of poly-unsaturated acids, which are more easily oxidized, provides an evidence for lipid peroxidation. Priestly and Leopold (1979) reported slight decrease in phospholipid, an increase in total lipids and no change in the proportion of fatty acids during lipid peroxidation in soybean seeds subjected to accelerated ageing for five days at 40 $^{\circ}$C and 100% Relative Humidity. Priestly and Leopold (1983) demonstrated a gradual shift in the proportion of polyunsaturated fatty acids towards saturated and monounsaturated acids that accompanied a decline in vigour and germinability under natural ageing. Wilson and McDonald (1986) reported evidence linking lipid peroxidation and storage deterioration in some species.

7.10.8 Detection of Free Radicals

The technique of electron spin resonance quantifies the content of unpaired electrons. Free radicals are very reactive, though their accumulation is likely, detection may not be possible. They possess a potential for short distance damage due to their short shelf life period. Whereas, the long-distance detrimental effects of lipid peroxidation is due to their conversion in to more stable chemical species. The fatty acid hydro peroxides can be further degraded or reduced into a wide array of stable secondary products like epoxides, hydroepoxides, hydroxyl fatty acids and aldehyde ketones. Spin traps are molecules, which react with common free radicals, and form unusually stable radicals that persist long enough to be detected later, (Borg, 1976). Therefore, by incorporating the spin traps in seeds before ageing or germination free radicals can be captured for subsequent electron spin resonance analysis.

7.10.9 Detection of Hydro-peroxides

In seeds, if lipid peroxidation occurs during storage, oxygenated fatty acids accumulate as non-volatile products, which can be detected by Peroxide value (AOAC, 1980) or Gas Chromatograph mass spectrometry and high performance lipid chromatography.

7.10.10 Detection of Secondary Products

Secondary products are nothing but the end products of lipid peroxidation like malandialdehyde and hydro-peroxides, formed from acyl chain of oxygenated fatty acids. Stotzky and Schenk (1976) reported that seeds of many species release volatile aldehydes like acetaldehyde during early stages of germination. Whereas, Woodstock and Taylorson (1981) reported an increase in non-mitochondrial respiration through increase in aldehydes larger than C_2 is suggestive of lipid peroxidation (Wilson, 1983). Similarly, presence of hexanal in aged seeds with or without hydration acts as an index of lipid peroxidation.

 ## 7.11 Strategies to Limit Lipid Peroxidation in Seeds

Application of exogenous factors like (a) lipid modification, (ii) Seed irradiation, (iii) Antioxidant treatment, (iv) Hydration treatments and (v) Regulation of oxygen pressure, helps in the investigation of the role of lipid peroxidation in seed ageing.

(i) Lipid modification

Rudrapal and Basu (1980) reported that the exposure of mungbean seeds with iodine vapour at 40 °C, 95% RH, 20 days and 16 hours a day renders them immune to accelerated ageing 100 % germination, increased root and shoot length and concluded that the addition of iodine to unsaturated fatty acids rendered the acyl chains less prone to peroxidation, improved membrane integrity and higher dehydrogenase and amylolytic enzyme activity during germination. Similarly, Dey *et al.,* (1983) concluded that hydroxyl radical formation initiates lipid peroxidation by abstracting hydrogen atoms from the polyunsaturated fatty acids.

Unsaturated fatty acids being more prone to lipid peroxidation, reducing the content of polyunsaturated fatty acids through breeding might result in increased resistance to lipid peroxidation. Similarly, development of lines with low lipoxigenase enzyme, active against unsaturated fatty acids in the presence of water and oxygen, is considered to be the next best approach.

(ii) Seed Irradiation

The chemicals that offer protection to seed tissues from radiation damage are also expected to enhance the shelf life of the seeds under natural as well as accelerated conditions. Solutions of KI, Na_2HPO_4, Cysteine, Para-aminobenzoic acid, oxalic acid and NaCl produced promotive effects in trace quantities in seeds. During accelerated ageing, Sun and Leopold, (1995) reported decrease in oligo- saccharides like stachyose in soybean seeds due to non-enzymatic degradation or due to hydrolysis by galactosidase.

(iii) Treatment with Antioxidants

Pretreatment of seeds with antioxidants like tocopherol, starchphosphate, ascorbic acid, cinnamamic acid, BHT (butylated hydroxyl touline) have been reported to increase seed vigour in rice, maize, mustard, sunflower, French bean, pea, lentil, millet and jute. Antioxidant treatments are basically aimed at inhibiting lipid peroxidation by reducing the level of active oxygen species like superoxide radicals, hydroxyl radicals and hydrogen peroxide by blocking free radical chain reactions. Contrary to this, some of the antioxidants have tying up with metals and act as potent chelating agents and hence capable of catalyzing lipid peroxidation. Electron Spin Resonance or Chemi-luminescence methods are capable of measuring radical scavenging activity of antioxidants against free radicals like DPPH radical, superoxide anion radical, hydroxyl radical and peroxyl radical. Some volatile antioxidants, like di-tert-butyl sulfide, being toxic, could mask the promoter effect of enhancing seed longevity.

(iv) Hydration dehydration treatments

The pre-sowing techniques are meant for improving the germination rate at sub optimal temperatures and hypoxia, wherein the seeds are incubated in solutions of low water potential like PEG and is called as osmopriming. This not only facilitates reduction in the level of lipid peroxidation but also helps in restoration of antioxidant defense systems. Hydration-dehydration treatments help even the aged seeds also to regain their initial ability to germinate by repairing the lipid peroxidation induced membrane damage.

(v) Regulation of oxygen pressure

Since oxidative reactions are considered to be the chief cause of seed deterioration, minimizing oxygen pressure around the seeds ensures both storability and longevity. This is clearly evident from the success of storing seeds for longer durations in hermitically sealed containers. Lesser diffusion of oxygen enhances the storage life of seeds with Hard seededness, albeit at low moisture levels. Contrary to this, the rate of loss of seed viability under accelerated ageing conditions at 44°C and 100 per cent RH, is independent of oxygen pressure and only indicates the non-oxidative manner in which seeds are damaged in accelerated ageing system. Vanangamudi and karivaratharaju (1986) reported that exogenous application of plant growth regulators like GA3, ABA, and IBA before storage have retarded seed deterioration and maintained the seed vigour and germination.

7.12 Ageing / Repair Hypothesis

This hypothesis is aimed at evaluating physiological age/ repair

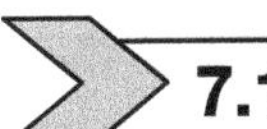

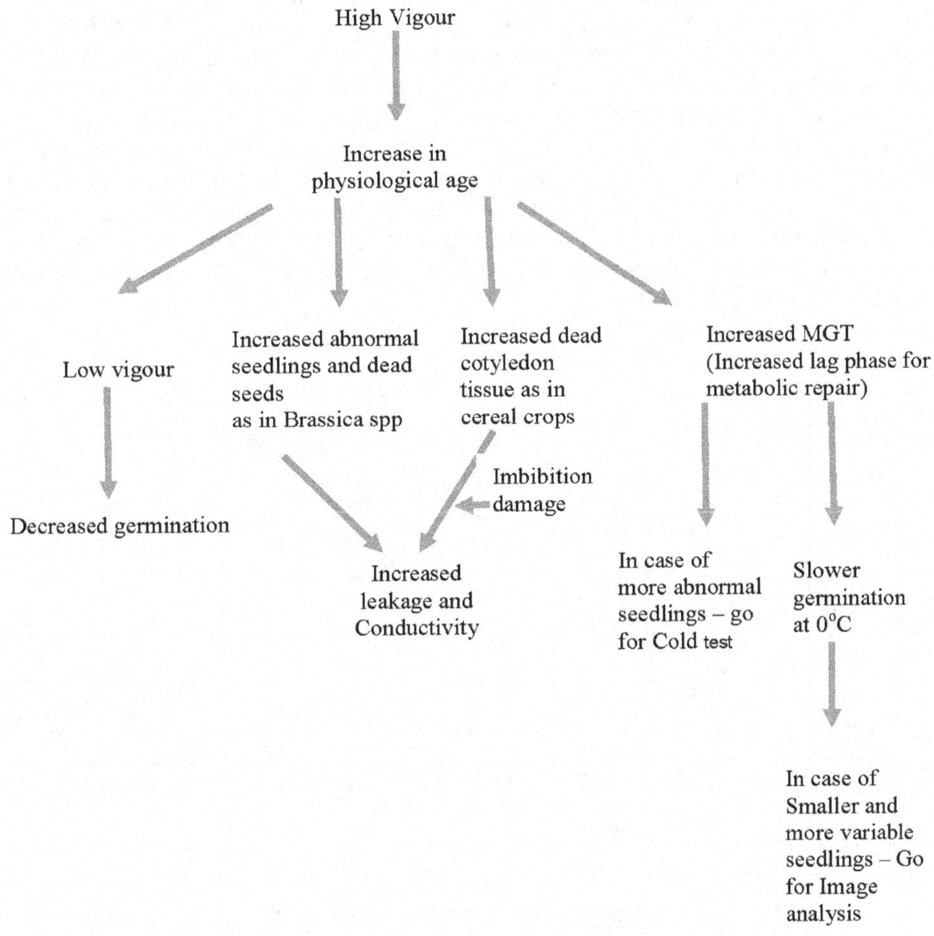

Interpretation of Various vigour tests on the basis of ageing repair hypothesis

Test	Interpretation
Tetrazolium test	Staining Pattern of living or dead tissues
Conductivity test	Leakage from dead seeds or damaged tissues after ageing and imbibitional damage
Controlled deterioration test	Positionof seed lot on the seed survival curve
Radical emergence Test	Time required for repair of ageing
Seedling size and uniformity	Differences in timing rate of emergence
Cold test at 10°C	Incomplete repair at an aerobic condition sresulting in more abnormal seedlings
Cool test at 18°C	Incomplete repair at lower temperatures resultingsmaller seedlings due to slowrate of emergence

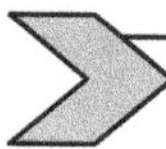

7.13 Vijay- Dadlani Model of Seed Ageing (2003)

Water content is another important factor determining the rate of seed deterioration. The role of enzymatic reactions in deterioration of dry seeds is very little because of the lack of active enzymatic metabolisin on dry seeds. Therefore, non-enzymatic reactions like Amadori and Millard reactions can play an important role under these dry seed conditions. (5).   In dry seeds the enzymatic reactions may play little role Wilson and McDomald (1986) demonstrated a relationship between the efficiency of hydration- dehydration treatments and amelioration of peroxidation induced detrimental changes in aged seeds, because water acts both as a deterrant and a protectant against peroxidative reactions and cause loss of seed viability. Basu (1993) demonstrated the role of water molecules in countering the free radicle induced chain reactions of hydration treatments in seed invigouration treatments for enhancing the seed longevity. Based on the integration of several factors various seed constituents, their hygroscopicity, efficiency of antioxient machinery etc., Vijay and Dadlani (2003) proposed a model of seed viability. (fig. 7.5).

Seed				
Physical structure	Micro-flora	Major Chemical Constituents		
		Oil	Carbohydrates	Proteins
		High PUFA	Slow release of moisture	Hygroscopic
		Anti-oxidants, low temperature		
		High Peroxidation		High EMC + Hysteresis
		Free radicals		Low moisture storage
			Radical quenching	High metabolic rate
			Macromolecular cross linkage	
Loss of viability		Invigouration or cellular repair mechanism	Protein denaturation and disruption	Biomembrane

Fig. 7.5 Model of Seed Ageing Proposed by Vijay and Dadlani (2003)

7.14 Summary

Several multipurpose species with great economic importance, simply suffer from their short shelf life that renders even genetic conservation a difficult task leave alone the reforestation programmes. A mature seed is designed to function only under favourable conditions. Till such conditions are made available, the seeds will remain in physiologically inactive state with reduced respiration rate. Depending on the prevailing conditions, the seeds will deteriorate and cause huge loss to agriculture and forestry in general and seed industry in specific. When compared to mature and large seed, it is still more difficult to store a seed lot with immature and small seed. Considerable amount of variation exists between different species in respect of seed longevity and forms the basis for classification of seeds. Seed longevity depends on both environmental and genetic factors.

Seed deterioration is a natural phenomenon with progressive, irreversible, deleterious and universal changes occurring before death, is in fact an inevitable consequence of life. Ageing related deterioration causes damage to various molecules, cells and organelles of seeds. Yet providing optimum storage conditions can retard the rate of deterioration. Some varieties exhibit less deterioration than others and individual seeds of different seed lots vary significantly in respect of their storage potential. It means seed deterioration is nothing but the falling of seed quality from higher level to a lower level with a set pattern of specific symptoms at each stage. Deterioration being a catabolic process, once initiated cannot be reversed. Therefore, seed deterioration has been aptly defined as an irreversible, inexorable process of seed populations and exerts direct impact on viability.

A good majority of seeds though equipped uniquely to survive till a right place and time are available for beginning a succeeding generation does not mean the seeds will retain their viability indefinitely. It is purely the physical condition and physiological status of the seeds that determine the shelf life of seeds in addition to the genetic constitution. In the post fertilization era, the translocate carbohydrates condense and store in different storage tissues as starch, amylose, amylopecting and hemicellulose. Similarly the proteins and fatty acids are stored in the form of polypeptides and triglycerides respectively. In addition to these, a number of auxins, inhibitors, vitamins, minerals, enzymes etc., exert direct influence on the seed deterioration. Longevity of seed depends on the destruction of some of the essential components required for triggering some biochemical steps needed for initiation of germination.

Each and every seed lot gets exposed to a specified period of storage before being planted. In order to achieve high germination, storage under optimum conditions become mandatory. Seeds with low moisture and high quality stored under cool/dry conditions perform better than those with high moisture and low quality seeds under hot humid conditions. Kind, value, volume, class and climate of the storage area determine the duration of storage required. Breeding material and germplasm have to be stored on long-term basis because of their value. Both long term experiments and accelerated ageing testing revealed that longevity of seeds is under the control of seed moisture, temperature in addition to genetic factors. Symptoms of deterioration at physiological level start appearing in a set pattern the moment seeds attain physiological maturity. Seed deterioration appears to be the consequence of a lethal damage caused to various cellular components such as membranes, proteins, nucleic acids and cytoplasm. Though several mechanisms have been proposed from time to time, they are mostly species specific and vary with ageing conditions. Hence, none of these mechanisms are fully capable of giving a detailed and exclusive account on seed deterioration. However, (i) Lipid peroxidation model involving Free radical attack, (ii) Amadori and Millard reactions, (iiii) loss of glossy state of cytoplasm can be considered as the chief mechanisms. Assuming that seed deterioration as a multifactorial event, each component is critical on its own weight. Therefore, any approach capable of minimizing lipid peroxidation would obviously help in improving seed vigour and viability.

7.15 Exercise

I. Fill in the blanks with suitable words.

1. Individual seeds of a seed lot vary greatly in respect of their --------.

2. The translocated carbohydrates are stored in different seed tissues in the form of --------. and -------- in addition to starch and amylose.

3. The terms orthodox and recalcitrant seeds was proposed by --------.

4. KNO_3, Thiourea, Kinetin are examples of --------.

5. Coumarin, ammonia and absicic acid are examples of --------.

6. Remarkably short lived seeds are also known as --------.

7. Survival during dry storage is the property of -------- seeds.

8. The low moisure content of orthodox seeds are bound in -------- is mostly immobile.

9. Neem and papaya are examples of -------- type of seeds.

10. The classification of AS-1, AS-2 and AS-3 was given by --------.

11. In addition to genetic factors, seed longevity is also geverned by -------- and --------.

12. Most seeds attain equilibrium moisture content (5%) when stored at -------- RH and -------- °C.

13. Usually seeds stored at high moisture content cannot tolerate --------.

14. High storage temperature, in the presence of high moisture content increases the rate of --------.

15. -------- of seeds in storage facilitate invasion of insects and pathogens.

16. The lower the moisture content, the lower the temperature, then the slower the process of --------.

17. To inactivate most of the seed borne fungi and insects, the storage temperature should be --------.

18. Life span of seeds in storage will be halved by every one per cent increase in --------.

19. Temperature and moisture stress during flowering, seed filling and maturation results in --------

20. The success of germination quality is determined by the quality of -------- stored in seed during embryo maturation.

21. DNA plays a key role in germination by regulating -------- pathways.

22. -------- reactions are dependent on the availability of moisture and temperature.

23. Raffinose sugars that retard -------- may promote seed longevity.

24. Physical reactions are --------, when the seeds are removed from storage.

25. -------- reactions occur in seed stored at high relative humidity.

26. Oxidative degradation is caused by -------- mechanisms.

27. The chief purpose of oxidative reactions is to retrive energy from glycolysis through ------.

28. The by products of oxidative reactions are referred to as -------- or --------.

29. -------- is the known strongest oxidant.

30. Incresed leakiness of membranes is the result of -------- of fatty acids.

31. The inability of non viable seeds to resist penetration of heavy metal salts is an expression of loss of -------- integrity.

32. The accumulation of toxic chemicals during germination or ageing activated enzymatically results in --------.

Answers

1. storage potentials
2. Amylopectin and Hemicellulose
3. Roberts(1973)
4. Germination promoters
5. Germinatin inhibitors
6. Recalcitrant seeds
7. Orthodox
8. Macromolecules
9. Intermediate
10. Justice and Bass (1978)
11. Moisture and Temperature
12. 30%RH, $15\,^{\circ}C$
13. Low oxygen pressure
14. Metabolic processess
15. High moisture content
16. Deterioration
17. Less than $8\,^{\circ}C$
18. seed moisture content
19. Reduced source sink relations
20. m-RNA,
21. Harmonal signalling,
22. Physical
23. Crystallization
24. Reversible
25. Hydrolytic
26. Non-enzymatic
27. B- oxidation of fatty acids
28. ROS or free radicals
29. Hydroxyl radical
30. Oxygenation
31. membrane,
32. Seed death

II.   Define and explin the following

1. Seed Deterioration
2. Recalcitrant Seeds
3. Orthodox seeds
4. Physical Reactions
5. Amadori reaction
6. Spin traps
7. Glassy state

III. Answer the following in brief

1. List out the consequences of physiological seed deterioration.
2. How the translocated carbohydrates are stored in different storage tissues of seed?
3. List out the factors that determine type and duration of storage.
4. List out the types of seed retained for more than two seasons in storage.
5. List out the changes associated with ultrastructural degenerration of cell.

6. List out the by-products formed during lipid peroxidation.

7. List out the functional changes occuring during mitochondrial degradation leading to seed deterioration.

8. List out the logical conclusions based on which one can infer that seed is deteriorated.

9. List out the exogenous factors and their role in lipid peroxidation during seed ageing.

IV. Answer the following in detail

1. Explain the changes occuring at physiological level leading to deterioration of seeds.

2. Explain the changes occuring at biochemical and molecular levels leading to deterioration of seeds.

3. Explain in detail the lipid peroxidation model of seed deterioration.

4. Explain various factors responsible for exogenous modulation of peroxidation.

V. Write short Notes on the following

1. Free radicals

2. Autooxidation

3. Depletion of Food Reserves Theory of seed deterioration

4. Accumulation and repair of Cellular Lesions Model.

5. Monitoring of lipid loss.

6. Formation of cytotoxic aldehydes.

7. Biological effects of lipid peroxidation

7.16　References

Abdul- Baki, A.A., (1969). Relationship of glucose metabolism to germinability and vigour in barley and wheat seeds. Crop. Sci. 9 : 732 – 737.

Abdul – Baki, A.A. and Anderson, J.D. (1970). Viability and leaching of sugars from germinating barley. Crop Science. 10: 31- 34.

Abdul - Baki, A. A and James Baker, E., (1973). Are changes in cellular organelles or membranes related to vigour loss in seeds. Seed Sci and Technology. 1 : 89-125.

Abu-Shakra, S.S., and Ching, T.M., (1967). Mitochondrial activity in germinating new and old soybean seeds. Crop Sci. 7 : 115 -118.

Agrawal, P.K., (1980). Relative storability of seeds of tens species under ambient conditions. Seed Res.8: 94-99.

Anderson, JD., and Baker, JE.,(1983). Deterioration of seeds during ageing. Symposium Deterioration Mechanisms in seeds. American Phytopathological Soc. Vol. 73(2) 321-330.

Anderson, J.D., (1977). Adenylate metabolism of embryonic axes from deteriorated soybean seeds. Plant Physiology 59:610-614.

Bailly, C ., Benamar,A., Corbineau, F., and Come D., (1996). Changes in the malondialdehyde content and in superoxide dismutase, catalase and glutathione reductase activities in sunflower seeds as related to deterioration during   accelerated aging. Physiol. Plantarum. 97: 104-110.

Barton, L.V.,   (1961). Seed preservation and Longevity. Plant Science Monographs. Leonard Hill. Ltd. London, U.K.

Baker, E.H., and Bradford,K.J., (1994). The fluorescence assay for Millard product accumulation does not correlate with seed viability. Seed Sci. Res. 4 : 103-108.

Banarjee, A., Choudhuri, M.M., and Ghosh, B., (1981). Changes in nucleotide content and histone phosphorylation of ageing seeds.Zeitschrift fur Pflanzen physiology. 102: 33-36.

Basavarajappa BS., Shetty HS., and Prakash, (1991). Membrane deterioration and other biochemical changes associated with accelerated ageing of maize seeds. Seed Science and Technology 19: 279-286.

Basu, R.N., (1993). Seed invigouration in extended storability. Seed Res. 1: 216-219.

Baudry, A., Heim,M.A., Dubreucq,B., Caboche, M.,WeissharrB., and Liprince,L., (2004.) TT2, TT8 and TTG1 synergistically specifythe expression of BANYLUS and pro-anthocyanidin biosynthesis in *Aradopsis thaliana.* The Plant J., 39: 366-380.

Betty, M., Finch- Savage, W.E., King, G.J., Lynn, JR., (2000). Quantative Genetic analysis of seed vigour and pre-emergence seedling growth traitsin *Brassic Oleracia.* New Phytol.148: 227-286.

Bentsink, L.C., Alonso-Blanco, D., Verugdenhil, K., Grost, S.P.C., and Koorneef, M (2000). Genetic analysis of seed soluble oligosaccharides in relation to seed storability of Aradapsis. Plant Physiology. 124: 1595-1604.

Berjak, P., and Villers, T.A., (1972). Ageing in plant embryos. II Age induced damage and its repair during germination. New Phytol. 71: 135-144.

Bernal lugo, I., and Leopold, A.C., (1998). The dynamics of seed mortality. Journal of Experimental. Botany.49: 1455-1461.

Bernal Lugo, I., and Leopold, A.C., (1992). Changes in soluble carbohydrates during seed storage. Plant Physiology. 98:1207-1210.

Bhaskaran, M., Vanangamudi, K., Natesan, P and Bharathi, A., (2004). In: V. Krishnasami, AS, Ponnusami, Balamurugan, P., Srimati, P., natarajan,N and Raveendran, TS (eds.). Compendium on seed science and Technology.pp 352-359. TNAU, Coimbattore, India.

Black,M., and Pitchard,H.W., (2002). (eds.) Desiccation and survival in plants; Drying without dying. Wallingford, UK., CAB Internatinal.

Boubriak, I., Naumenko,V., Lyne,L., Osborn, D.J., (2000). Loss of viability in rye embryos at different levels of hydration. Senescence with apoptotic nucleosome cleavage on death with random DNA fragmentation. In Black, M., Bradford, K.J., and Vazquez- Ramos. (eds.) Seed Biology Advances and Applications. Pp 205-214. CABI. Publishing, Cambridge.

Buitink, J., Hoekstra, F.A., Hemminga, M.A., (2000). A critical assessment of the role of oligosaccharides in intracellular glass stability.Pp 461-466. In Black, M., Bradford, K.J., and Vazquez- Ramos . (eds.) Seed Biology Advances and Applications. CABI. Publishing, Wallingford, Cambridge.

Buitink, J., Hoekstra, F.A., and Leprince, O. (2002). Biochemistry & Bio physics of tolearne systems .In. Black, M and Pritchard H.M (Eds). Dessication and surviaval in Plants: drying without dyeing, Wallingford. CABI Publishing, Wallingford, U.K. Pp: 293-318

Bray, C.M. and Dasgupta, J., (1976) Ribonucleic acid synthesis and loss of viability in pea seed. Planta. 132: 103-108.

Bray, C.M., Chow., T.Y., (1976a). Lesions in post-ribosomal supernatant fractions associated with loss of viability in pea (Pisum arvense) seed. Biochim. Biophys. Acta 442: 1-13.

Bray, C.M. and Chow, T.Y., (1976b). Lesions in the ribosomes of non-viable pea (Pisum   arvense) embryonic axis tissue. Biochim. Biophys. Acta. 442: 14-23.

Bruni, F., and Leopold, A.C., (1991). Glass transitions in soybean seed: relevance to anhydrous biology. Plant Physiology. 96: 660-663.

Bruni, F and Leopold, A.C.,, (1992a). Pools of water in anhydrobiotic organisms: a thermally stimulated depolarization current study.. Biophysics Journal. 63: 663-672.

Casthilo, R.F. Meinicke, A.R. Almeida, A.M. Hermes, -lima, M. & Vercesi, A.E.,(1994) Oxidative damage of mitochondria induced by Fe(II) citrate is potentiated by Ca^{2+} and includes lipid peroxidation and alteration imembrane proteins. Arch. Biochem. Biophys. 308: 158-163.

Chin, H.F. 1988. Recalcitrant Seeds - A Status Report. International Board for Plant Genetic Resources, Rome. 28p.

Ching, T.M. and Schoolcraft, I., (1968). Physiological and chemical differences in aged seeds. Crop Sci. 8: 407-409.

Cheah, K.S.E. and Osborne, D.J., (1978). DNA lesions occur with loss of viability in embryos of ageing rye seed. Nature. 272: 593-599.

Clerkx, EJM., Blankestijin De Vries, ., Ruys, h., Groot SPC., and Koornmeet, M., (2004). Genetic differences in seed longevity of various Arabidopsis mutants. Physiol.   Plant.   121: 448-461.

Copeland, L.O., and McDonald, M.B., (2001). Principles of Seed Science. and Technology. pp. 192-229. 4 [th] editon . Kluwer Academic Publishers. M.A. U.S.A.

Gutteridge, J.M.C., and Halliwell, B., (1990). The measurement and mechanisms of lipid peroxidation   in biological systems. Trends .Biochem. Sci. 15: 129-135.

Dandoy, E., Schyns, R., Deltour, L. and Verly, W.G. (1987). Appearance and repair of a purine / pyrimidine sites in DNA during early germination. Mutat. Res. 181: 57-60.

D'Amato, F., (1952). The problem of the origin of spontaneous mutations. Cytologia. 5: 1-13.

Daneehy, J.P. (1986). Maillard reactions: non-enzymatic browning in food systems with special reference to the development of flavour. Advances in Food Research   30: 77-138.

Dell' Aquilla, A., and Spada, P., (1994). Effect of low and high temperatures on protein synthesis patterns of germinating wheat embryos. Plant Physiology and Biochemistry.   32:65-73.

Devaiah, S.P., Pan, X., Hong, Y., Roth, M., Welti, R., Wang, X., (2007). Enhancing seed quality and viability by suppressing phospholipase Din Arabidopsis. The plant J. 50 (6): 950-957.

Dey, P.M. (1985). D-Galactose-containing oligosaccharides In: Biochemistry of storage carbohydrates in green plants,ed. Dey, PM and Dixon, R pp.53-129. Academic press, New York.

Dzuba, S.A., Golovina Y.A., and Tsvetkov., Y.D., (1993) Echo-induced EPR spectra of spin probes as a method for identification of glassy states in biological objects. Journal of Magnetic Resonance, B 101: 134-138

Delouche, J.C. (1973). Percepts of seed Storage. Proc. Mississippi state seed processors short course. Marcel Dekker, Inc. NY.

Delouche, J.C., (1969). Planting seed quality. Journal Paper Number 1721. MAES, MSU, Belt wide Cotton production Mechanisation Conf. New Orleans, LA, pp 16-18.

Delouche, J.C., and Baskin, C.C. (1973). Accelerated ageing techniques for predicting the relative storability of seed lots. Seed Science and Technology. 1: 427-452.

Dey, A., Chipalkatty, K.S., and Aiyar, A.S., (1983). Effects of chronic irradiation on age related biochemical chenages in mice. Radiation research 95: 6370645.

Dey, B., Circar, P.K., Circar, S.M., (1967). Phenolics in relation to non availability of rice seeds. Proc. Inter. Symp. Pl growth substances. Calcutta.

Dopont, J., Moreau, F., Lance, C., Jacob, JL., (1976). Phospholipid composition of the membranes of lutoids from *Hevea brasiliensis* latex. Phytochemistry. 15(8): 1215-1217.

Dussert, S., Daveyb, M.W., Laffarguea, A., Doulbeaua, S., Swennenb, R., and Etiennea., (2006).

Oxidative stress, phospholipid loss and lipid hydrolysis during drying and   storage of   intermediate seeds. Physiol. Plant. 127: 192-204.

Duvel, J W T, (1905). The viability of buried seeds. USDA. Bulletin 83. Washington, D.C.,

Ellis, R.H., Hong, T. D., and Roberts, E.H., (1990). An intermediate category of storage behaviour? Coffee, J. Expt. Bot. 41: 1167-1174.

Ellis, R.H., T.D. Hong and E.H. Roberts. 1988. A low-moisture-content limit to logarithmi crelations between seed moisture content and longevity. Ann. Bot. 61: 405-408.

Ellis, R.H. and Roberts, E.H., (1981). The quantification of ageing and survival in orthodox seeds.Seed Science and Technology. 15: 1- 17.

Ellis, R.H. and E.H. Roberts. 1980. Improved equations for the prediction of seed longevity. Ann. Bot. 45:13-30.

Elders, R.H., and Osborne, D.J., (1993). Function of DNA synthesis and DNA repair, in the   survival   of embryos during early germination and dormancy. Seed Sci. Res. 3: 43-53.

Earnshaw, M.J., Truelove, B., and Butter, R.D., (1970).Swelling of *Phaseolus vulgaris* mitochondria in relation to free fatty acid levels. Plant Physiology. 45: 318-321.

Esashi, Y., Kamataki, M., and Zhang,M., (1997). The molecular mechanisms of seed deterioration in relation to accumulation of protein aldehyde abduct. In: Ellis, R.H., Black, M., Murdoc, A.J., and Hond, T.D., (eds.), Basic and Applied Aspects of seed Biology. Pp 489- 498. Kluwer Academic Publishers,

Esterbauer, H., Jugens, G., Quehenberger,O., and koller , E., (1987). Autooxidation of human low density lipo protein . Loss of polyunsaturated fatty acids and Vitamin E and generation of aldehydes. J. Lipid Res. 29: 495-509.

Ewart, (1908). On the longevity of seeds. Proc. Roy. Soc. Of Victoria, Vol XXL: 1-211.

Farrant, J.M., N.W. Pammenter and P. Berjak. 1986. The increasing desiccation sensitivity of recalcitrant *Avicennia marina* seeds with storage time. Physiol. Plant. 67: 291-298.

Farrant, J.M., P. Berjak and N.W. Pammenter. 1985. The effect of drying rate on viability retention of recalcitrant propagules of *Avicennia marina.* S. Afr. J. Bot. 51: 432-438.

Fielding, J.L and Goldsworthy, A (1982). The evolution of volatiles in relation to ageing in dry wheat seed. Seed Science and Technology. 10: 277-282.

Franks, F., Hatley, RHM., and Mathias, S.F., (1991) Materials science and production of shelf-stable biologicals. Pharmaceuticals Technology International 3: 24-34.

Galletti, P., Introsso, D., Manna, C., Clemente, G and Zappia, ,V., (1995) Protein damage and methylation-mediated repair in the erythrocytes. Biochemical Journal   306: 313-325.

Ghosh, B., Adhakari, J., and Banerjee, N.C., (1981). Changes some metabolites in rice seeds during ageing. Seed Sci. Tech. 9: 469- 473.

Gille, J.J., and Joenje, H., (1992).Cell culture models for oxidative stress superoxide and hydrogen peroxideversusnormo-baric hyperoxia. MutantRes. 275(3-6): 405-414.

Harrington, J.F.,(1973). Problems of seed storage. In W. Heydecker(ed.) Seed Ecology   Proceedings.pp 251-263. PSU Press, Univerity Park.

Harrington,J.F.,(1972). seed Storage and Longevity. In: TTKozlowski (ed.) Seed Biology, Vol. III. 145-245. NY.

Halliwell, B., and Gutteridge, J.M.C., (1989). (eds.) Free radicles in biology and medicine. Clare dome Press. Oxford. ISBN: 0- 1985-5294-7.

Hoekstra, F.A.,(2005). Differential longevities in dessicated anhydrobiotic plant systems.   Integrative Comparative Biology. 45: 725-733.

Hoekstra, F.A., Golovina, E.A., and Buitink, J., (2001). Mechanism of plant desiccation tolerance.Trends in Plant Sciences. 6 (9) : 431- 438.

Hailstones, M.D., and Smith, M.T., (1988).Lipid peroxidation in relation to declining vigour in seeds of soya (Glycine max L.) and cabbage (*Brassica oleracia* L). J. plant Physiology.133: 452-456.

Herman, D., (1969). Prolongation of life. Role of free radical reactions in ageing. J. American Geriabies. Society. 17: 721-735.

Herman, D., (1956). Ageing : A theory based on free radical and radiochemistry. Joural of Gerontology. 11: 298-300.

Harman, G.E., and Mattick, L.R., (1976). Association of lipid oxidation with seed ageing and death. Nature. 260 : 323-324.

Heydecker, W., (1972). Vigour. In: Roberts, E.H., (ed.) Viability of seeds. pp209-252. Chapman and Hall. London.

Hong, T.D. and R.H. Ellis. 1995. Interspecific variation in seed storage behaviour within two genera - *Coffea* and *Citrus*. Seed Sci. Technol. 23:165-181.

Hong, T.D. and R.H. Ellis. 1992a. The survival of germinating orthodox seeds after desiccation and hermetic storage. J. Exp. Bot. 43:239-247.

Hong, T.D. and R.H. Ellis. 1992b. Optimum air-dry seed storage environments for arabica coffee. Seed Sci. Technol. 20:547-560.

Hong, T.D. and R.H. Ellis. 1990. A comparison of maturation drying, germination, and desiccation tolerance between developing seeds of *Acer pseudoplatanus* L. and *Acer platanoides* L. New Phytol. 116:598-596.

Hunt, L., Holdsworth, M.J. and Gray, J.E. (2007). Nicotinamidase activity is important for germination. Plant J. 51, 341–351

James, E., (1961). An annotated Bibliography on seed storage and deterioration. USDA, ARS, 34: 14-1.

James, E., (1960). Seed deterioration. Proc. 5th fatm Seed Research Conf. pp 31-39. Americal seed trade association. Washington, D.C.

James, E., (1967). Presservation of seed stocks. Adv. Agron. 19: 87-106.

Jones, H.A., (1920). Physiological study of Maple seeds. Bot. Gaz, 69(2): 127-152.

Justice, O.L. and Bass, L.N., (1978). Principles and practices of seed storage. USDA, Hand Book, pp . 506, Washington, D.C.

Kaloyears, S.A., (1958). Rancidity as a factor in the loss of viability of pine and other seeds. J. Ame. Oil. Chem. Soc. 35: 176-179.

Kalpana, R. and Madhava Rao. KV,1994. Absence of the role of lipid peroxidation during accelerated ageing of seeds of pigeonpea (*Cajanus cajan* (L.) Millsp.). Seed Sci. Technol. 22: 253-2

King, M.W. and E.H. Roberts. 1979. The Storage of Recalcitrant Seeds: Achievements and Possible Approaches. International Board for Plant Genetic Resources, Rome.

Kotak, S.,Vierling, E., Baumlein, H., and Von Koskull, D.P., (2007). A novel transcriptional cascade regulating expression of Hat stress proteins during seed development of Arabidopsis. The plant Cell.19: 182-195.

Karmas, R., Buera, M.P., and Karel., M., (1992). Effect of glass transition on rate of non-enzymatic browning in food systems. J. Agric. Food Chem.40: 302-304.

Leprince, O., Hendry, GAF, and McKersie, BD., (1993) The mechanism of desiccation tolerance in developing seeds. Seed Science Research 3: 231-246.

Lanteri, S., and Belleti, P., (1990). Frequency of chromosomal aberrations induced during storage in Pinusnigra. Arnold Seeds. J. of Genetics and Breeding. 44: 281-290.

Larson, R.A. (1997). Naturally occurring Antioxidants. Lewis Publication, Boca Raton.

Leopold, A.C., and Musgrave., (1979). Respiratory changes with chilling injury in soybeans. Plant Physiology. 64: 702-705.

Leprince, O., Harren, FJM., Buitink, J., Alberda, M., Hoekstra, F.A., (2000). Metabolic disfunction and unabated respiration precede the loss of membrane integrity during dehydration of germinating radicles. Plant Physiology. 122: 597- 608.

Linnane, A.W., and Crowfoot, P.D., (1975). Biogenosis of yeast mitochondrial membranes. In Membrane Biogenesis. (ed.) Tzagoloff, A.) 113-136. Plenum Press New York.

Loic Rajjou., Manuel Duval., Karine Gallardo., Julie Catusse., Julia Bally., Clandette, Job., and Dominique job., (2012). Seed germination and vigour.Annual Review Plant Biology. 63: 507-533.

Macovei , A., Balestrazzi, A., Confalonieri, M., Fae,M., Carbonera,D., (2011). New insights on the barrel medic. Mt. OGG1, and MtFPG functions in relation to oxidative stress response in planta during seed imbibition. Plant Physiol. And Biochem. 49: 1040-1050.

Madhavarao, K.V., and Kalpana, R., (1994). Carbohydrates and the ageing process in seeds of pigeon pea (*Cajanuscajan*) cultivars. Seed Sci. and tech. 22: 495-501.

Maki, K.S., Bartsch, J.A., Pitt, R.E., and Leopold. A.C., (1994) Viscoelastic properties and the glassy state in soybeans. Seed Science Research 4:27-32.

McDonald,M.B., and Nelson, C.j., (eds.) (1986). Physiology of seed deterioration. Crop Sci.Soc. Of America. Madison. WI.

Miura, K., Lin, S.Y., Yano,M., Nagamine, T., (2002). Mapping quantitative trait loci controlling seed longevity in rice. Theor. Applied. Genet.104: 981-986.

Moreau,F., Dupont, J and and LanceC., (1974). Mitochondria in higher plants, structure, function, and Biogenesis. In Roland Douce. Biochem. Biophysics. Acta. 345-295- 304.

Nautial, A.R., Thapliyal, A.P., and Purohit, A.N., (1985) Seed viability in sal IV. Protein changes accompanying loss of viability in Shorea robusta. Seed Science and Technology 13: 83-86.

Obendorf, R.L., (1997). Oligosaccharides and galactosylcyclitols in seed desiccation tolerance. Seed Sci. Res. 7: 63-74.

Ohga, I., (1926). A comparison of life activity of century old and recently harvested Inidan lotus fruits. Amer. J. Bot. 13 : 760-765.

Orlova, N.N. and Soldatova, O.P., (1980). Possibility of the estimation of the mutability critical level in stored seeds. ISTA Congress. Vienna. Preprint. 100.

Osborne, D.J., Boubriak, I., Pitchard, H.W., and Smith , R.D., (2003). Seed lifespan and telomeres..In : G. Nicholas, K.J. Bradford, D.Come, and H.W. Pitchard (eds.) The biology of seeds: Recent Research Advances.pp- 301-308. CABI Publishing, Wallingford,

Osborne, D.J., and Boubriak, I., (2002). Telomeres and their relevance to the life and death of seeds. Crit. Rev. Pant Sci. 21 (2): 127-141..

Pal, P., and Basu, R.N., (1989). Volatile aldehyde production in relation to seed vigour of rice, (*Oryza sativa L.*) Indian Agriculturist. 33 (4): 255-258.

Pammenter, N. W., and Berjak, P., (1999). A review of recalcitrant seed physiology in relation to desiccation tolerance mechanisms.Seed Science Research.9: 13-37.

Patil,V.N., and Andrews, C.H., (1985). Cotton seeds resistant to water absorption and seed deterioration. Seed Science and Technology. 13: 193-199.

Priestly, D.A., (1986).Seed ageing implications for seed storage and persistencein soil. Comstock publishing, Ithaca, New York.

Priestly, D.A.,(1985).Hugo de Vries and the development of seed ageing theory.Ann. Bot.56: 267-269.

Priestley, D.A., and Leopold, A.C.,(1983). Lipid changes during natural ageing of soybeanseeds. Plant Physiol. 63: 726-729.

Priestley, D.A., and Leopold, A.C., (1979). Absence of lipid oxidation during accelerated ageing of soybeans seeds. Plant Physiology. 63: 726-729.

Pritchard, H.W. 1991. Water potential and embryonic axis viability in recalcitrant seeds of *Quercus rubra*. Ann. Bot. 67: 43-49.

Pritchard, H.W. and F.G. Prendergast. 1986. Effects of desiccation and cryopreservation on the *in vitro* viability of embryos of the recalcitrant seed species *Araucaria hunsteinii* K. Schum. J. Exp. Bot. 37: 1388-1397.

Puntarulo, S., Galleano, M., Sanchez, RA., and Boveris, A., (1991) Superoxide anion and hydrogen peroxide metabolism in soybean embryonic axes during germination.   Biochem. Biophys. Acta. 1074: 277-283.

Rajjou, L.Y., Lovigny, SPC., Groot, M., Belghazi, M., Job, C., Job, D., (2008). Proteome wide characterization of seed ageing in *Arabidopsis*; A comparison between artificial and natural ageing protocols. Plant Physiol. 148: 620-641.

Roberts, E.H., (1972). Viability of seeds. Chapman and Hall. London.

Roberts, E.H., (1973). Predicting the storage life of the seed. Seed Science and technology. 1 :pp 499-514.

Roberts, E.H., (1972). Storage environment and the control of Viability. In: E .H. Roberts (ed.) Viability of Seed. pp 14-58. Chapman & Hall London, U.K.,

Roberts, E.H., and Abdalla, F. H., (1968). The influence of temperature, moisture and oxygen on the period of seed viability in barley, broad beans and peas. Annl. Bot. 32: 97-117.

Roberts, E.H. and R.H. Ellis. (1982). Physiological, ultra-structural and metabolic aspects of seed viability. Pp. 465-485 *in* The Physiology and Biochemistry of Seed Development, Dormancy and Germination (A.A. Khan, (ed.). Elsevier Biomedical Press, Amsterdam

Roberts, E.H. and R.H. Ellis. 1989. Water and seed survival. Annl. Bot. 63: 39-52.

Roberts, B.E and Osborne, DJ., (1973) Protein synthesis and loss of viablity in rye embryos - the lability of transferase enzymes during senescence. Biochem.   J 135: 405-410.

Rosnoblet, C., Aubry, C., Leprince, O., Ly Vu,B., Rogniaux, H., Buitink, J., (2007). The regilatory gamma subunitSNF4b of the sucrose non –fermenting related kinase   complex is involved in longevity and stachyose accumulation during maturation of Medicago truncatula seeds. The Plant J. 51 (10): 47-59.

Ross C.R., Sartorelli, L., Tato, L., Siliprandi, N., (1964). Relationship between oxidative phosphorylation efficiency in phospholipid content in rat liver mitichondria. Archives of Biochemistry and Biophysics.107: 170-175.

Shen-Miler, J., Schopf, J, W., Harbottle, G., Cao, R.J., Ouyang, S., Zhou, K. S., Southan, J.R., and Liu, G.H., (2002). Long living lotus: Germination and soil gamma- irradiation of centuries old fruits,   and cultivation, growth and phenotypic abnormalities of Offspring. Amer. J. Bot. 89 (2): 236- 247.

Sircar, S.M., and Biswas, M., (1960). Viability and germination inhibitor in rice seed. Nature London. 73: 377-420.

Smith,M.T., and Berjack, P., (1995). Deteriorative changes associated with the loss of viability of stored desiccation tolerant and desiccation sensitive seeds. Seeds development and germination. In: Klgel, J., and Galili, (eds.) Marcel, Dekker, New York. 701.

Smith IK.,. Polle, A., and Rennerberg, H., (1990). Glutathione. In: Stress responses in plants: Adaptation and Acclimation Mechanisms (R. G. Alscher and J. Cumming (.eds), pp. 201-215. Wiley-Liss Inc, New York, NY. ISBN 0-471-6810-4.

Stedman, K.J., Pritchard, H.W., and Dey P.M., (1996). Tissue specific soluble sugars in seeds as indicators of seed storage capacity. Ann. Bot. 77: 667-674.

Stotzky, G., and Schenk, S., (1976). Observations on organic volatiles from germinating seeds and seedlings. Amer. J. Bot. 63(6): 798-805.

Sun, W.Q., (1997). Glassy state and seed storage stability.The WLF kinetics of seed viability loss at TTg and the plasticization effect of water on storage stability. Annals of Botany.79 : 291-297.

Sun, W.Q. and Leopold, A.C. (1995). The Maillard reaction and oxidative stress during ageing of soybean seeds.Physiologia Plantarum 94: 94-105.

Sun, W.Q. and Leopold, A.C., (1993). the glassy state and accelerated ageing of soybeans. Physiologia Plantarum 89:767-774.

Teng, Y.T. and Y.L. Hor. (1976). Storage of tropical fruit seeds. Pp. 135-146 *in* Seed Technology in The Tropics (H.F. Chin, I.C. Enoch and R.M. Raja Harun, eds.). Universiti Pertanian, Malaysia.

Tompsett, P.B. (1994). Capture of genetic resources by collection and storage of seed: a physiological approach. Pp. 61-71 *in* Tropical Trees: The Potential for Domestication and the Rebuilding of Forest Resources (R.R.B. Leakey and A.C. Newton, eds.). ITE Symposium No. 29, ECTF Symposium No. 1. HMSO, London.

Vanangamundi, K and Karivaratharaju, T.V. (1986). Effect of pre- storage chemical formulation of seeds on shelf life of redgram, blackgram, and green gram seeds. Seed Science and Technology 14:477-482.

Villers,T.A., and Edgcumbe, D.J., (1975). On the cause of seed deterioration in dry storage. Seed Science and technology. 3: 761.

Vijay,D., Dadlani,M., Anand kumar, P., and Sivakumar Panguluri.,(2009). Molecular marker analysis of differentially aged seeds of soybean and safflower. Plant Mol. Biol. Rep. online.

Waterworth, W.M., Kozak, J., Provost, C.M., Bray, C.M., Angelis, K.J. and West, C.E. (2009). DNA ligase 1 deficient plants display severe growth defects and delayed repair of both DNA single and double strand breaks. BMC plant biol. 9, 79.

Williams R.J., and Leopold. A.C (1989)2.The glassy state in corn embryos. Plant Physiology. 89: 977-981.

Wilson,D.O., and Mc Donald, M.B.,(1986). The lipid peroxidation model of seed ageing. Seed Sci.and Technology. 10: 269- 300.

Woodstock, L.W., and Feeley, J., (1965). Early seedling grwoth and initial respiration rates as a potntial indicator of seed vifourin corn. Proc.AOSA. 55: 131-133.

Woodstock, L.W. and Taylorson, R.B., (1981). Ethanol and acetaldehyde in imbibing soybean seeds in relation to deterioration. Plant Physiology. 67: 424-428.

Woodstock, L.W., Furrman, K., and Solomos, T., (1984). Changes respiratory metabolism during ageing in seeds and isolated axis of soybeans. Plant Physiol. 25: 1071-1076.

Zanakis, G.N., R.H. Ellis and R.J. Summerfield. (1993). Response of seed longevity to moisture content in three genotypes of soybean *(Glycine max).* Exp. Agric. 29: 449-459.

Zhang,M.,Nakamaru, Y.,Tsuda, S.,Nagashima,T., Esashi,Y., (1995). Enzymatic conversion of volatile metabolites in dry seeds during storage. Plant and Cell. Physiology. 36:157-164.

Zhang,M.,Yajima,H.,Umezawa,Y.,Nakagawa,Y.,and Esashi, Y.,(1994). GC-MS identificcation of volatile compounds evolved by dry seeds in relation to storage conditions. Seed Science and Technology. 23: 59- 68.

Zhang,M., Liu,Y., Torii,I., Sasaki, H., and Esashi, Y., (1993). Evolution of volatile compounds by seedsduring storage periods. Seed Science and Technology. 21: 359-373.

CHAPTER 8

Requirements and Procedures for Standardization of Seed Vigour Testing

(A fault once denied is twice committed)

 ## 8.1 Introduction

What do we expect from our seeds of field-sown crops? The answer to this simple question may not be that simple. A mere germination is not the solution. Be it in the field sown crops or in a transplant production system, the requirement is the uniform field emergence in the former and high emergence and uniform seedlings in the later. Testing of seeds before sowing has become a common practice among the seed producers to make sure that they derive maximum returns out of their investment on seeds. A germination test usually provides information on seed's ability to develop into a healthy seedling under almost ideal conditions. The test criterion and conditions under which germination tests are conducted have been standardized across the seed industry long back and hence expected to convene the requirements as stipulated under the Seeds Act. Seed producers in general and end users in specific have started showing increased dissatisfaction over the past few decades in the absence of an alternative to replace the germination test, that became the sole criterion for presentation of potential seed performance on a seed label.

- Need for uniformity, reproducibility and standardization of vigour tests.
- Reasons for lack of standardization in vigour tests among ISTA laboratories
- Efforts put forth for development of vigour testing methodology
- Changes occurred over a period of time in testing for vigour
- Social and technological challenges encountered during vigour testing.
- Requirements and variables in seed vigour testing.

Even today, seed lots with high germination differ substantially in their field emergence when sown at the same time in the same field and or may also differ in performance after storage in the same environment or transport to the same destination. The two most pertinent questions before us are (i) Are the results of germination test wrong? (ii) Why difference in performance? Since the answer to the first question is a clearly 'no'. The answer for the second question is very simple that germination test is not sensitive enough to indicate subtle but significant quality differences among germinating seed lots, caused by another component of seed quality referred to as seed vigour. This is so because depending on the genetic background and production environment, biochemical changes in seeds can occur either quickly in days or more slowly in years. As a result, seeds lose vigour much before they lose the ability to germinate. Therefore, seed lots with similar high germination results are bound to differ in their physiological age, storage potential or extent of deterioration and so differ in seed vigour and ultimately their ability to perform. Seed vigour can also be regarded as an additional aspect of the physiological quality of germinable seeds. Differences in the physiological age of the seed are the chief cause of vigour differences and could be due to the prevailing conditions during seed production (weather conditions, delayed harvest, improper processing and transport) and storage (temperature, seed moisture content, relative humidity in storage). Seed lots of agricultural, horticultural and silvi-cultural species exhibit differences in seed vigour.

Seeds are tested for vigour to know how they will perform in the field because a seed lot is composed of a mixture of individual seeds each is capable of giving a mature individual plant. When conducted under standardized conditions, this test enables a seed producer to compare and determine the vigour of a seed lot well before it is marketed, McDonald, (1988). In addition to this, a vigour test also provides exact standards required for testing so as to reproduce the results among the laboratories testing the same seed lot, a process known as standardization.

A seed vigour test is only a measure of the seed's ability to emerge and develop under known stressful seeding conditions such as wet soils, salinity, cold soils, meant for deep planting etc. In order to gain acceptability by the seed testing community, a seed vigour test must fulfill three basic criteria viz., (i) should provide a more stable index of seed quality than germination test, (ii) capable of ranking the seed lots consistently in terms of potential performance in field and storage and (iii) objective, inexpensive, uncomplicated, rapid, reproducible, precisely interpretable and ultimately related to the seedling field emergence. In the absence of a clear working definition for decades on one hand and the disproportionate variability in the number of vigour tests, test parameters and protocols, vigour test results and inability in interpretation of these results on the other hand not only lead to great confusion in testing seed lots for vigour, but also ended up in still greater controversy because of the hope and fear between some that vigour tests will become more widely accepted as a means of establishing the planting value of the seed lots. Even after a century of research, neither seed vigour tests are standardized across the industry nor the results of vigour tests find place in the mandated labels attached with the seed bags.

Basically, both the seed analysts and consumers must understand that seed vigour tests are not designed to predict the exact number of seedlings that are likely to emerge and survive in the field. The primary purpose of these tests is to indicate whether or not trouble is anticipated from a high germinating seed lot when placed under adverse environmental conditions in the field, storage or transportation. In order to be meaningful, the vigour tests must be standardized in respect of both testing methodology as well as interpretation of test results. In the event of lack of pertinent information on standardization of seed vigour tests, it is going to seriously impede the use as well as acceptance of seed vigour testing.

 ## 8.2 Aspects of Performance Associated with Seed Vigour

Seed vigour is a characteristic of seed lots with high germination and exerts influence on rate, uniformity and final level of emergence in the field and glasshouse besides storage potential.

- Emergence ability of seeds under unfavourable environmental conditions,
- Rate of uniformity of seed germination and seedling growth
- Performance after storage, especially the ability to germinate.

8.2.1 At the Time of Sowing

Seed vigour is not an important factor, when seedbed and environmental conditions are close to ideal because field emergence will correlate well with the germination of the seed lot. But in reality finding such an optimum field conditions is a rarity. As a consequence of extreme values of soil moisture, temperature and also depending on the vigour status of the seed lot, varying degrees of field performance in respect of rate or level of emergence, differences in uniformity of crop growth including differences in both vegetative and reproductive yield in some species can be expected. Even though there is no difference in the laboratory germination, seed lots with high vigour will perform better under stressed seedbed conditions than low vigour seeds.

8.2.2 At and During Storage

The vigour status of seed lots at the time of entering storage obviously determines the storage potential. Stress in any form exerted by the storage environment in respect of changes either in temperature or relative humidity in uncontrolled storage, seed lots with high vigour will be able to withstand these stresses and decline in quality at a slower rate than low vigour seed lots. The performance of a seed lot even after storage under controlled conditions of low temperature and low moisture content is decided by the vigour status of the seed lot.

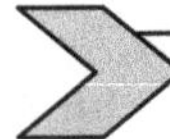

8.3 Standardization of Vigour Test Procedures

Standardization of seed lots is basically a very complex and crucial process aimed at determining their performance well before their actual sowing in the field by assigning precise standards such that (i) The results should be comparable and reproducible both between and within seed testing laboratories testing the same seed lot and (ii) the results should be reliable and help in the settlement of disputes if any, if at all arises, amicably by all parties involved in seed quality evaluation. For this purpose previous tests for seed quality should have been completed in a recognized seed-testing laboratory in a standard manner acceptable to all involved in quality evaluation. In the initial stages, the requirements for precision and standardization of test methods to ensure repeatability and reproducibility of the results have been seen as a problem for seed vigour testing even by AOSA, (1983), McDonald, (1995) and Hampton (1995b).

In order to make the vigour test results reproducible, uniformity in testing procedures/ protocols across the laboratories becomes a must. No doubt, the organizations working at International level like ISTA and AOSA, together with seed trade associations of USA, Canada and Europe have initiated efforts since their inception in this direction through their respective vigour test committees. The aims of the Vigour Committee are (i) to develop and validate vigour tests for introduction in to ISTA Rules, (ii) to encourage the use of vigour tests, and (iii) to ensure that these are carried out to achieve repeatability between and reproducibility within laboratories. These organizations have already introduced the concepts of comparative and referee testing and conducted extensive referee tests. As a result, seed vigour testing laboratories were able to reproduce the results in their own laboratory within acceptable confidence limits with the same sample of seed, but the amount of variability between different laboratories was found to be in an unacceptable level and resulted in disputes. This once again proves beyond doubt the necessity of identification of uniform test procedures across laboratories and rectifies the loopholes towards standardization of testing procedures.

Standardization is badly needed to day keeping in view the requirements of plant breeders, seed production, quality control and marketing in addition to strengthening seed vigour research and consumers besides the relatively unimportant wholesalers, retailors and carry forwarding agents etc.

During the course of development of a standard germination test, initially numerous variants were proposed, tested, developed, refined and ultimately appropriate method was standardized together with tolerance limits and assigned for each seed type. This level of robustness in planning and execution of standard germination test had made it remain the sole criterion for the presentation of potential performance of seed lots for a long time. Law mandates the information pertaining to germination percentage need to be printed on seed labels that are to be attached with bags offered for commercial sale. Because, under optimum planting conditions, the result of germination test expressed as percentage normal seeds germinated, becomes an effective predictive measure of plant stand establishment. In the absence of any alternate or substitute, this method has lead to dissatisfaction among seed the producers and the consumers alike. Even $1^{\circ}C$ increase or decrease in temperature or 1 per cent increase or decrease in moisture content of seed as well as substrate is likely to exert influence on the rate of germination. The subsequent germination evaluation in to normal and abnormal seedlings is a subjective assessment and has created a necessity for formulating a new seed testing techniques, with a better seed quality index than germination percentage.

In contrast to the standard germination testing, typically, the seed growers in general and the farmers in specific, encounter stressful germination environments in the field in which even the seed lots with recommended standard germination value failed to establish the desired plant population. Under such circumstances, measurements of seed such as vigour became necessary as predictive tool. A seed vigour test is simply an analytical procedure meant for evaluating seed vigour under standardized conditions in such a way that these results permit the seed producer to determine and compare the vigour of various seed lots before offering for sale.

Vigour testing not only measures the percentage of viable seed in a sample, it also reflects the ability of those seeds to produce normal seedlings under less than optimum or adverse growing conditions. Seeds may be classified as viable in a germination test, which provides optimum temperature, moisture and light conditions to the growing seedlings. Despite this, the seedlings may not be capable of continuing their growth and will only be completing their life cycle under a wide range of field conditions. Though the seed vigour test provides a better index than standard germination test, still this particular information is neither required nor printed on the seed labels. But, when the AOSA has conducted a survey in 1983, the respondents have not permitted any reference to be made pertaining to seed vigour on seed labels, for one reason or the other, even after the identification of test procedures and quantification of results. Even when the vigour information has to be presented on the seed labels, regulations obviously become a necessary to ensure fairness to both seed producers and users.

By this time several seed testing associations have developed variable cold test methods using 100 per cent soil, sand with a thin layer of soil, a mixture of sand and soil, with varying levels of microbial concentrations and compositions in tune with soil sources. Obviously the results from these tests varied significantly between and within laboratories. In principle, the sole objective of vigour tests besides being useful they must also provide accurate, unbiased, reliable and meaningful information to the seed grower, seed testing laboratory and the ultimate end user.

Therefore, AOSA, perceived the idea that, precision of a vigour test is difficult in the absence of a standard, against which a comparison can be made. Taking clue from seed germination test, the first essential step would be to develop more variants of vigour tests and finally accept one or more tests, which can reliably measure the vigour of seed lots. For example, problems associated with the effects of genotype, seed size or pesticide, lack of standard concentrations etc. Similarly, McDonald (1995) and Hampton (1995b) also reported that repeatability and reproducibility of the vigour test results have been a problem in the past in the absence of precision and standardization of treatments. Problems associated with standardization like sampling, subjective nature of several assessments, moisture content of the seed, fluctuations in the environmental conditions like temperature and moisture content of seed and substrate have been identified.

Accordingly both ISTA and AOSA proposed a series of vigour tests. At the same time, identification of the above-mentioned factors responsible for influencing the test results have also acted as a catalyst in R&D and helped in the documentation of further variables that are most likely to exert impact on the test results and assisted in the development of vigour test protocols. The efforts resulted in a number of single seed vigour tests for specific crops duly minimizing the variability and permitting accurate and reliable vigour data for supporting marketing and regulatory efforts. Therefore, these procedures in turn were evaluated through their respective Vigour Test Committees over a long period of time, to assess variability within and between laboratories. In addition to this, they have also designed comparative tests to evaluate the sharpness of test procedures and proposed the protocols for the most useful ones and published the same in the form of their respective Hand Books duly defining the limitations for each test to meet the ultimate aim of supplying high quality seed to facilitate better crop production.

Standardization of seed lots is a process, aimed at determining their performance well before their actual sowing in the field by assigning precise standards as detailed below.

(i)   The results should be reproducible between seed testing laboratories testing the same seed lot and

(ii)   the results should be reliable and help in the settlement of disputes if any, if at all arises, amicably by all parties involved in seed quality evaluation.

Simultaneously, it should also be kept in mind that, the critical goal of seed production can only be realized by minimizing the rate of deterioration process. Further, it is also better to remember the fact, that vigour tests are conducted in several capacities with the ultimate aim of supplying quality seed to the growers. On the other hand, despite the existence of unwritten agreements in the seed industry, vigour testing became a much-needed component of seed quality. But in reality, still, nowhere this information is printed or utilized on the label attached with the seed bag. It took more than 100 years for publishing the Rules for Testing Seeds after First Seed Legislation was passed in Berne, Switzerland (1816). Similarly, it is true with the evolution of seed vigour too; it took more than 100 years for deriving a working definition acceptable for all parties involved in seed business by AOSA and ISTA (ISTA, 1977). Its much-delayed acceptance could be attributed to several (i) social and (ii) technical challenges it has encountered.

8.3.1 How far a "Vigour Test" is Worthwhile?

(i)   Whether or not the scientific background of the test explained is biologically sensible;

(ii) Whether or not vigour test is conducted with commercial seed lots having high germination;

(iii) Whether or not the test results can be related to the expression of seed vigour in respect of field emergence capacity and seed storage potential;

(iv) whether or not the test is reproducible, repeatable and validated;

(v)   Whether or not clear cut evidence is existing.

A vigour test is considered to be worthwhile, if answers to the above five questions are 'Yes."

8.3.2   Need of Uniform Seed Testing

Nobbie (1877) initiated first comparative testing duly involving the participation of 17 laboratories to test *Poa pratensis* (Kentucky Blue grass) for germination testing (Ediam, 1878) and stressed that, the ultimate goal of seed testing lies in its uniform application of approved methods prescribed by rules and advocated worldwide uniformity in seed testing should become a basis and ultimate prerequisite for reliability of seed testing results.

8.3.3   Need of Reproducibility of Vigour Test Results

The seeds should be made available at the doorstep of end users well in advance of the sowing season. At this particular point of peak sowing season, the seeds will be crossing state and country boundaries very fast. This obviously necessitates testing and retesting at both exporting and importing points of entry. Discrepancy in result between different testing stations is bound to cause mistrust and leads to litigation and ultimately ends up with legal suites. This is only one side of the coin. On the other side is the loss of one valuable production season. This stubbornly proves that these vigour test results are of little use and points out a necessity that these results should be reproducible, if at all were of any help.

ISTA vigour test committee after conducting comparative tests over a long period of time concluded that vigour tests which rank the seed lots consistently for their potential performance in field or in storage can be standardized. Despite the complexity and variability of factors involved in planting/storage/transportation environment, it is not possible to produce a planting or storage index. Single tests based on some aspect of germination like accelerated ageing and cold tests along with conductivity test is showing some promise.

Keeping in view the fact that seed vigour tests are bound to play a pivotal role in seed business including development and production of various trait specific strains. Seed vigour test information is useful to seed companies in several ways, be it in monitoring quality control during seed production, or post harvest handling including storage till planting, inventory management, seed treatment, quality enhancement and rectification measures, carryover seed decisions, etc., resulted in rapid acceptance of seed vigour tests as an essential quality control measure. Despite these advantages, seed industry has adopted the attitude that seed vigour tests are not sufficiently standardized among laboratories (AOSA Vigour Testing Hand Book-pp 34-35) for obtaining reproducible results. For several obvious reasons listed below, standardization of vigour tests is badly in need.

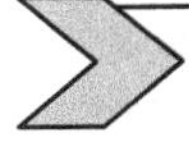 ## 8.4 Reasons for Lack of Standardization among Laboratories identified by AOSA

1. Most of the vigour tests, including ISTA approved cold test and accelerated ageing test, together with seedling vigour classification test and even the age old tetrazolium test classifies seeds and seedlings into normal and abnormal and various other sub-classes even to date and use subjective interpretation for reporting the results purely based on characters that are difficult to describe precisely. The vigour tests are expected to provide a quantitative method of assessment to avoid subjective assessments and hence make standardization easier. This kind of testing is used for identification of seed performance either after or during stress, the adverse conditions are then commonly followed by a germination test.

2. According to one broad classification, vigour tests could be broadly grouped into those that use measurements either during the process or at the end of the process. The tests included in the former group like recording of observations as in seedling growth rate measurement while in case of later, as in standard germination test the observations are measured at the completion of the test.

3. Variation is possible in respect of both test materials and testing conditions like media, substrate, consumables, equipment, initial moisture and quality of the test seed material on one hand and moisture, temperature and other environmental conditions on the other hand directly exerts influence on end results of tests in which the rate of growth or rate of biochemical process is measured than for those such as standard germination test that measures the completion of the process. For example, $1^{\circ}C$ difference in temperature in a standard germination test has little or no effect, but the same $1^{\circ}C$ difference in temperature exerts significant difference in accelerated ageing test measuring seed deterioration rate or seedling growth rate test. Similarly minor fluctuations in respect of moisture content of testing seed material and substrate, or relative humidity of air etc., makes the equipment of standard germination test unsuitable for conducting most seed vigour tests.

4. Variation is also possible in the members of Malvaceae and Leguminosae families, with their hard seed coat, slowly permeable, partially permeable or impermeable to water and exchange of gasses and thereby affect both percentage germination and length of seedling growth and ultimately causes a biased index of vigour result.

5. In case of seeds bestowed with dormancy also the interpretation of vigour test results are biased and hence necessitate conformation with a warm germination or tetrazolium test.

6. The cold test for cotton though considered being an ISTA approved test, standardization is very difficult in respect of creation of testing conditions, because seed is exposed to soil microorganisms. Standardization of soil microflora in the laboratory is not only difficult, predicting the behaviour of pathogens in a population is altogether different and hence the results are difficult to explain. Standardization of cold test will be a difficult task if not impossible, because the test procedure, the substrate and the microorganisms have to be managed simultaneously. Though use of sterile media with specific microorganism inoculation has been suggested as a simple way than to culture and maintain certain level of pathogenicity. Byrum and Copeland (1995) demonstrated the differential behaviour of pathogens and potential for standardization of vigour tests.

7. Electrical Conductivity Test is considered as a fast and practical procedure offering objective oriented information, Vieria *et al.,* (2004), which can be used in most of the seed testing laboratories. This test does not require expensive equipment and skilled personnel, hence can be used as supplement or a quick alternative to a germination test. This test though does not include germination phase. Since, intensive rainfall and fast water uptake by dry seed after sowing in the field leads to imbibitional damage, seed evaluated with this test also exposes to similar conditions and hence imbibitional damage in tested seeds could be expected. Lisjak *et al.,* (2009). Dijana *et al.,* (2014) reported a strong negative correlation of Electrical Conductivity test with most maize seed and seedling vigour parameters in the applied germination tests like germinability, germination rate, root and coleoptile length and seedling mass and these results will help greatly in standardization of test for maize and also suggested for adoption as a routine vigour test for maize seed production and quality control.

8.5   Development and Validation of Vigour Tests

8.5.1 Stages in Test Development

There are several ways to evaluate seed vigour. This has created confusion among seed companies, growers and end users as to what these laboratory tests mean and how they should be interpreted. Research, development and validation are the three important stages in the development of vigour test. By 1980, a seed vigour definition had been approved by every major organization involved in commerce and testing of seed. The study of referee and research results by the vigour test committee revealed that seven useful vigour tests were available and the procedures were published in the Vigour Test Hand Book of AOSA in 1983.

8.5.2 Research

Establishment of the principle of the idea is the first step in research wherein the basis of new idea whether or not works. For this purpose, selected lots may be used for comparison and usually this idea is tested only in one laboratory.

8.5.3 Development

Development involves the extension of the principles using the commercial lots. During this phase it is determine whether or not the new test / equipment is fit for the purpose. If so, determine the influence of various possible biological and experimental variables. Based on the data generated establish the basic repeatability and reproducibility and this provides the basis for submission for Method validation. The test variables have been comprehensively evaluated, Loeffler, TeKrony and Egli (1986) and revised test procedures provided. Vigour test referees have consistently demonstrated that conductivity test results are reproducible within and among seed laboratories.

8.5.4 Validation

The method validation programme of ISTA ensures rigorous evaluation of the method for repeatability within and between laboratories and confirms relationship to expression of vigour.

Typically, method validation involves six laboratories wherein six commercial lots are tested using a specified test method. The results are analyzed for repeatability and reproducibility and three experts critically review this analysis report. Based on the recommendation of the reviewers the test can be validated and such validated methods can be proposed for publication in ISTA Rules.

8.5.4.1 Comparative Testing

To be useful, any quality test must first provide reproducible results. Nobbe (1877) initiated the first comparative test. Since 1950, comparative testing became the responsibility of the Rules Committee, and after 1953, technical committees continued to arrange comparative tests. Since inception, these tests were designed for measuring the two important aspects viz., monitoring uniformity in testing and method validation and was also popularly known as Experimental Sample Testing (EST). It is a means by which uniformity in seed testing can be monitored and comprises of partitioning of a sample into several subsamples and testing takes place at different laboratories simultaneously with a prescribed method. The pooled results are statistically analyzed and variation worked out for uniformity. These tests were used both for monitoring uniformity in testing and method validation. The decision of Association of American Seed Control officials resolved in 1974 that vigour test committee of AOSA has to develop standardized vigour test procedures. The AOSA Special News letter of 1976 became a significant milestone by way of providing specific guidelines for the conduct of eight proposed vigour tests using a referee format. Comparative testing covered all ISTA member laboratories, while method validation became the task of technical committees, Steiner (2011) and organized voluntarily by technical committees or even by individual laboratories.

8.5.4.2　Referee Testing

With a view to reinforce comparative testing, the term was replaced in 1962 as referee testing. The first Referee Test Committee was constituted with experienced M.J.F. Koopman in the Chair without any change in the nature of test. Usually comparative testing programmes involve a large number of laboratories that assess the variation both between and within laboratories. These tests can also be designed to regulate the ruggedness of a test procedure, i.e. to determine the importance of differences in procedure at possible critical stages of the test to the test results. Organization of referee tests through trade channel was considered but not implemented; Participation in referee testing is made obligatory for laboratories issuing ISTA certificate. To evaluate the reproducibility of various vigour tests, both ISTA and AOSA have conducted extensive referee tests and demonstrated that (i) seed testing laboratories can reproduce their own results on the same sample of seed within acceptable confidence limits. (ii) the same tests when conducted by different laboratories the variability is often unacceptable.

8.6　Recent Classification of Vigour Tests

In an attempt to speed up the process of vigour test adaptation, ISTA has divided the vigour tests into two categories: Suggested vigour tests and Recommended (Validated) vigour tests. The former are those that correlate with seed vigour and offer promise as standard vigour tests while the later are those that have been critically evaluated for determining seed vigour and are considered standardized (eg. Accelerated Ageing Test). Vigour Testing took a major step forward within ISTA in 2001 with its members voting for incorporation of vigour chapter in the Rules for seed Testing. As per ISTA (2001), a seed vigour chapter was included in Rules for Seed testing and published in the Rules, Amendments, 2001. It gives the details of precise methods to be used for carrying out ISTA validated tests. The validation procedures for the new test of seed quality, seed vigour, by ISTA have been formally laid out in the Hand Book of Method Validation, (ISTA, 2005). The two main components of method validation applicable to vigour tests are, (i) the description of the method should be clear and complete, so as to give reliable and reproducible results, and (ii) the relationship between the results of the vigour tests and a practical expression of vigour should be confirmed. So far two tests, (i) the Conductivity Test for Pea (*Pisum sativum*) and Accelerated Ageing test for soybean (*Glycine max*) have been validated by ISTA by inclusion in the ISTA Rules. The third vigour test proposed for validation is the Controlled Deterioration test for small seeded vegetable crops, Powel and Matthews (2005) has been accepted for Brassica species and added to the Rules in 2010. Whereas, the fourth and latest one to be added to the Rules for vigour testing is The Radicle Emergence Test for *Zea mays*. But research within this committee has provided evidence that this test provides an assessment of rate of germination and vigour for a range of species such as canola, cotton, pepper, water melon, cucumber, melon, viola, etc.

The well-known and established conductivity test is based on the leakage of solutes from the seed into water, now can also be applied to *Glycine max and Phaseolus vulgaris*. The fundamental basis of conductivity test involves changes that occur during the ageing of grain legumes and hence the test could be expected to apply to all such species.

8.6.1 Validated (Recommended) Vigour Tests

The term validate indicates that these tests have under gone extensive international comparative testing to establish the reproducibility and repeatability of the described methods. The data from comparative tests forms the basis for tolerance tables for the results of the test. In addition to this relationship between test results and an expression of vigour has also been established very clearly. In case of Conductivity test, the result is related to the field emergence of peas and the result can be used to rank the seed lots by vigour level as follows, Matthews and Powell (1981).

<25 uScm^{-1}g^{-1}	Nothing to indicate, seed is unsuitable for early planting or sowing in adverse conditions
25-29 uScm^{-1}g^{-1}	Seed may be suitable for early sowing but involves some risk of poor performance under adverse conditions.
30-43 uScm^{-1}g^{-1}	Seed not suitable for early sowing especially under adverse conditions
> 43 uScm^{-1}g^{-1}	Seed not suitable for sowing

Whereas the accelerated ageing test is related to both field emergence as well as storage potential of soybean. Accelerated ageing test was initially developed as a test to estimate the longevity of seed in commercial storage, Delouche and Baskin (1973) and hence has been utilized in the prediction of life span of a number of different species. Subsequently, this test has been evaluated as an indicator of seed vigour in a wide range of crop species and has been related to field emergence and stand establishment and hence has been accepted as one of the two ISTA recommended vigour tests. In accelerated ageing test seeds are exposed to the variables moisture and humidity that are known to cause rapid seed deterioration. Hence, seed lots high vigour should be able to withstand these conditions and deteriorate at a slower rate than lots those with poor vigour.

8.6.2 Suggested Vigour Tests

Suggested tests are those tests that have revealed the potential for identifying differences in seed vigour but have not under gone extensive comparative testing in a validation procedure. The Hand Book of Vigour Testing by AOSA has described about seven suggested vigour tests. For this purpose the two validation procedures developed and followed are (i) Multi-laboratory validation which involves the participation of 6-8 laboratories and (ii) performance based validation that aims at evaluating the performance of proprietary test methods such as those used for seed health testing. Acceptance of a test by vigour test committee leads to its description as an ISTA approved method. The acceptance is subjected to ISTA membership voting. After voting only it becomes an ISTA official method.

8.6.3   Proficiency Testing

An ISTA working group operating between 1992 to1995 has developed ISTA Accreditation standard for seed testing laboratories and this standard was approved in the general meeting of 1995. Consequent to the approval of ISTA Laboratory Accreditation standard in 1995, it has once again updated "Referee testing" in the name of "Proficiency Testing" without making any changes in the nature of testing (Austin, 2002). The already existing Referee tests were modified, extended and adopted to become International Proficiency tests and the responsibility was entrusted to an internationally operating accreditation body at the ISTA secretariat. ISTA Vigour Committee and Statistics Committee have jointly established proficiency tests for laboratories to demonstrate their competence in carrying out these two tests and hence included in ISTA accreditation. The responsibility of preparation of samples in consultation with ISTA Secretariat lies with Proficiency Test committee. Proficiency testing programme has been made compulsory prerequisite since 2004 for accreditation of laboratories including authorization for issuing of ISTA certificates (cf. Muschick, 2010). Therefore, performance in proficiency testing eventually became decisive. The Proficiency Test committee of ISTA, Congress, has decided to organize 11 test rounds in the triennium 2014-2016 with a view to cover every crop group one proficiency test round. In order to bring balance in Proficiency Test Programme, crops from all parts of the globe should be considered. Since identification of foreign seeds is not an easy task for inexperienced, the Proficiency Test Committee organizes one voluntary training test round on seed identification every year. All laboratories participating in ISTA Proficiency Testing Programme receives an individual and confidential detailed statistical analysis results. The name and numbers of ISTA Accredited Laboratories indicate successful performance in proficiency testing. Though the term comparative testing is used still widely in technical areas of Product Testing services, Certification Agencies and International Organization for Standardization (ISO) uses it in addition to Proficiency test. As per ISTA Secretariat, Zurich, the procedure of ISTA Proficiency test flow chart is presented here under.

8.6.3.1   The ISTA Vigour Test Committee

Consequent to the establishment of ISTA Biochemical and Seedling Vigour Committee, the concept of vigour testing got much impetus. Over the past 60 years, more than sixty vigour test methods have been proposed, studied intensively and used by the seed technologists. As far as research refinement is concerned, vigour testing has become excessively diversified and created a necessity to redirect the goals since only few methods are used internationally. As a result, ISTA Vigour Testing Committee has concentrated its efforts on the standardization of nine vigour tests, Hampton and TeKrony, (1995). Contrary to this, AOSA has suggested or recommended procedures for seven test methods since 1983. The methods considered efficient by ISTA and AOSA are almost the same. Some vigour tests were designed to identify seed performance / germination after exposure to different stresses such as low temperature, moist substratum (cold test); high humidity and high relative humidity (Accelerated ageing test); high seed water content(Controlled deterioration test) are followed by germination test. On the other hand, seed vigour per se can also be evaluated either through enzyme action or seedling growth rate or even measuring the changes in cell membranes organization etc., that measures the expression of physiological potential of seeds in addition to physical and health status of seeds. Indirect evaluation of cell membrane integrity through electrical conductivity test has been accepted internationally as standardization has almost been achieved. But still, some refinements in methodology still constitute a challenge in seed technology research.

The Focus of ISTA Vigour Test Committee during 2013-2016 periods was on current validated vigour tests in particular and extending the scope of radicle emergence and conductivity tests and shortening the controlled deterioration test. In case of radicle emergence test, the focus was on a wide range of species including vegetables (Onion, Cabbage and leek) grasses,(*Lolium perenne, Elymus nutans, Festuca arundinacea*) and field crops (Soybean, wheat, alfalfa and sunflower). Consequent to the development of initial work, that has determined the suitable conditions and timing for obtaining a count of radicle emergence before test results are compared to emergence and storage trials. Small- scale comparative tests would determine whether the test protocols are repeatable and reproducible. This is followed by comparative tests that would form a part of method validation.

At present the controlled deterioration test will take 10-14 days for completion. It is proposed to reduce or shortening the controlled deterioration test to 3 days. For this purpose two alternative approaches have been developed. The first approach involves rising of seed moisture content whereas the second approach comprises of adopting a conductivity measurement as an alternative to the germination test that is currently performed afer the deterioration step of the test. Extension of the scope of the conductivity test will be limited to the small-seeded, coloured Desi-type chickpea (Cicer *arieti-num).* In this case, the validated conductivity test in the ISTA Rules will be applied and the result compared to the results of field emergence trials.

Procedure of ISTA Proficiency Test Programme Flow Chart Effective from 15-08-2013

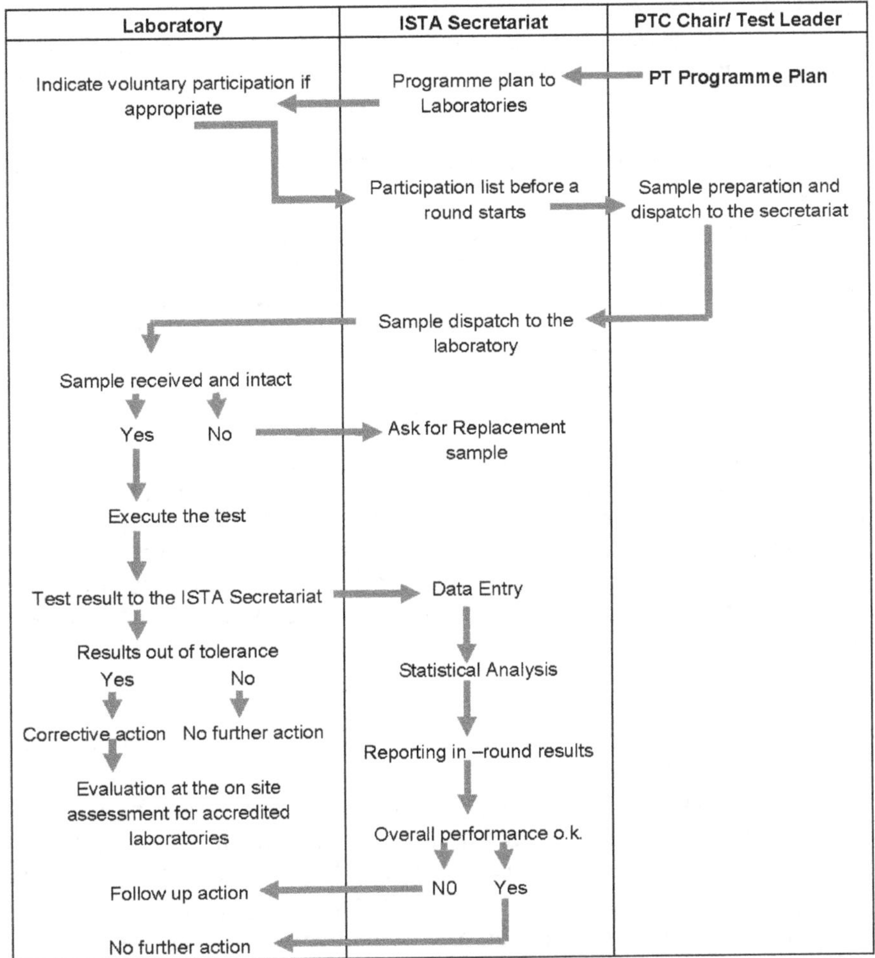

8.7 Standardization of Vigour Test Protocols

Simultaneously, it should also be kept in mind that, the critical goal of seed production can only be realized by minimizing the rate of deterioration process. Further, it is also better to remember the fact, that vigour tests are conducted in several capacities with the ultimate aim of supplying quality seed to the growers. On the other hand, despite the existence of unwritten agreements in the seed industry, vigour testing became a much-needed component of seed quality. But in reality, still, nowhere this information is printed or utilized on the label attached with the seed bag. It took more than 100 years for publishing the Rules for Testing Seeds after First Seed Legislation was passed in Berne, Switzerland (1816). Similarly, it is true with the evolution of seed vigour too; it took more than 100 years for deriving a working definition acceptable for all parties involved in seed business by AOSA and ISTA (ISTA, 1977). Its much-delayed acceptance could be attributed to several (i) social and (ii) technical challenges it has encountered.

8.7.1   Social and Attitudinal Challenges for Seed Vigour Testing

During vigour testing, challenges encountered in respect of social and attitudinal needs of ultimate seed users, seed growers and seed industry needs to be answered. Therefore, in the first instance, in order to derive maximum benefit out of vigour testing, it becomes very much necessary to answer some of the most important questions.

8.7.1.1   Who are going to derive the real benefit from the seed vigour testing information?

Obviously the major /minor seed companies of the developed countries working in both public and private sector with built in Research and Development facility are obviously the real end users of this technology. They have already developed their own protocols, and conducting their own vigour tests on a routine basis as a part of their own "in-house quality control" programmes.

8.7.1.2   How this information on seed vigour is useful?

The information on seed vigour is useful to different stakeholders in different ways. For example, both seed growers or seed producers and end users or farmers derive the benefit of different kind. Similarly, a person engaged in seed research derives an altogether different benefit from that of person(s) engaged in either marketing or handling the seed in storage godowns. Since, the seed vigour has been designed to monitor the quality of seed lots, the information generated through these tests is useful in several ways to different stake holders as detailed below.

8.7.1.2.1   Seed Producers

The seed producers have readily accepted and involved the following attributes of seed vigour as essential measure of quality control.

1. The information derived from Vigour tests assist in seed growers in monitoring seed quality at every stage of seed production starting from seeding emergence through marketing, conditioning and treatment, storage and bagging to till planting in the next season.

2. The information derived from Vigour tests is useful in detecting adverse practices, if any, and hence facilitates initiation of timely corrective action so as to prevent further deterioration.

3. The information derived from Vigour tests renders seed producers in the establishment of minimal seed quality standards to seeds offered for sales in the competitive market.

4. The information derived from vigour tests, at times, may result in identification and recommendation of additional measures required to be taken into consideration like seed treatments with fungicides, insecticides or seed enhancements etc.

5. The information on seed vigour can also be helpful to the seed producer with respect to inventory management like allocation of optimum space for storage, handling of valuable hybrid seed material, reducing their operational costs, etc., in such a way duly providing quality seed acceptable to the end user at appropriate time.

6. Vigour tests are capable of discriminating seed lots that qualify for long-term storage.

8.7.1.2.2   End users of the Seed

Unlike the seed industry, which is quick enough to absorb the seed vigour testing into their quality control programmes, the level of acceptance of vigour information by the farmers is yet to be accomplished and is useful to the farmers as detailed below.

1. The farmer, in the first step, as a contract seed grower, can expect better price for the highly vigour seed he has produced.

2. A commercial farmer, as an end user of seed for grain production purpose, can plan for

    (i)   The quantity or seed rate to be used for obtaining satisfactory emergence and

    (ii) How early he can plant the seed material in the given set of or existing field conditions.

3. Facilitates farmers in predicting the uniformity in plant stand establishment and probable competition.

4. Facilitates a farmer to plan and take timely decisions on requirements for secondary tillage, application of fertilizers and pesticides *etc.*

5. A farmer can foresee and match the conditions of environmental stresses like compaction of soil, drought and cold etc., which the seed with specified level of vigour can only tolerate.

6. Ultimately, the end user gets a rough idea regarding how plentiful or uniform yield he is anticipating before hand from his field sown with seed of specified vigour level.

When both seed producers and seed end users are deriving clear-cut advantage from this seed vigour information, why it is not presented "on the labels along with other information" as mandated by law?

May be the probable answer to this question lies in the fact that (1) either the vigour tests are not sufficiently standardized so as to generate reliable and reproducible results between seed testing stations or because, in the interstate and international seed trade, seed lot consignments are liable for testing both at dispatching and receiving points. It means, in the absence of specified standards for testing, variation in the results between testing laboratories will be a common phenomenon which in turn is likely to create mistrust, makes marketing a difficult task which automatically leads to litigations between the different parties involved in seed trade transactions.

8.7.1.3   Research Leaders

Research leaders are of the strong feeling that adoption of vigour tests in haste is leading to challenges in the courts subsequently. This is more detrimental to the credibility of vigour testing and do more harm than good and hence demands standardization of vigour tests very carefully.

8.7.1.4   Seed Analysts and Research Managers

The attitude of most seed analysts and seed research managers is to conduct vigour tests with the same equipment, testing methodology and evaluation criterion as employed for germination testing. This will definitely create a problem because, vigour testing being a more sensitive index of seed quality obviously demands a more rigorous testing of control variables besides the criteria of interpretation.

8.7.2 Technical Challenges Encountered during Seed Vigour Testing

In addition to the above mentioned few social and attitudinal challenges, the following technical challenges in respect of (i) vigour test requirements, (ii) variables in vigour testing and (iii) presentation and interpretation of vigour test results also need to be taken in to consideration at this juncture.

8.7.2.1 Vigour Test Requirements

1. Vigour test, being a more sensitive index of seed quality, cannot afford the luxury of first count; second count and a final count etc., duly providing the safe time for each seed to germinate leisurely as in a standard germination test. Most of the seed vigour tests have been designed to monitor one or the other aspects of seedling or biochemical growth parameters.

2. A standard germination test, no doubt, offers an excellent flexibility to the analyst with respect to timing of analysis. Contrary to this, a vigour test has to be conducted and evaluated at the earliest moment precisely possible. Because, inordinate delays beyond the prescribed protocol period makes the test invalid or liable for rejection.

3. Availability of moisture within the vicinity of seeding zone as an environmental factor exerts immense pressure and impact not only on imbibition speed, but also on the subsequent seedling growth rate and on various other biochemical processes also. Even the minor variations in moisture content of germination substrata or relative humidity of the air shows considerable effect on seedling growth rates. Despite this, none of the vigour tests, including that using seedling growth as a parameter never defines neither the quality nor the quantity of moisture used in the test.

4. Similarly, it is a well-established fact that the temperature, another important environmental parameter, directly exerts influences on the seedling growth rate. In germination testing, either direct or alternating temperatures with $\pm$ 1°C difference are permitted. Whereas in case of vigour tests, maintenance of uniform temperature throughout the test duration is highly essential. That too, especially in case of tests like accelerated ageing; the level of tolerance should not be more than $\pm$ 0.3°C. Therefore, ISTA (1995) specifically recommends for the use of specific equipment with excellent descriptions to meet vigour testing protocols.

5. Therefore, the conditions and equipment that are recommended / suitable for standard germination test may not be suitable for conducting vigour tests. Hence, use of still more sophisticated equipment is required to produce a more reliable seed vigour index than germination test (ISTA, 1995).

8.7.2.2.1 Skills required

For successful conducting of seed vigour tests, required skills include the ability to

  (i) Plan and execution of work in a sequential manner

  (ii) Perform the calibration checks routinely

  (iii) Carry out the procedures pertaining to step wise sample preparation.

  (iv) Designating the components of the test to other staff members.

  (v) Prepare and treat the seeds for testing.

  (vi) Strictly adopt guidelines and Standard Operating Procedures (SOPs) of the organization.

(vii) Maintain the integrity, safety, security and quick traceability of samples, subsamples, test data and test results and their documentation.

(viii) Prepare solutions with required concentrations for titration purpose to carryout chemical reactions

(ix) Calculate the means and percentages.

(ix) Calculate and balance equations duly including appropriate units of measurements.

(x) Interpret the test methods and procedures.

(xi) Evaluate and report the results to the superior immediately.

(xii) Record and communicate the results in the prescribed proformae of the enterprise.

8.7.2.2.2 Knowledge Required

For successful conducting of seed vigour tests, required knowledge include the ability to

(i) Understand the biology of the seed submitted for analysis

(ii) Thorough understanding of the Rules for Seed Testing and role of ISTA.

(iii) Identify the seed correctly

(iv) Discriminate between a range of vigour testing procedures,

(v) Materials and apparatus used in seed vigour testing.

(vi) Enterprise and /or legal traceability requirements.

(vii) Understand the Standard Operating Procedures of enterprise.

(viii) Sound knowledge on relevant health and safety of surrounding environments.

(ix) Calculations, tolerances and rounding off the results

8.7.2.3 Variables in Vigour Testing

No doubt, the development of an acceptable vigour test is a challenging task, that too in the absence of standards, there is no vigour test which is universally accepted, unlike in germination test, against which the merits can be compared. Most of the available vigour tests being crop specific, it appears that the main idea behind this should be to confine the focus on standardization rather than wider application in the first step by reducing the number of tests followed by development of optimum procedures for each crop. In addition to this, the following variables that affect seed vigour tests also need to be taken into consideration while interpreting the results.

8.7.2.3.1 Seed Size

The following examples are sufficient to demonstrate that seed size modifies the vigour test results. A seed lot is a population of seeds each one is unique in size and shape and hence exerts effect on vigour test results directly. Small seeds, by virtue of their greater surface area to volume ratio, absorb water at a greater rate when compared to larger seeds and tend to complete imbibition and initiate radicle protrusion much earlier than larger seeds. Contrary to this, the larger seeds leak more electrolytes than smaller seeds even though they are more mature and vigorous and the results of a conductivity test are presented on seed weight basis rather than a per seed basis. Therefore, even in accelerated ageing test, a specified weight of seed is placed in the inner chamber rather than the specified number of seeds. Ensure that size of the seed does not bias any vigour test results.

8.7.2.3.2   Seed Treatments

Usually, the seeds are treated to offer protection from pathogens after sowing in the field. Treated seed lots definitely affect the result of vigour tests especially when treated and untreated seed lots are to be compared. No doubt, the Cold test was initially designed to test the efficacy of the seed treatment only. Since, one of the components of vigour testing is to know the stand establishment potential of a seed lot, arguments both for and against testing treated/untreated condition have been advocated. Conductivity test, fortunately provide guidelines on seed treatment, and recommends the use of untreated seed or seed treatments should be removed prior to testing. On the other hand, accelerated ageing test advocates the use of seed as it is received in the laboratory for testing. Keeping in view the future requirements, where most of the seeds will be offered for commerce only after treatments/ enhancements like pelleting, film coating, physiological seed enhancements like priming and bio-control etc., all vigour tests to be proposed in future must provide guide lines pertaining to treatments.

8.7.2.3.3   Seed Dormancy

Keeping in view the detrimental nature of dormancy with regard to stand establishment, this trait has been eliminated in several cultivated crop species through breeding procedures. Despite this, dormancy remains a serious problem during seed testing in several crop species such as *Oryza sativa, Arachis hypogaea*, etc. If dormancy is suspected of affecting vigour results, the seed should be checked by Standard warm germination test or a Tetrazolium test. Several species of legume seeds are impermeable or slowly permeable to water. This can affect the percentage of seeds that germinate and the length of seedling growth in stress tests and can cause a biased index of vigour in a seedling growth rate test. This does not necessarily mean that dormant seeds are less vigorous in any way. Further, it can also affect the leaching of electrolytes from seeds in a conductivity test. Irrespective of the method used, guidelines must be provided for treatment of seed dormancy while testing for vigour. Since, most vigour tests use seedling growth as a measure, the two basic questions: How to evaluate a dormant seed? What will be the level of vigour if dormancy is relieved off?, still needs to be worked out.

8.7.2.3.4   Vigour Test Design

Several seed vigour tests have been designed to evaluate seeds / seedlings and provide a composite value like percentage of the seed lot. In case of Conductivity test, though all seeds are treated alike, but still the results are reported as average of all seeds. Since this design is more rapid, less expensive and offers individual seed analysis, can be considered good but the results should be interpreted with great caution, because presence of even one or two very bad leakers would alter the result despite the excellent quality. On the other hand, conducting a series of vigour tests like accelerated ageing for storage life, conductivity test for membrane integrity etc., and summarizing their results as a single vigour index though sounds good, is in fact difficult to implement successfully because different tests are designed to evaluate different parameters of seed quality and ultimately the result also varies both in respect of degree and intensity. For example, cold test gives more sensitive information than conductivity test result.

The cost benefit ratio also needs to be taken into consideration because the additional information provided by a test, at times, may not be sufficient either with respect to the increased cost and or time required to generate the data. For several quality tests to be useful, they must provide reproducible results. To evaluate the reproducibility of various vigour tests, both AOSA and ISTA have conducted extensive referees. These referees have shown that seed testing laboratories should have their own results on same sample of seed within acceptable confidence limits, but the amount of variability between different laboratories can be unacceptable. Lack of standardization among testing stations could be attributed to degree of subjectivity, i.e., seedling classification in TZ test, Cold and AA test are different, seeds and seedlings are separated into categories based on characteristics that are difficult to describe precisely.

8.7.2.3.5   Validity of Results

It is very easy to define the validity rather than to demonstrate it conclusively. This is so because the concept of validity is purely relative rather than being absolute. Therefore, the validity of a test will vary according to the purpose for which its results are utilized and hence, may demand retesting till a conclusive body of evidence is accumulated. Seeds in open storage must be retested for germination usually after five months and 24 months in case of hermitically sealed. Since, vigour test is considered to be a still more sensitive index of seed quality; obviously, the retest intervals would be shorter than the germination testing. If so, by how early? What would be the influence of seed storage environment on the validity of test results? These questions need to be answered in due course of time.

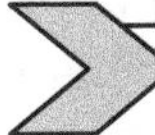

8.8   Summary

A standard Germination test, being robust, remained the sole criterion for several decades for presentation of potential performance of seed lots under optimal conditions and facilitated printing of germination percentage on seed labels as mandated by law. Contrary to this, seed vigour test, supposed to be a still more sophisticated analytical procedure offering a more precise index than germination test, is neither required not printed on seed labels offered for sale. Even for presenting this information on labels regulations becomes necessary to ensure fairness to seed producers and users.

Seed vigour testing information has several useful applications in the seed industry. Despite this, AOSA has identified several reasons for lack of standardization in procedures between different seed testing laboratories. Discrepancies in test results between different seed testing laboratories are likely to create legal hurdles and also a loss of one fruitful production season. This obviously creates a necessity for establishment and standardization of procedures. A complex process of predicting the performance of seed lots well before their sowing in the field is referred to as standardization.

Variable cold test methods developed obviously resulted in significant differences between and within laboratories. The objective of Vigour test is to provide accurate, unbiased, reliable and meaningful information to users. In the absence of standardization, repeatability and reproducibility of vigour tests itself became a problem and facilitated identification of other variables associated with standardization like sampling, subjective assessment, fluctuations in moisture and temperature etc. This resulted in development of single seed vigour tests for specific crops, minimized variability and provided accurate data for marketing and regulatory efforts duly defining limitations of each test and to meet ultimate aim of supplying high quality seed. These procedures were further tested in the respective vigour test committees of AOSA / ISTA for further sharpening and refinement in testing protocols either comparative testing or its revised form referee testing and its still sophisticated form of proficiency testing.

The acceptance of definition for vigour satisfactory to all engaged in seed trade could be due to several social and technical challenges. Among social challenges, answering questions like who are the real users of this information? In what way seed vigour information is useful to seed producers and farmers, why vigour information is not presented on seed labels etc. needs to be answered. Since the vigour tests have not been sufficiently standardized to produce reliable and reproducible information, is likely to cause variation in test results and ultimately mistrust and litigations in seed transactions. Research leaders have divided vigour tests into suggested and recommended tests. Contrary to this seed analysts and research managers suggested more rigorous testing of control variables in addition to providing interpretation criteria. On the other hand, technical challenges in respect of vigour test requirements, variables in vigour testing together with interpretation and presentation of results needs to address. Vigour tests have been designed variously to evaluate various parameters of quality results in variation in respect of degree and intensity, cold test is more sensitive than conductivity test. Changes in seed vigour take place very rapidly. Therefore even minor variations in testing equipment, testing materials, protocols etc., are bound to cause variation in test results. Cost benefit ratio needs to be considered. Use of standard check sample is a must for vigour testing because these tests demand precise moisture and temperature control in a specified time frame for in-house quality control programmes and determination of discrepancies in test protocols. Selection of control seed lot together with its maintenance and preservation is more important. Control samples needs to be equilibrated at room temperature before every use. Similarly periodical measurement of control samples for moisture is also highly essential.

In future, the requirement for vigour testing will always be there as an important component of seed quality. Seed testing laboratories will perform vigour tests only on the request of their clients. The results of vigour tests will differ depending on their use between species and countries. Before the vigour test results are interpreted and understood effectively, the ultimate user must be educated thoroughly on this subject of seed vigour.

8.9 Exercise

I. Answer the following questions.

1. Explain why necessity arose to have a better quality index than germination test?

2. Why standard germination test remained the sole criteria for a long time for presentation of potential performance of seed lots?

3. The information generated through vigour test is neither required nor presented on seed labels. Why?

4. Repeatability and reproducibility is a problem with vigour tests. Explain why?

5. Who are the real users of the vigour information?

6. Explain in detail how and why vigour information is useful to various stakeholders?

7. What is the response of seed analysts and research managers in respect of testing and use of vigour test results?

8. A vigour test has to be conducted and evaluated at the earliest moment possible. Why?

9. In accelerated ageing test a specified weight of sample is placed rather than specified number of seed. Why?

10. Cold test gives more sensitive information than conductivity test. Why?

11. Why it is more difficult to standardize a cold test?

12. Why sufficient standardization is lacking between seed testing laboratories?

13. The control seed lot selected should be high in germination and moderate in vigour. Why so?

14. Why control samples are allowed to equilibrate at room temperature before every use?

15. Periodical checking of moisture content of control seed lot is a must. Why?

II. Write short notes on the following.

1. Vigour test requirements
2. Variables in vigour testing
3. Validity of results
4. Use of standards
5. Use of control samples
6. Purpose of control samples
7. Selection of control seed lot
8. Maintenance and preservation of control samples
9. Maintenance of moisture in control samples
10. Standardization of test procedure

III. Define the following.

1. Standardization of vigour test
2. Suggested and recommended tests
3. Referee and Proficiency testing

IV. Fill in the blanks with suitable words.

1. Vigour information is useful to a commercial farmer in deciding ----------

2. Based on the field conditions and available vigour information, a commercial farmer can decide ----------

3. The information of vigour facilitates seed producers in identification of adverse practices and helps in initiation of ---------- measures.

4. The information of vigour facilitates seed producers in establishment of minimal quality standards to seeds offered for sale in the ---------- market.

5. Seed vigour information facilitates seed growers to expect ---------- price for the high quality seed he has produced.

6. A commercial farmer, based on vigour information, can predict uniformity in plant population and probable ----------.

7. Initially cold test is designed to test the efficacy of ----------.

8. The control samples enable seed technologists to discover discrepancies in the test ----------

9. Comparative tests are meant for evaluating the sharpness of ----------

10. Several species of legume seeds are ---------- or ---------- to water.

Answers

1. Seed rate
2. Early planting date
3. Corrective
4. Competitive
5. Premium price
6. Competition
7. Seed treatment
8. Protocols
9. test procedures.
10. Impermeable or slowly permeable.

8.10 References

AOSA (1983). Seed Vigour testing Hand Book. Contribution No. 32 to the Hand Book of Seed Testing. (ed.) Clark, B.E.., McDonald, M.B., and Joo, P.K., pp 88., Lincoln. NE. AOSA

Ashton, D. (2002). Important notice – Name change from "Referee" to "Proficiency". ISTA News Bulletin, 124, 22.

Byrum, J.R. and Copeland, L.O. (1995). Variability in vigour testing in maize seed. Seed science and Technology. 23: pp 543-549

Dijana, Ocvirk., Marija Spoljarevic., Sanja Spoljaric Markovic., Miroslav Lisjaak, Renata Hanzer and Tihana Teklic., (2014). Seed germinability after imbibition in electrical conductivity test and relations among maize seed vigour parameters. Journal of Food, Agriculture and Environment.12 (1): 140145. WFL Publisher Science and Technology.

Eidam, E. (1878). Dritte Versammlung von Vorständen der Samencontrol-Stationen zu München am 17. September 1877. Landwirt schaftliche Versuchs-Stationen, 21, 473-476.

Ferguson, A.J., (1993). The agronomic significance of seed quality in combining peas. University of Aberdeen, U.K.

Ferguson Spears, J., (1995). Introduction to seed vigour testing. In H A Van de Venter (ed) Seed Vigour Testing seminar. Pp 1 - 9. ISTA. Zurich.

Hampton, J.G., (1995a). Conductivity test.In: H A Van de Venter, (ed.) Seed Vigour Testing Seminar. pp 10-28. ISTA, Zurich.

Hampton, J..G., (1995b). Methods of Viability and Vigour Testing: A Critical Apprisal . In: AS.Basra (ed.) Seed Quality, basic mechanisms and agricultural implications.pp 81-118. Food Products Press, Binghamton. NY.

ISTA, (1995). In: Hand Book of Vigour Test Methods, 3rd Edition. (eds.). Hampton, J. G., and D. M. TeKrony, ISTA, Zurich. Switzerland.

ISTA, (2013). Procedure of ISTA Proficiency Test Programme flow chart effective 15-08-2013, ISTA Secretariat, Zurich Strasse, 50. 8303, Basserdorf, CH: Switzerland.

Lisjak, M., Wilson, I. D., Civale, L., Hancock, J. T. and Teklić, T., (2009). Lipid peroxidation levels in soybean (Glycine *max* (L.) Merr.) seed parts as a consequence of imbibitions stress. Poljoprivreda (Agriculture) 15: 32-37.

Loeffler, T.M., TeKrony, D.M., Egli, D. B., (1986). The bulk conductivity test as an indicator of soybean seed quality. Jour. of Seed Technology. 12: 37-53.

Marcos- Filho, J., (1998). New approaches to seed vigour testing. Scientia. Agricola. 55; pecial issue.

McDonald, M.B., (1995). Standardization of vigour Tests. In H.A. Van de Venter (ed.) Seed Vigour Testing Seminar. Pp. 34 – 52. ISTA, Zurich.

McDonald, M. B., (1994).The History of seed vigour testing. Jour. of Seed Tech.17: 93 -101.

McDonald, M.B. (1988). Challenges in seed technology. pp. 11-31. In Proceedings 10th Seed Technology Conference. ed. J.S. Burris. Ames, IA.

Muschick, M. (2010). The evolution of seed testing. Seed Testing International, 139, 3-7.

Nobbe, Fredric., (1876). Handbuch der Samen Kunde, Wiegandt- Hempel, Parey, Berlin.

Reusche, G.A., (1987). Comparison of the AOSA recommended conductivity analysis and the alternative single seed procedure. AOSA. News Letter, 61(1): 79-97.

Steiner, A.M. Kurse,M., Leist, N., (2011). Anobligation from the beginning: Comparative, Referee and now Proficiency testing. Seed Testing International. 141: pp 20-21.

Steiner, A.M., Kruse, M. and Leist, N. (2008). ISTA method validation 2007: a historical retrospect. Seed Testing International,136: 32-35.

Tao, K.J. (1980). The vigor "referee" test for soybean and corn. Association of Official Seed Analysis Newsletter, 54(3); 53-68.

Vieria R. D., Neto, A. S., Mudrovitsch de Bittencourt, S. R. and Panobianco, M. (2004). Electrical conductivity of the seed soaking solution and soybean seedling emergence. Scientia Agricola. 61: 164-168.

Precision of Seed Vigour Testing and Reporting of Seed Vigour Test Results

(Efficiency is the capacity to bring proficiency in to expression)

9.1 Introduction

LEARNING OBJECTIVES

- Need and Importance of precession in Seed Vigour Testing,
- Ways and means of acquiring precision in vigour tests,
- Presentation and Interpretation of Vigour Test Results,
- Use of Standards in Seed Vigour Testing
- Selection of Control Seed Lot
- Use of Control Samples
- Maintenance and Preservation of Control Samples.
- Usefulness of seed vigour tests,

Seeds have been undoubtedly the most important and most crucial means of production since mankind is working in plant production and agriculture, clearly indicates the real value of seeds, Muschick, (2010). Vigour testing is a more sensitive measure of seed quality and hence requires more sophistication in respect of both protocol as well as interpretation by the seed analyst. This obviously demands in the first instance that the vigour test results must be standardized among the seed testing laboratories. Further, this has to be achieved in such a way that the vigour testing has to be made acceptable to the consumers by way of testing the material in well defined and equipped vigour test laboratories under the supervision of proficient seed analysts duly following all the technical aspects.

One should always remember that, high seed vigour does not always necessarily mean higher yields. Unfortunately, we have only fewer comparisons of seed vigour to final yield than to field emergence. This is so because, yield comparisons are variable depending on the crop planted, stand achieved, original level of vigour, growing season etc. In addition to these, genotypes of certain crop species have the capacity to compensate for the minor loss in stand differences which ultimately result in little differences in yield. Therefore, it appears that educating the consumers is essential to make them understand the vigour test results and also to be able to interpret the vigour test results. This could be attributed to the fact that, neither the ISTA nor AOSA have the responsibility to interpret the results as their role is confined only in standardization of test methods rather than test results. Hence, Mc Donald (1995) strongly emphasized that interpretation of vigour test results is the role of the consumer but not that of

Seed Analyst's. The Seed Analyst's role is to accurately present the test data and allow the consumer the responsibility of interpreting whether the seed lot is of acceptable quality based on his own standards, a time honoured process known as a caveat emptor.

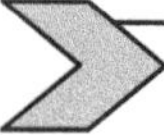

9.2 Precision is important in Seed Vigour Testing. Why?

In seed technology, seed vigour is the most researched subject. The knowledge that started pouring in worldwide pertaining to several seed vigour tests have simultaneously resulted improvement in respect of sensitivity, efficiency and standardization.

9.2.1   The Need

Since the role of seed analyst is continued only to present the vigour test results accurately, precision assumes much significance. Evaluation of seed physiological potential demands the use of both standardized procedures as well as the correct interpretation of the meaning and scope of the derived results. As a consequence of much progress achieved in respect of standardization, some tests have now reached a stage for recommendation with greater level of confidence.

9.2.2   Why Precission ?

During seed vigour testing precision plays an important role at least on two counts as detailed below.

On one hand,

1. Changes in seed vigour occur very rapidly.
2. Even minor differences in testing materials and environments like temperature, initial moisture content of seed, test duration, fluctuations in temperature and moisture in test chambers during testing, etc., can cause minor changes in test results.
3. Use of improper equipment and substandard supplies are also likely to exert influence on the result

Therefore, keeping in view the speed with which the seed is likely to deteriorate, conducting of a vigour test becomes more challenging and demands more precision in respect of both consumables and test conditions.

On the other hand,

First of all, both the seed analyst and the consumer must understand essentially that almost all seed vigour tests are not designed for predicting the exact number of seedlings that are most likely to emerge and survive in the field. No doubt, several seed vigour tests do correlate well with field emergence. Further, the primary objective of a vigour testing is to predict whether or not any trouble is anticipated from a high germinating seed lot particularly when placed under adverse environmental conditions in the field, or in storage or even in transport and this in turn demands reliable information from seed vigour tests so as to rank the seed lots in tune with their quality level and to discard those seed lots that fall below the standards prescribed by the company's vigor test procedures, Ferguson, (1993). Simultaneously, it also becomes essential that how long a seed lot will be in a position to maintain the required germination percentage, say "x" while in storage for a period of "Y" months. This obviously demands correct precision in interpretation of test results so as to facilitate appropriate selection of better seed lots of required quantity for sowing as well as to predict the probable success achievable under diverse field conditions.

9.2.3 How to achieve precision while conducting vigour tests?

1. Precision is required both within and among seed testing laboratories.
2. This can be achieved by use of control supplies for all vigour tests and species to be tested.
3. Measurement of seed moisture and adjustment as and when needed.
4. Maintenance of seed storage at appropriate conditions.

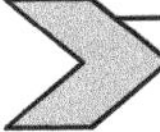

9.3 ISTA Vigour Tests

Seed vigour, being an important parameter of seeds quality, plays an important role in the commercial production of agri-horticultural crops. Since the quality of seed is expected to decline during field weathering, harvesting and or storage and hence is bound to result in a serious economic loss. This in turn necessitates an understanding on the source and mechanism of seed loss either to reduce or slow down the process of deterioration if not stopped completely. A vigour test basically is expected to assess, the physiological and physical basis of a potential seed lot quality and performance, be it directly or indirectly and thereby facilitates a more sensitive differentiation among seed lots than does the standard germination test. No doubt, the seed industry is widely using the seed vigour tests for quite some time. Yet only recently, vigour testing by ISTA seed testing laboratories was incorporated after approval in ISTA Rules for seed testing. The recognition, that greater precision is required for vigour testing than is required for standard germination testing has led to improved standardization and uniformity in comparative tests despite much more difficulty encountered n achieving the standardization of vigour tests. Moreover, most of the vigour tests have not experienced widespread use either owing to their subjectivity or high cost and even a good amount of variability in test results between seed testing laboratories. The vigour test committee of ISTA aims (1) to encourage the use of vigour tests, (2) to develop and validate vigour tests for introduction in ISTA Rules and (3) ensures that these are carried out to achieve repeatability between and reproducibility within the seed testing laboratories.

9.3.1 Electrical Conductivity Test

This is the first test included by ISTA in the International Rules for Seed testing in 2001 for garden peas, (*Pisum sativum*), field beans (*Phaseolus vulgaris L*), Soybeans, (*Glycine max L*). This biochemical test was proposed by Fick and Hibbard (1925) for predicting the germinability of timothy seeds. This idea was further developed with a detailed methodology by Matthews and Bradnock (1967) to estimate the percentage seedling emergence in peas. The test is based on the principle that less vigorous more deteriorated seeds exhibit relatively lower speed of membrane repair during imbibition and release greater amounts of solutes in to the external environment. Under real field conditions, the exudates thus leached from seeds after sowing results in impairment of seedling emergence. This is the most widely used vigour test, Mracos Filho (2015).

9.3.2 Cold Test

This is the oldest and most widely used test for seed vigour assessment in hybrid corn, barley, carrot, egg plant, field beans, lettuce, onion, rice, sorghum soybean and others. The concept for this test as developed by Alberts (1927) involves testing the efficiency of post fungicides applied in humid media after sowing the seeds. Further, the test procedure involves evaluation of seed samples using soil as substrate and exposed to a combination of low temperature, high substrate moisture coupled with the presence of pathogens. Suboptimal temperatures, on hand, facilitate solute leakage during imbibition due to relatively slow and poor membrane repairing and on the other hand the low temperatures coupled with sugars leached into soil medium loaded with microorganisms enhances the damage to seed performance and results in reduction in the speed of germination and seedling emergence stimulates the discharged.

9.3.3 Accelerated Ageing Test

This test offers precious information on storage and seedling field emergence potentials and involves subjecting of seed samples of numerous species to high temperatures and high relative humidity followed by germination test. The concept for this test came from the observations of Crocker and Groves, (1915) and developed by Helmer *et al.,* (1962), Delouche and Baskin (1973) and the final protocol was designed by McDonald and Phaneendranath, (1978) and is based on the principle that high vigour seed lots tolerate ageing stress better than lower vigour seed lots. Standardization capability of this test rendered it one of the most globally recognized tests to evaluate seed vigour. However, Powell (1995) reported that, in respect of vegetable and small seeded crops absorb water more rapidly and result in large scale variation and lack of uniformity among the seed samples tested. In order to solve these problems, Jianhua and Mc Donald (1996) proposed SSAA, Saturated salt accelerated aging test, with a method more similar to natural ageing, duly substituting water with saturated salt solution. Markos – Filho (2015) reported that this test is suitable for more species.

9.3.4 Tetrazoluim Test

Lakon (1950) proposed the differentiation of low and high vigour seeds based on the location and extent of colour and texture of the staining tissue. More and Smith (1956) classified seeds in to Vigorous, viable but not vigorous and non viable. Detailed test procedures for evaluation of seed vigour by tetrazolium test were published by Vieria and Von Pinho (1999) for cotton. The test is also reported to be suitable for groundnut, maize, soybean and tropical forages.

9.3.5 Tests based on Seedling Growth

First count of the germination, speed of germination, seedling growth rate, and seedling vigour classification and emergence rate of primary root are based on seedling growth and designed to predict the uniformity and speed of seedling emergence directly affecting plant stand establishment. Matthews *et al.,* (2012) reported that the extent of deterioration is directly proportional to the length of phase–II of seed germination or lag phase (Matthews and Khajeh-Hossain (2007).

 ## 9.4 Reporting and Interpretation of Seed Vigour Test Results

Vigour testing, being a highly sensitive measure of seed quality, demands more sophistication in respect of both protocol and interpretation. In order to achieve this, first of all, the vigour test results must be standardized both within and among the seed testing laboratories, in such a way, that vigour testing is made acceptable to the consumer by means of testing the material in well-equipped vigour test laboratories under the supervision of proficient seed analysts.

Seed vigour test results are purely relative values, because the test is conducted on a number of seed lots simultaneously with the inclusion of a control seed lot to act as a reference point as well as to serve in arranging the seed lots in high to low ranking order. Just one has to keep it in mid that, it is not only the plant stand which will vary from field to field but also the results of these tests vary both between and among seed testing laboratories at times, thereby making it very difficult to establish cut off points between acceptable and unacceptable levels of seed vigour across several crop species.

For a moment, let us assume a situation, where in a seed lot with high germination is at the end of the slow decline stage of a seed survival curve. Probably, this will result in a rapid decline in a standard germination percentage, at least for a short period, which is completely unrelated to germination test result. Therefore, the germination test results are often repeated within three to six months. Thus, the ISTA vigour test committee has recommended that, since vigour falls much before germination, repeated vigour tests should be completed more frequently than germination tests. Further, on the other hand, the dynamic environmental conditions will be changing from field to field, day to day and year to year giving scope for a given seed lot to behave differently with respect to varying emergence percentages under each set of conditions. Therefore, those interested in purchasing the seeds should be extremely cautious in recognizing that the seed vigour test does not predict the emergence percentage.

9.4.1 Presentation of Vigour Test Results

How best the vigour test results can be presented is still not yet resolved, because, variation in the results of various vigour tests conducted on a specified seed lot itself is considered as the first problem identified. A seed analyst should always remember that the potential physiological performance of a seed or seed lot is the culmination of a set of characteristics that can be evaluated either directly or indirectly by ether assessing the present metabolic state of the seed or the ability to tolerate stresses and thereby a relationship seedling emergence and storability can be established. Moreover, a vigour test is usually conducted in conjunction with a germination test because the latter is required before the seed can be sold commercially. The results of accelerated ageing test and cold test are given as percentage germination and hence acceptable by the consumer immediately because of their similarity with germination percentage. Similarly, the Tetrazolium test's considered being one of the most useful vigour tests. However, presentation of the results of this test in colourful staining patterns to the uneducated consumer in a more meaningful way appears to be remotely difficult. Similarly, in brick grit/gravel test, the results are reported as the total of normal emerged seedlings either as ...% vigour or as % brick grit value. On the other hand, in conductivity test, an important test of vigour for several crops, the results are expressed as a measure of electrical conductivity of seed leachate and presented as μS cm-1 per gram of seed, a value particularly applicable to field emergence. Moreover the relationship with seed quality is inverse, i.e. higher reading correlates with poor quality. For example, readings below 20 μS/cm/g are acceptable, readings above 40 μS/cm/g are unacceptable and those between 20-40 μS/cm/g are considered marginal. It means a conclusion opposite of what most people would expect and that too majority of the end users are uneducated. It means, depending on the species under consideration, the nature of tests differs and hence the way in which the results are expressed or reported also varies accordingly.

Thus, the responsibility of understanding the nature of test conducted based on the result lies with the consumer for a particular species in which he is interested. Further, AOSA (1983), in addition to the above also perceived that, the results of vigour for a sample tested is only the expression of vigour as it is at a particular point of time and hence the results of vigour have nothing to do with or not valid if at all the seeds were shifted subsequently to a place with poor storage conditions, because the vigour of whole lot subsequently may feel very rapidly or under good storage conditions, however, there would not be any change in the vigour test results. Since, the same would be true in case of germination results also.

Several seed companies are conducting a series of vigour tests and combine the information to form a seed vigour index and establish an acceptable level of vigour for a specific production season and seed lots are evaluated before conditioning, processing and marketing. Those seed lots exhibiting a specified vigour level after conditioning are marketed as vigour proven, vigour rated or high vigour seed, without giving the details of how many tests are conducted or criteria used for assessing the vigour potential.

9.4.2 Use of Standards while Presenting Seed Vigour Test Results

In order to encourage the quality control process, a seed testing laboratory should have a standard seed lot with known values of vigour and by employing standards while comparing the results between seed testing laboratories. It appears that it is difficult to standardize the cold test because this test exposes seeds to soil microorganisms and their ability to resist the attack under cold or wet conditions and their consequent measurements. As such use of local soil in cold test makes the comparison difficult between seed testing laboratories because the soils vary in their moisture holding capacity, pathogen load etc., that directly exert influence on cold test results. This obviously demands the standardization of cold test procedure with respect to substrate, micro flora etc., that would be difficult if not impossible. Though the use of sterile media with some specific microorganisms has been suggested, it is very simple to suggest an approach, but it is very difficult to culture and maintain a constant level of pathogenicity. Further, one should not forget that the behaviour of microorganisms change drastically when they are in soil as a population and how they behave in the absence of an interacting population needs to be worked out. Despite these associated problems, Byrum and Copeland (1995) have clearly indicated that cold test could be repeatable just like standard germination test.

Contrary to this, Conductivity test, became a widely accepted test of seed vigour, not because this test has been studied thoroughly and correlated well with seed vigour, but because the test variables have been evaluated comprehensively, Loeffler *et al.*, (1986) and revised test procedures provided. According to Tao (1980) and Reusche (1987) vigour test referees have consistently demonstrated that conductivity test results are reproducible both within and among seed testing laboratories.

Under these circumstances, McDonald (1994) pointed out that incorporation of conductivity test in to "Rules" does not necessarily mean that seed lots must be tested for vigour and result should appear on the label. It only means that, for testing vigour using conductivity test, a specified procedure is available and that must be followed because this procedure not only ensures standardization in testing but also ensures the appropriate interpretation of results in addition to maintenance of confidence and credibility of testing protocols.

 ## 9.5 Control Samples in Seed Vigour Testing

Use of Standard check samples or control samples becomes mandatory for all laboratory based seed testing programmes. Because, unlike in germination testing, seed vigour tests during testing demands a specific time frame, precise moisture, and temperature management. Even small fluctuations/ variations in these test components are likely to produce drastic changes in the test results. This automatically creates a necessity for all seed vigour testing laboratories to obviously opt for selection, maintenance and use of check samples for each species tested. The basic use of a check sample in seed testing laboratories is aimed at providing the 'in-house quality control' and is expected to serve two basic purposes, viz., (i) providing internal quality control and (ii) enables the seed technologists and seed analysts to discover even minor discrepancies in the testing protocols.

9.5.1   Selection of Control Seed Lots

For selecting a control seed lot, every year several seed lots in each crop species need to be screened for germination and vigour tests. The lots thus selected should be high in germination, moderate in vigour and relatively free from physical injury or disease. It is always better not to select a seed lot with too high vigour, because it may not permit detection of minor differences in methodologies and hence could not become a god check sample.

9.5.2 Use of Control Samples

The seed vigour tests need to be standardized because they play an important role in plant breeding, seed production, marketing and quality control. Seed vigour testing even today, has not yet achieved even the same level of standardization possessed by a standard germination test for all species and regions. Both the International Seed Testing Associations like AOSA and ISTA and the entire seed trade are merrily looking for and focusing much on bringing improvement in the reproducibility of vigour tests. Control samples are included just like routine samples in every vigour test run in a coded form. The seed technologist or Seed Analyst only will determine whether the check sample falls outside the prescribed parameters or not, only at the end of the test run. The test should be repeated after identification and solving the problem. The vigour test results of control samples summarized at regular intervals help in evaluating the uniformity of samples and testing procedures.

9.5.3 Maintenance and Preservation of Seed of Control Samples

Precise and careful planning is required so as to have sufficient quantity of check sample required for the entire year or testing season. As far as possible pack those small quantities for use as working samples on day-to-day basis and store them under highly controlled conditions such that the vigour and germination capacity are perfectly preserved. It is always preferable and advisable to draw the working sample sufficient for a week from the stores. The control seed lot is divided into several sub samples of around each 500 seeds and packed in moisture proof packets or stored in a moisture proof container at -20°C before testing. Allow the control samples to equilibrate at room temperature before every use.

9.5.4 Maintenance of Moisture in Control Samples

A seed lot with too high vigour could not become a good check sample, as the same may not permit detection of minor differences in the testing methodology. In addition to precise estimation of initial moisture content, it is better to ensure through periodical checks, the gain or loss of moisture content of the control sample. The initial moisture content should be precisely estimated at the time of selection of the lot meant for use as a control sample. If necessary, adjust moisture content at initial level itself because the extremely dry seed is likely to cause imbibitional damage and ultimately gives inaccurate readings during testing.

9.6 Interpretation of Vigour Test Results

Today, being very much sophisticated, vigour testing reveals the true quality of the seed lots and is perhaps one of the most important tools in the kit of an end user of seed. The farmer should use it as a guide in avoiding planting into an unfavourable set of conditions. Usually, a vigour test is conducted in conjunction with a germination test because the later is required before the seed can be sold in the market. These two tests together will give much of the required information to most of the producers who are interested in knowing how they will perform at both ends of the spectrum.

The germination test usually and almost always, have the higher result because the evaluation parameters are more forgiving than those used in the vigour test. Therefore, the germination test tells the producers how their seed will perform under optimum conditions. Contrary to this, the result of vigour test will always be a lower number because the deliberately applied stress will suppress germination. Hence, the result of a vigour test is likely to be more consistent with field emergence and tells the producers how their seed will perform if it is subjected to stressful conditions. This can be best explained with the following example. Assume that a seed lot has registered 90% germination and 65% cold test result. This clearly indicates that the seed lot would produce 90 per cent plant stand under ideal conditions and 65 per cent under stressful conditions. Therefore, in order to derive better results, early and accurate evaluation of seed quality is essential for planting and production of a high quality crop. No doubt, having overall information pertaining to seed quality definitely render help in taking important crop input decisions like use of seed treatments and other inputs including herbicides and ensures a healthy and vigorous plant stand establishment. If seed bed conditions are favorable then the vigour will be closer to germination.

A consumer is supposed to know that a vigour test neither predicts nor forecasts the emergence, but at the most, it allows the consumer an opportunity to determine that one seed lot is superior to another (Ferguson, 1995). The degree of superiority or how much better is determined in the field at the time the environmental stress is experienced in the particular year. It means, in some years the stress may be so severe that no seeds germinate even from the most vigorous seed lot. As such the seed analyst should avoid predicting or forecasting field performance of a seed lot based on vigour test results. Therefore, it is the consumer who should consider the conditions in to which the seed will be sown, i.e. high vigour seed, no doubt will increase the potential for good emergence in poor conditions but there is no guarantee for it. As such, AOSA (1983) advises the consumers to compare vigour along with standard germination test results for more than one lot before buying, if possible. This automatically facilitates the consumer to select a lot having higher vigour. The consumer has to exercise his judgment regarding the vigour status of the seed lot *vis-a vis* the conditions under which it is sown because the vigour information is used to judge the element of risk at sowing for emergence.

The germination test result is reported as percentages of normal seedlings, abnormal seedlings, dormant seed and dead seeds. Therefore, the seed analyst has to impart great care and carefully look at this breakdown so as to list out any issues that are limiting the seed in having a higher percentage of normal seedlings. Germination test is conducted at 20-25 ^{0}C and seed quality is estimated under optimum conditions where as very rarely a seed grower plant his seed at 20 ^{0}C soil temperature. Most farmers use the difference between the germination result and vigour result as the expected per cent marked and accordingly determine the seed rate or else a low vigour seed is planted at a late when the soil condition become warmer. It is not possible to establish a seed vigour-testing laboratory over night. It means expertise in analysis is acquired only over a period of time. At the same time precision in equipment and testing methodologies also play a more important and crucial role than conducting the germination test, because in case of vigour testing it is planned to measure more minor changes in deterioration, which are not obvious in the germination testing.

Early and accurate evaluation of seed quality is essential in the planting and production of a high quality crop. Knowing the overall seed quality will help with important crop input decisions, such as the use of seed treatments, and maximize use of crop inputs, such as herbicides, and ensure a healthy vigorous plant stand. Therefore, organizations interested in seed storage potential and seedling emergence like seed growers, officers from Seed Certification, Seed Processing Companies, and Agricultural Extension Officers and NGOs may join together and develop guidelines necessary for educating the consumer and guide him in taking/ selecting appropriate decisions while purchasing the seed.

 ## 9.7 Summary

Reliable information is desirable from a vigour test for ranking the seed lots in tune with their level of quality. Because, the very purpose of a vigour test is to indicate whether or not any trouble is anticipated from a high germinating seed lot when placed under adverse environmental conditions either during storage or in the field. Precession in vigour testing is a must because keeping in view the speed with which loss in vigour occurs very rapidly and even minor changes in testing materials and test conditions, environments, equipments, substandard supplies are all capable of bring in variation and alters the test results and hence demands precision both between and within seed testing laboratories. Vigour testing being a more sensitive index of quality demands more sophistication in terms of both protocol and interpretation of results by the seed analyst. In order to make the test more reliable and acceptable to the consumer, conducting the test in well-equipped laboratories under the guidance of proficient seed analysts is the only solution.

Reporting the results itself is considered to be the first problem in vigour testing. Several vigour tests proposed thus far have their own merits and demerits connected with their respective background principle. Further, the results of different tests are reported in different units. This only adds to the confusion to the end users. The results of cold test and accelerated ageing test because of their similarity with germination test are nearly acceptable to the end users. However, the problem with TZ test lies in the explanation offered to the consumer with respect to colorful staining patterns in a meaningful way. The unit of expression of result also varies greatly depending on the test conducted. How best the vigour test results can be presented are not yet resolved because either the dynamic field conditions or seed lot at the end of slow decline stage of survival curve behave differently in respect of their emergence percentages. Hence, those interested in purchasing the seeds should be extremely cautious in recognizing that, the seed vigour test does not predict the emergence percentage.

At international level both ISTA and AOSA have conducted extensive referees and concluded that seed testing laboratories are capable of reproducing the results repeatedly within the same acceptable limits of confidence with same sample. The extent of variability achieved between laboratories is not within the confidence limits, which indicate that sufficient standardization is lacking. Even the procedure of cold test needs standardization in respect of substrate, micro flora etc., and is difficult if not impossible. Dormancy character present in legumes with hard seed coat offers a biased index of seed vigour and only confounds the interpretation of vigour test results.

Use and maintenance of standard check samples becomes mandatory for all laboratories for in-house quality control as well as to discover minimum discrepancies in test protocols. The control lot selected should be high in germination, medium in vigour and relatively free of physical injury or disease.

On the other hand, several seed companies are conducting a number of vigour tests duly combining the information to form a seed vigour index with acceptable level of vigour for a specific season. Seed vigour test results are purely relative values at a particular point of time, hence invalid if seeds are shifted to a place with poor storage conditions subsequently and are likely to deteriorate very fast. Therefore, the responsibility of understanding the result lies with the end user only, because a vigour test neither predicts nor forecasts the emergence. Hence, educating the end user appears to be essential to make them understand and interpret the vigour test results properly as it is his business and the seed analyst has nothing to do with the test result. The superiority of a seed lot is determined only in the field when exposed to environmental stress prevailing in that particular point of time. Having information on the overall seed quality obviously helps in taking important crop input decisions. Seed vigour falls much before the germination, it is advisable to conduct vigour tests more frequently than germination tests. Further, frequent changes in environmental conditions from field to field and day to day, a given seed lot behaves differently in respect of germination. On the other hand, seed companies are marketing seed lots after conditioning as vigour proven without providing the details or criteria used for assessing the vigour potential. Expression of vigour test results in different units on one hand, the information provided by a single vigour test is only a comparative value and one test cannot detect all possible facets of seed performance render vigour testing and field seedling emergence estimation really a difficult task.,

9.8 Exercise

I. Answer the following questions.

1. Briefly explain why precision is impotant in seed vigour testing.

2.   How precision is achieved during vigour testing.

3.   List out the problems associated with reporting the results of seed vigour tests.

4.   Interpretation of vigour test results is the role of consumer. Justify.

5.   The role of seed analyst is confined to presentation of vigour test data. Coment.

6.   Vigour test will neithr predict not forecast the emergence, but only offers the consumer an opportunity to determine how one seed lot is superior to another. Coment.

7.   List out the usefulness of seed vigour tests.

8.   Summarise the limitations of seed vigour testing.

9.   Suggest precautionary measures to be taken while conducting vigour tests.

10.   Why a vigour test is conducted in connjunction with germination test.

11.   Explain why the results of accelerated ageing and cold test are accepted immediately.

12.   Why reporting results of a tetrazolium test is difficult.

13.   Explain why the results of a vigour test are not valid if the seeds are shifted to a poor storage subsequently.

14.   Explain why the germination test results are often repeated within 3 to 6 months.

15.   When percent 95 and 65 per cent are the respective results of germination and cold tests of a seed lot. What inference would you draw.

16.   In which way knowing the overall quality of a seed lot helps you as an end user of seed?

II. Write short notes on

1. Presentation of seed vigour test results   2. Reporting seed vigour test results

3. Usefulness of seed vigour tests   4. Limitations of seed vigour tests

III. Fill in the balnks with a word or phrase

1.   Changes in seed vigour occur very --------------.

2.   The result of a vigour test is influencd by -------------- equipment and -------- supplies.

3.   Precision is required both ---------------- and -------------- seed testing laboratories.

4.   The result of conductivity test is expressed as --------------

5.   The result of conductuvuty test expressed as µS cm-1 per gram of seed is applicable to---.

6.   The results of a brick gravel test are reported as -------------.

7.   Seed lots exhibiting specified vigour level after conditioning are referred to as ----.

8.   Vigour information is used to judge the element of -------- at sowing for ----------.

9.   Seed vigour tests help in ranking seed lots based on their-----------------.

10.   Seed vigour tests provide an indication of seed ---------- well before planting.

11.   Seed vigour tests are great tools in the hands of seed industry for ------------------ .

12.   The values obtained in a seed vigour test is only ---------- and not --------------.

13. Compensated seed rate can be used in case of -------------- crops.

14. Precision and standardization of vigour testing methods can be achieved by ------.

15. Educating farmers on vigour testing may not be of any help where farmers are ---.

Answers

1. Rapidly
2. improper, substandard
3. Within, among
4. µS cm-1 per gram of seed
5. Field emergence
6. % vigour or % brick grit value
7. Vigour proven, vigour rated or high vigour seed
8. risk, emergence
9. Physiological quality
10. Deterioration
11. in-house quality control
12. Relative, absolute
13. Transplanted
14. Referee testing
15. Illiterate

9.9 References

Alberts, H.W. (1927). Effects of pericarp injury on moisture absorption, fungus attack, and viability of corn. Journal of the American Society of Agronomy 19: 1021-1030

AOSA (1983). Seed Vigour testing Hand Book. Contribution No. 32 to the Hand Book of Seed Testing. (ed.) Clark, B.E.., McDonald, M.B., and Joo, P.K., pp 88., Lincoln. NE. AOSA.

Baalbaki, R.; Elias, S.; Marcos-Filho, J.; McDonald, M.B. (2009). Seed vigor testing handbook, AOSA, Ithaca, NY, USA. (Contribution to the Handbook on Seed Testing, 32).

Basavarajappa, B.; Shetty, H.S.; Prakash, H.S. (1991). Membrane deterioration and other biochemical changes associated with accelerated ageing of maize seeds. Seed Science and Technology 19: 279-286.

Byrum, J.R. and Copeland, L.O. (1995). Variability in vigour testing in maize seed. Seed Science and Technology. 23 : pp 543-549

Crocker, W.; Groves, J.F. (1915). A method for prophesying the life duration of seeds. Proceedings of the National Academic Sciences USA 1: 152-155.

Ferguson Spears, J., (1995). Introduction to seed vigour testing. In H A Vande Venter (ed.) Seed Vigour Testing seminar. Pp 1 - 9. ISTA. Zurich.

Hampton, J.G., (1995). Methods of viability and vigour testing- A critical appraisal. In A.S. Basra (ed.). Seed Quality 81-118.Food Products Press, Bringhamton.

França-Neto, J.B.; Krzyzanowski, F.C. (2009). The tetrazolium test for seed vigor determination, p. 100-104. In: Baalbaki, R.; Elias, S.; Marcos-Filho, J.; McDonald, M.B., eds. Seed vigor testing handbook, OASA, Ithaca, NY, USA. (Contribution to the Handbook on Seed Testing, 32).

Ferguson, A.J., (1993). The agronomic significance of seed quality in combining peas. University of Aberdeen, U.K.

ISTA, (1995). In : Hand Book of Vigour Test Methods, 3rd Edition. (eds).Hampton, J. G., and D. M. TeKrony, ISTA, Zurich. Switzerland.

Jianhua, Z.; McDonald, M.B. (1996). The saturated salt accelerated aging test for small seeded crops. Seed Science and Technology 25: 123-131.

Loeffler, T.M., TeKrony, D.M., Egli, D. B., (1986). The bulk conductivity test as an indicator of soybean seed quality. Jour. of Seed Technology. 12: 37-53.

Marcos-Filho, J. 2015. Seed Physiology of Cultivated Plants Fisiologia de Sementes de Plantas Cultivadas. 2 ed. ABRATES, Londrina, PR, Brazil (in Portuguese).

McDonald, M.B., (1995). Standardization of vigour Tests. In H.A. Van de Venter (ed.) Seed Vigour Testing Seminar. Pp. 34-52. ISTA, Zurich.

McDonald, M. B., (1994).The History of seed vigour testing. Jour. of Seed Tech.17: 93 -101.

McDonald, M.B.; Phannendranath, B.R. 1978. A modified accelerated aging vigor test procedure. Journal of Seed Technology 3: 27-37.

Muschick, M. (2010). The evolution of seed testing. Seed Testing International,139, 3-7.

Reusche, G.A., (1987). Comparison of the AOSA recommended conductivity analysis and the alternative single seed procedure. AOSA. News Letter, 61(1): 79-97.

Tao, K.J. (1980). The (1980). Vigor "referee" test for soybean and corn. Association of Official Seed Analysis Newsletter, 54(3): 53-68.

Vieira, M.G.G.C.; Von Pinho, E.V.R. 1999. Methodology of the tetrazolium test on cotton seeds Metodologia do teste de tetrazólio em sementes de algodão, p. 8.1.1-8.1.12. In: Krzyzanowski, F.C.; Vieira, R.D.; França-Neto, J.B., eds. Vigor seeds: concepts and testing Vigor de Sementes: Conceitos e Testes. ABRATES, Londrina, PR, Brazil (in Portuguese).

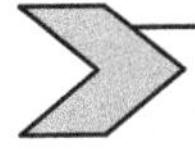

CHAPTER 10

Seed Vigour Testing

(If quality is definite, the benefits are infinite)

10.1 Introduction

The conceptual knowledge gained by us thus far is confined only to how the seed vigour is acquired during the post fertilization seed development on the mother plant till harvested and how the same is lost subsequently through the two crucial processes of seed ageing and imbibition and their interactions. But, in reality, the process of seed ageing also occurs consequent to attaining of physiological maturity while the seed is still intact on the mother plant in a good majority of species and continues even during the post harvest handling including storage in some other species and even after sowing in the field upon consequent exposure to the severities of environmental conditions in some other species and this may ultimately result in the decrease or reduction of seed vigour.

- Major and Minor areas where seed Vigour Information can be utilized
- Major and Minor players engaged in the utilization of vigour information.
- How Plant Breeders use this information in Varietal Development
- Which characters are used for identification of varieties for rain fed ecosystems?
- List of most desirable Vigour Traits
- Usefulness of Seed Vigour Tests
- Limitations of Seed Vigour Tests
- Precautions to be taken while conducting the Seed Vigour Tests.

Therefore, it is possible, that reliable vigour information thus generated needs to be utilized confidently by both the seed producers including plant breeders and seed growers in planning and organizing the seed production programmes to meet the demand. At the same time the same information can also be utilized as a predictive tool again by both the seed growers including export / import managers, seed store house managers and marketing managers for making effective decisions for the forth coming sowing season. Several authors like Hampton and Coolbear, (1990); Dornbos, (1995); and Ferguson Spears, (1995) have described the best utilization of results of seed vigour test results.

10.2 Applications of Seed Vigour Testing

A vigour test is one, which reveals a seed lot's ability to withstand a variety of different stress factors. As such, usually vigour tests are designed to mimic poor seeding conditions to find out how a seed lot will perform under stressful situations. As of now we have only few methods for vigour tests approved in ISTA Rules. This is so because vigour tests will only be relevant where there is a problem of seed vigour within a species. At the same time not all species show differences in vigour for a number of reasons. Therefore, it may not be relevant to apply vigour tests to all species without identifying and describing a problem in emergence or storage potential in that species. At present, potentiality exists for two types of test for more general application. For example, Conductivity test, because of its rapidity and potentiality for applying to several species, it appears to be attractive to many people. When the result of germination test is taken together with vigour offers a complete performance profile of seed lot for a wide range of field conditions. In addition, this information also facilitates timely help in guiding important seeding decisions like when to take up sowing? Is it necessary to wait for warmer soil temperatures? Survival capability of seed already purchased to an early planting into the cooler soils?

Loss of seed vigour arising as a consequence of ageing during storage and imbibition after the seeds have been sown in the field are the two major problems. The twin processes of ageing and imbibition, coupled with their varied interactions, have great relevance in at least in five broad areas. The major areas included are plant breeding, quality control during seed production, and the reasonable efforts while dealing with the marketing comprising of seed quality enhancement, storage, export and import. The major players in these respective fields are plant breeders, seed growers, seed producers, persons engaged in seed certification and other quality enforcement agencies, farmers, marketing personnel, store house managers, exporters and importers and so on including small players like carry forwarding agents, wholesalers and retailers involved in the marketing chain.

10.2.1 Plant Breeders

Seed vigour is a quantitatively inherited trait in part, genetically controlled and is governed by many genes and hence exerts influence on yield and stability. Obviously, it is the genetic composition of the seed that interacts with the surrounding environmental conditions of the production environment in many ways determines the phenotype, its precise rate of emergence, the subsequent growth rate and ultimately the yield level.

Polygenic control of several seed characters is expected to either directly or indirectly exert their impact on seed germination ability and vigour and the same has been reported in plenty in published literature. Therefore, real potential exists for plant breeders to incorporate and improve seed vigour characters both directly and indirectly in the varieties they plan to develop. Though the breeding stocks exhibit considerable amount of variability with respect to vigour, it is indeed very difficult to establish the cause of the low vigour because it could be either due to the genotype and / or the production environment to which it is exposed. The potential difference existing with varietal stocks may or may not be in a position to affect the final plant stand and the yield.

During the course of development, evaluation, and release of strains, plant breeders will be generating enormous amount of technical information pertaining to their suitability and / or adoptability to a particular stress situation or location. This is clearly evident from the fact that different varieties released by plant breeders in the same crop differing significantly in respect of their various characteristics. Depending on the situation, need for incorporation of that trait to a specific situation, a number of tedious methods and techniques will be adopted by the plant breeders to achieve and ensure highest quality seed to the end users. During this time they scrupulously plan and control both pre and post harvest procedures. The same information can be passed on to the agronomists for effective utilization during the subsequent seed multiplication. The very information and knowledge can also be shared with certification agency for taking up succeeding production cycles in the seed supply chain system.

10.2.1.1   Objective

The basic objective of all breeding programmes are aimed at ensuring either high yield directly or indirectly by way of achieving optimum plant stand establishment or population density dynamics, maturity and reaction to various forms and kinds of stresses. The seed vigour is though known to directly exert influence on yield, it was never considered as a breeding objective directly in any of the plant breeding programmes so far. However, the same was given due weightage by both plant breeders and agronomists alike by indirectly banking upon the seedling vigour rating, referring to the capacity of a variety to emerge, establish rapidly and grow vigorously.

1. Several morphological characters like split coleoptile, husk coverage, susceptibility of corn to ear rot and hypocotyl length of soybeans etc. are some of the quantitative traits that have contributed much to the seed vigour.

2. Hard seededness, a typical character of legumes, is associated with maintenance of vigour duly withstanding field weathering after maturation.

3. Seed coat etching, an indicator of direct vigour, associated with large seeded legumes is governed by a single incompletely dominant gene for susceptibility with 25-50 percent penetrance.

4. Existence of genetic differences due to seed coat thickness and seed density lead to identification of lines resistant to mechanical injury, (Barriga, 1961).

5. Similarly, Smith and Millett (1964) identified two tomato varieties with improved germination capacity at 10°C.

6. Seed maturity on the other hand exerts influence indirectly on the seed vigour. For example, early maturing soybean varieties during their maturation are exposed to hot and dry weather culminating in reduced seed quality. This, obviously, necessitates the selection of appropriate locations for seed production as well as date and time of planting and harvesting that can minimize the potential liability of maturity on vigour.

10.2.1.2   Practice in Plant Breeding

1. A plant breeder can initiate selection for seed size in species where the character is associated with high vigour.

2. In case of crops like soybean, selection of genotypes can, be practiced based on morphological characters associated with vigour like hypocotyls growth.

3. Genotypes can also be selected in tune with the maturity under specific environmental conditions for good seed production.

10.2.1.3   Examples

For example, in case of rice, seed vigour has been identified long back as the chief factor responsible for poor germination and seedling growth and establishment, (Ellis, 1992). Even prior to this, Pandey *et al.,* (1989), and Shenoy, (1990) reported that, under rainfed ecosystems, characters like early vigour, drought tolerance and other traits can also be considered for identification of varieties. This is so because, seedling vigour has been found to be associated mostly with seed size and density together with other parameters of germinating seed. Recently, a close relationship between seed vigour and seed yield was reported in West African rice varieties by Okelola, (2005), who reported that, speed of germination index, seedling vigour index, seedling emergence and seedling establishment are the most desirable vigour traits. Similarly, in case of wheat, seed quality parameters like germination per cent, seedling length, seedling dry weight, Vigour Index-1 and II are considered best.

10.2.2 Consumers

When high quality seed is offered for sale in the market, the consumer has nothing to enquire about the tests conducted to ensure the quality. The farmer has to make his judgment only based on the percentage of germination printed on the tag accompanying the bag. This germination percentage is most probably contributing to a false sense of security among the farmers. Assuming a situation, wherein the seed is wrongly placed in a stressful environment or field. At the most the seed may not give germination as expected because as all of us are aware that the standard germination percentage is only a poor predictor of field emergence capacity of a seed lot under suboptimal conditions.

Therefore, for an ultimate consumer, the farmer, it would always be advantageous to know the vigour status of each high germinating seed lots before making any decision as to which one to buy. Because farmers require vigour information to quickly determine the expected rapidity and uniformity of seedling emergence. No doubt, this information will not provide an expected field emergence value, but will only indicate whether a seed lot is of high or low in vigour and therefore indicate which seed lot will be more likely to perform under sowing conditions which provide some form of stress. It means, depending on the level of stress present in his field, a farmer has to select and purchase a seed lot for immediate planting.

10.2.3 Quality Seed Production

Keeping in view the age old saying, 'Prevention is better than cure', application of various vigour tests during different stages of seed production is bound to help the seed producer in identification of stages where actually loss in vigour in specific and seed quality in general is occurring during the production process. This is one of the most important areas in which application of vigour tests shows most immediate impact. The loss in seed vigour could be occurring either when the seed is still intact on the mother plant itself, referred to as weathering, or during harvesting, may be manual or mechanical, or during threshing, again may be manual or mechanical and processing and conditioning including during storage. Seed production companies in order to establish their own 'in house quality control standards' on one hand and also to monitor the seed quality during various phases of seed production and processing on the other hand commonly use vigour tests.

Seeds men require vigour information because pretty well they are aware of the fact that loss in vigour precedes loss in viability and that vigour tests could aid in monitoring seed quality. Application of vigour tests during different stages of seed production, viz., pre sowing to post harvest processing operations will definitely facilitate the seed producers in general and seed certification officers in particular to identify the area(s) where reduction in seed quality is occurring during the production system i.e. whether the loss is occurring either when the seed is still intact on the mother plant, (pre-harvesting stage) referred to as weathering, or during harvesting, may be manual or mechanical, and after maturity i.e., post harvest handling operations like during threshing, processing and conditioning including during storage may be manual or mechanical, McDonald (1975).

In order to minimize reduction in seed vigour loss during various crop growth stages, seed production fields will be subjected to tolerances in respect of planting ratios of female and male lines, border rows, pollen shedders, off type plants, volunteers, inseparable other crop species, other genetic contaminants, quality history of the parental lines used in hybrid seed production, average of seed females, etc.,

10.2.4 Seed Quality Control

Although research considers all characteristics of seed quality, literature shows that the physiological component has received more attention with an emphasis on studies involving the identification of the relationship between seed vigor and seedling emergence and the development of reliable methods for seed vigor evaluation. Seed companies could then take preventive measures by improving specific procedures to guarantee the best levels of seed quality instead of just being satisfied when seed lots only attain the minimum quality standards.

The assessment of seed vigour has several important implications for seed industry. Seed production companies in order to establish their own 'in house quality control standards' on one hand and also to monitor the seed quality during various phases of seed production and processing on the other hand, commonly use vigour tests. Perhaps, this is one of the most important areas in which application of vigour tests shows most immediate impact. Seeds men require vigour information because pretty well they are aware of the fact that loss in vigour precedes loss in viability and that vigour tests could aid in monitoring seed quality. Application of vigour tests during different stages of seed production, viz., pre sowing to post harvest processing operations will definitely facilitate the seed producers in general and seed certification officers in particular to identify the area(s) where reduction in seed quality is occurring during the production system i.e. whether the loss is occurring either when the seed is still intact on the mother plant, (pre-harvesting stage) referred to as weathering, or during harvesting, may be manual or mechanical, and after maturity i.e., post harvest handling operations like during threshing, processing and conditioning including during storage may be manual or mechanical, McDonald (1975).

In order to ensure the delivery of highest quality seed to the end users, postharvest processing procedures such as maintenance of appropriate duration of drying time, drying temperature, rate of drying for each entry needed to be managed carefully so as to derive maximum benefit out of standard germination ability and seed vigour. In addition to this, a perfect and conservative population dynamics is adopted to encourage development and production of healthy, robust and high-density seed. For this purpose, tolerances or standards have already been prescribed for different classes of seed offered for certification. Threshold levels of insects and incidence of diseases will also be properly monitored and managed from time to time.

In order to maintain the precise integrity and identity of the seed lot, they are tested for quality parameters in accordance with the standards prescribed for that crop keeping in view the policy of the company together with the seed laws prevailing in that area. May be it becomes necessary to print the information pertaining to germination percentage on the labels to be accompanied with the bags as mandated by Law. But it is not necessary for vigour test results. Because, according to AOSA, (1983), the referee testing of seed lots for vigour among different seed testing laboratories are not sufficiently reproducible within acceptable limits. Further, not even a single vigour test is available that could meet the broad-spectrum requirements of the seed industry.

Contrary to this, in order to meet the in-house quality control programmes, different public/private companies engaged in seed production is obviously undertaking some form of vigour testing so as to allow the seed lots for distribution into the commercial market. Thus, it definitely involves the preventive measures taken up by the company by not offering the low vigour seed for sale or else it may suggest suitable corrective measures for its seed storehouse managers for maintaining the potential conditions required by the seed lots during their storage. In other words, the companies cannot afford to store seed lots indefinitely as they are likely to lose their vigour rapidly at any time and thereby lose their ability to germinate during storage itself. Two probable situations prevail upon are, the seed companies have to either opt for better storage conditions or else store the seed lots with better storage potential, either way it only adds up to the cost of maintenance.

Further, vigour tests offer a clear area for application by identifying critical limitations during seed production and thereby aims at improving the quality of seeds. Seed companies engaged in seed production and marketing may wish to determine and use vigour information in the improvement of seed quality through the seed treatments such as priming. The vigour information pertaining to the initial seed quality becomes absolutely necessary for the seed producing companies to take a decision on whether or not the priming treatment of a seed lot is worthwhile economically. Treatment of aged seed or low vigour seed exerts greater impact on subsequent seed performance during storage.

Under the present situation, Cold Test is the most commonly used method for in-house quality control by several seed testing laboratories across Europe and Americas. Further, the standards and test methods prescribed vary significantly with seed supplying companies. Therefore, as pointed out earlier, since it is not possible to estimate the vigour with a single test for all crops, different companies, obviously, in tune with their own internal policies often combine standard germination test results with vigour test results. At times, when freeze / mechanical injury is suspected, usually tetrazolium test is used as a diagnostic test regularly by the members of the seed industry for assured supply of high quality of seed to the farmers or the end users of the seed. May be it is not an exaggeration to say that even today, majority of seed growers are not aware of the fact that vigour tests that are conducted are only a part of in-house quality assurance programme of the seed production companies.

The very technical information and knowledge generated by Plant Breeders and Agronomists from research when shared with Seed Certification Agencies and Seed Producers for taking up succeeding foundation and certified production cycles in the seed supply chain system. The seed production fields are also needed to be subjected to tolerances with respect to planting ratios of female and male lines, border rows, pollen shedders, off type plants, volunteers, inseparable other crop species, other genetic contaminants, quality history of the parental lines used in hybrid seed production average yielding ability of seed female parent or seed parent, etc., during various crop phonological growth stages so as to minimize reduction in seed vigour loss. In addition to this, threshold levels in respect of insects and incidence of diseases will also be properly monitored and managed from time to time.

10.2.5 Seed Storage

The storage potential of a seed lot is related the vigour status or in other words the stage of deterioration at the time of entering the storehouse. If a particular storage environment exerts any kind / form of stress like changes in temperature, relative humidity etc., the seed lots with high vigour obviously withstand such stress and will decline in quality at a slower rate than low vigour seed lots. Even after storage under controlled conditions, i.e., low temperature and low seed moisture, post storage performance of seed lot is dependent on the vigour status of seed only. Therefore, the information regarding the conditions of storage and duration of storage also needs to be taken into consideration. Powell et al., (2000) reported that both Polyethylene glycol (PEG) and aerated treatments have enhanced the longevity of low vigour seeds where as the longevity of high vigour seeds were reduced markedly after the treatment.

The seed storage managers may use vigour test results to make better knowledgeable decisions pertaining to the suitability of seed lots for storage, the possible length of storage and storage conditions required. Similarly seed exporters can also use this information to decide which seed lots can withstand the rigidities of transport and thus be expected to arrive in the port of importing country with the quality unimpaired.

10.2.6 Seed Marketing and Promotion

The need to meet the day-to-day requirements of the fast emerging consumer demands for uniform sized products from supermarkets have obviously created a necessity among the transplant producers and vegetable growers alike to demand information from seed companies regarding the quality of seed offered for sale. Since, only a high vigour seed is capable of meeting the requirements that the seed germinates uniformly and rapidly will produce a crop that can meet the requirements of these supermarket chains. Contrary to this, the seeds with low vigour can only give a slow and asynchronous germination and ultimately may not be in a position to meet the requirements of the market.

Seed companies may use vigour results in their in-house quality control and thereby not only prevent the sale of low vigour seed automatically, but also facilitate formulation of guidelines pertaining to storage potential of seed lots. As a consequence, the seed companies involved in production and marketing of the seed can avoid unnecessary storage of seeds that will rapidly lose their vigour. On the other hand, this will also facilitate the consumers to select the seed for storage that have good storage potential.

In reality, nowhere, the seed is either marketed or promoted based on the results of the vigour tests. However big or multinational may be the origin of the companies, they try to promote the sale based on 'High Quality Seed" without specifically referring to the type of test(s) conducted and the results achieved. Once, even the American Seed Trade Association has advised against the use of seed vigour tests until they are standardized and the use of vigour test has been resolved. According to Hampton and Coolbear (1990), though this caution is understandable, seed growers will demand more as the markets become more competitive. This leads to demand for more vigour test information protocols. Contrary to this several growers may not be aware that vigour tests are conducted as a part of the quality control program. Probably this could be attributed to the quantitative differences exhibited between different seed lots. According to AOSA, 1983, the quantitative test results are not reproducible when compared between different laboratories because, different laboratories, for obvious reasons, use a variety of methods for the same test for want of logistics and preference reasons ultimately contributing only towards variability in the test results.

Under the present situation, Cold Test is the most commonly used method adopted for in-house quality control by several seed companies and seed testing laboratories across Europe and America. Further, the standards and test methods prescribed vary significantly with seed supplying companies. Therefore, as pointed out earlier, since it is not possible to estimate the vigour with a single test for all crops, different companies, obviously, in tune with their own internal policies often combine standard germination test results with vigour test results. At times, when freeze / mechanical injury is suspected, usually tetrazolium test is used as a diagnostic test regularly by the members of the seed industry for assured supply of high quality of seed to the farmers or the end users of the seed. May be it is not an exaggeration to say that even today, majority of seed growers are not aware of the fact that vigour tests that are conducted are only a part of in-house quality assurance programme of the seed production companies.

In a good majority of crop species, seed germination and early seedling growth are considered as most sensitive especially under stress conditions. When the stress is due to salinity, it delays the onset of germination, reduces the rate of germination, and increases the dispersion events of germination leading to reduction in plant growth and ultimately loss in final yield. The consequential adverse affects due to salt stress can be minimized by pre –sowing seed treatments like priming. Seeds are partially allowed to hydrate with restricted amounts of water to a point where the process of germination has just begun but not completed and the seeds are dried back before using. This facilitates rapid and uniform emergence.

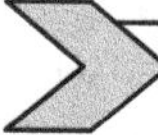

10.3 Plug Production in Nurseries for Transplant Production Systems

Sprout the seeds in trays and transplant the fragile seedlings into larger packs or pots is the traditional way to raise a large number of seedlings. Transplants are widely used in commercial production systems. This method is not only labour intensive, but also results in considerable mortality due to root loss of transplant shock. In order to minimize such losses plug trays are gaining prominence since 1980s for raising seedlings of annuals. A plug is a containerized transplant with self –enclosed root system and offers several advantages such as (i) growing quality plants under protection before setting out in the field, (ii) more efficient use of expensive seeds, (iii) providing optimal conditions for plant growth that is not possible in the field, (iv) facilitates growing of specific varieties that are not locally available, (v) eliminates inport of diseases that are not locally available, (v) ensures ideal plant population, in the field that may be difficult to achieve in the field from seeding (vi) less time and labour requirement for transplanting, (vii) reduced root loss, (viii) more uniform growth, (ix) faster crop establishment, and (x) increased production. On the other hand, this system demands more attention towards cultural practices, increased mechanization and additional well trained worker force.

Fig. 10.1 Plug production system

This practice obviously places a high reliance on high quality of seeds for maximal seedling emergence and uniformity. Specialization has led to increased capital investment in modern greenhouses, automated seeders and sophisticated transplanting robots (Styer and Koranski, 1997). This has challenged the seed industry to provide seeds that perform under these demanding production systems (Karlovich, 1998).

In most of the horticultural crops, barring certain root vegetables such as radishes, carrots, beets etc., raising seedlings for transplanting is extensively practiced around the world. The relatively high initial cost of horticultural crop seeds has led the growers to employ precision seeding and transplant production systems to maximize seedling stands. In several parts of the world, the practice of transplanting is extensively adopted in crops like rice and sweet corn. However, individual growers on their own farms raise a large proportion of the transplanted rice and maize. Now-a-days, commercial production of seedlings has been rapidly increasing in several countries. The relatively initial high cost of flower and vegetable seeds led the seed growers to use precision seeding and transplant production systems for maximizing seedling stands, McDonald and Kwong (2005). In order to derive maximum efficiency in plug production, optimum seedling emergence and uniformity demands the use of high quality seeds. The germination tests used to evaluate the production of normal seedlings under optimal conditions, Geneve, (2008) often do not always reflect the emergence potential of a seed lot under greenhouse or field conditions. Similarly conditions in the transplant production environment are also not always ideal and are likely to vary depending on the season of sowing like winter planting or summer planting. The vigour tests help to determine the failures in emergence and stand establishment and identify the levels of physiological aging of a seed lot and its potential for the development of normal seedlings under a wide range of sowing conditions, McDonald, (1975).

According to Lee, (2003), Lee and Oda, (2003), plug seedlings are typical examples of commercial transplants. The use of a sizable proportion of the grafted seedlings in vegetable production though occurred as early as 1930`s in Japan for watermelon it`s commercial application did not happen until 1960`s (Sataka *et al*, 2008). Grafting is a method of propagation where two pieces of living plant tissues are joined together to develop as a single plant. This facilitates serious crop losses caused by soil borne diseases caused by successive cropping. It is an ecofriendly technique for sustainable vegetable production using resistant root stocks and reduces dependence on agrochemicals. Revard *et al.,* 2008, Pradeep Kumar, *et al.,* 2015 and induces resistance against low and high temperature. Venema, 2008.   A sizable proportion of the grafted seedlings are routinely produced in cell trays. Even rice seedlings are produced in modern greenhouses for healthy and vigorous transplants for mechanized planting. In some cold countries, even sweet corn seeds are commonly sown in green houses till the seedlings attain a height of 30-40 cm at which stage these seedlings are transplanted in the main field to avoid the danger from early spring frost.

Use of transplants facilitates (i) intimate protection to seedlings from climatic and biological stress, (ii) permits early harvest and ultimately enhancement in yield, (iii) facilitates better root system development, (iv) promotes or prevents flower bud differentiation and germination rate, (v) enhanced opportunities for grafting operations, (vi) reduction in labour wages during early crop growth stages, (viii) Warm season crops can be grown in areas with too cold with direct seeding. (viii) Enhanced and efficient land use.

On the other hand transplants are most like to be subjected to premature bolting, decreased drought resistance, delayed fruit formation arising out of transplantation shock, formation and development of abnormal fruits, early senescence of seedlings caused by delay in transplanting. The quality of seedlings gets influenced by a number of factors both directly and indirectly as follows. (i) Seeds meant for transplanting should possess good genetic traits in order to ensure high purity, (ii) they should be fully ripe and free from mechanical defects and diseases including seed borne insects and viruses and (iii) capable of demonstrating high germination percentage, germination vigour in order to produce uniform and vigorous seedlings in shortest possible period.

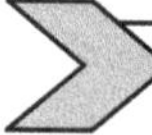

10.4 Uses of Seed Vigour Information

Seed vigour information is useful in several ways as listed below.

1. Determination of the true value of a seed lot is the most underlying reason for vigour testing.

2. By employing seed vigour information in regular plant breeding programmes, cultivars with improved seed performance can be developed.

3. Seed producers use vigour data to monitor seed quality during different stages of production like harvesting, drying, conditioning and storage.

4. Vigour information is useful especially in International Seed Trade and

5. Vigour test information is also useful in inventory management.

6. Vigour test information is mostly useful in carryover decisions.

10.5 Usefulness of Seed Vigour Tests

The usefulness of vigour tests can be briefly summarized as follows.

1. Seed vigour tests are useful in ranking seed lots for their physiological quality and to provide an indication of seed deterioration, well before germination test results.

2. Seed Vigour tests are used to identify seed lots that exhibit acceptable germination test results but are unlikely to store well or perform well under suboptimal conditions.

3. Seed Vigour test results provide information for future, which can be used in planning inventory, carryover seed and marketing strategies, distribution and sowing of seed.

4. Seed Vigour tests provide information, which can be used by the seed industry to answer customer inquiries about the seed lot performance or to prevent litigation.

5. Seed Vigour tests are great tools in the hands of seed companies for in-house quality control.

6. Routine vigor tests help a lot in making quality assurance decisions during production, conditioning, storage and marketing. The physical and physiological quality of seed may get damaged early even during the process of production. Drought during flowering and grain filling or extreme moisture during grain filling, improper drying and storage conditions, mechanical damage during processing and conditioning reduces germinability.

7. Seed vigour tests play an important role in crop improvement programmes of public and private sector research organizations aimed at determining and eliminating the inbuilt weaknesses of elite breeding material and thereby facilitates selection of superior lines with better capacity of emergence of seedlings even under stressful field   conditions.

8. Seed vigour tests for agri-horticultural crop seeds become necessary to comply with State / National and International laws dealing with seed trading.

10.6   Limitations of Seed Vigour Tests

1. It is not possible to prove with sufficient evidence that a strong relationship exists between seed vigour and yield in the absence of stand differences.

2. The values obtained in seed vigour tests are not absolute values, they are only relative and as such it is very difficult to compare the results of vigour tests at least in certain cases, because the results are often expressed in different units.

3. The services of trained and experienced analysts are a must not only for interpreting the results but also for educating the seed producers as well as seed consumers on seed vigour.

4. Neither the seed vigour tests nor the germination test predict the percentage field emergence.

5. Standard cut off points for acceptable and non-acceptable levels of vigour have been identified and recommended for only few tests like Conductivity test for peas, therefore the same is required for other frequently used tests also.

6. The compensated seed rates can be used in crops grown in nurseries like rice and some vegetable crops.

7. Use of low vigour seeds in direct sown crops may not be in a position to maintain uniform plant stand on par with the high vigour seed.

8. Published literature is confined to studies focusing only on the relationship between laboratory germination and field emergence vigour, *i.e.,* information on other aspects of crop performance is scanty.

9. While standardizing vigour tests acceptable level of vigour for different production seasons, crops, and genotypes has to be worked out.

10. In certain crops and genotypes where the inherent capacity to compensate minor differences in plant stand following emergence with little less final yield still needs to be exploited.

11. The ability of high vigour seeds to perform well in adverse soil environment is confined only to a limited range, *i.e.,* in case when the post planting conditions become too severe, there is no guarantee that even the most vigorous seed gives satisfactory germination.

12. Referee testing among seed labs only can do the precision and standardization of seed vigour testing methods.

13. Even in the developed and under developed countries, where the majority of farmers are illiterate, educating the farmers regarding vigour may not be of any major help.

14. A good majority of vigour tests needs to be conducted on non-toxic substrates like filter paper either alone or in enclosed petri-plates.

10.7 Precautions to be taken while conducting Seed Vigour Tests

1. More care and precision is required to be maintained both during planning and execution of vigour tests, when compared to a standard germination test.

2. Fluctuations in temperature and moisture in respect of both the substrate as well as test chamber results in variation in biochemical processes, thereby affects seedling growth rate, on which usually the vigour is estimated.

3. The precession of the equipment should be verified and checked at regular and frequent intervals while conducting the seed vigour tests, because even $\pm$ 1°C difference in temperature considerably affects the seedling growth test or deterioration of seed in the accelerated ageing test and the same holds good with respect to other parameters like relative humidity and seed moisture etc.

4. At times, dormancy is suspected to interfere with the interpretation of vigour test results; this necessitates verification or conformation with the standard tetrazolium test for viability. The variation in field emergence of high germinating seed lots is low under favourable environmental and soil conditions. Contrary to this under unfavourable soil and environmental conditions, the variation becomes glaring and more significant.

5. Therefore, planting high vigour seeds in the fields with adverse conditions will only result in high   field emergence compared to low vigour seed and hence necessitates gap filling or associated delayed maturity. This ultimately results in the reduction of yield and thereby poor seed standards.

6. Seed ageing process and Agro-climatic condition needs to be taken into consideration simultaneously for better realization of vigour results. It is always better to select a suitable test, viz., Accelerated Ageing Test (AA) or Controlled Deterioration Test (CD) for tropics and subtropics; Cold test for temperate areas; PEG test for drought prone areas; and NaCl test for crops grown on saline soils etc., may be adopted for obtaining better results of seed vigour.

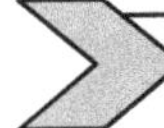 ## 10.8 Strategies to Improve Seed Vigour

First of all we have to accept the fact that the mechanisms underlying seed vigour are very poorly understood. With the limited knowledge, the approaches proposed or to be proposed in the future for bringing improvement in this crop trait is a challenging task.

1. **Ensure better maternal environment:** It has long been proved that the maternal environment in which the seed develops exerts greater impact on the subsequent generations of multiplication. First step in this direction will be to carefully select the areas for seed production based on their beneficial climatic requirements. In the second step, try to clearly demonstrate and establish a potential link between the maternal environment and the seed quality. Role of seed coat pigments in the control of dormancy and germination has been demonstrated by Debeaujion *et al.,* (2000). The existence of positive correlation between the depth of dormancy and quantitative expression of DOG1 QTL was demonstrated by Chiang et al (2011) and Kendell *et al.,* (2011). Similarly, the role of flowering time regulator was proposed by Chen *et al.,* (2014).The impact of maternal temperatures on the flavonoid production was demonstrated in mature seed coat by MacGregor *et al* .,(2015). Permeability of seed coat in soybean is under the control of a single transmembrane protein GmHa1-1 and concluded that during domestication of soybean, selection was practiced for high permeability of seed coat as quality property. Sun *et al., (*2015). Despite intangible mode of action of DOG 1, presence of this gene/protein in abundance and maternal control of dormancy still remains unknown. Dekkers and Bentsink, (2015).

2. **Ensure optimum time of harvest:** Seed development is not uniform with in a panicle/ spike/ inflorescence and hence seed bag or seed lot harvested at a time in several species, be it the members of Graminea or Brassicaceae or Umbelliferae. The individual seeds differ in their vigour and other characteristics and result in heterogeneity. Those seeds formed from early fertilization may be completely matured and ready for shattering and even begin to deteriorate while some others are under developed, developing and in different developmental stages that need to be sorted out by seed companies duly employing different strategies. Copeland and McDonald, 2001; Bewley et al., (2013). One can understand the shedding of seeds of wild species at different times as and when they mature, despite getting exposed to different environmental conditions during development. In several plant species this adaptive behavior could be attributed to the selection of characters during domestication with a view to derive maximum yield. Contrary to this in certain species plants senesce fully before harvest duly minimizing the ill effects of different developmental stages. Hence one should be very cautious while deciding the optimum time of harvesting for advancing the seed material for succeeding generation, which varies between different species. Deterioration in seeds starts once they attain physiological maturity both on and off the plant. Hence it is advised to harvest the commercial seed material just before harvest maturity and dry rapidly not only to maintain initial quality but also to reduce pre- harvest maturity deterioration. Bewley et al., (2013).   The enhancement in seed vigour that started consequent to detachment from the mother plant will get terminated owing to uninterrupted loss of moisture from seed during the course of seed drying.   Harvesting the seeds before harvest maturity and holding them under optimum conditions avoiding drying enhances seed lot uniformity by allowing less developed seeds to complete their development. In addition to this, a number of

techniques such as size and density grading during processing to improve seeds physiological performance were developed, Halmer, (2004).

3. **Ensure Continued Seed Development:** *Prunus avium*, a wild species domesticated for its cherry fruits. The seed being deeply dormant, exhibits differential performance between seeds due to either areas of production, extent of seed development, or age of mother plant, Finch – Savage et al., (2002). Though cherry fruits mature at different times on the mother plant, for practical reasons they are harvested at a time duly leaving several fruits as immature. Among these, matured seeds respond to simple dormancy breaking treatments, while those that are immature demand complex cold and wet temperatures treatments lasting for 26 weeks representing winter and summer seasons, Suszka *et al .,* (1996).

In order to overcome this problem, seeds after cleaning are prevented from drying by storing in wet sand at 15 ^{0}C for 8 weeks that facilitate aeration. This treatment renders characteristic development and maturity of all seeds, reduces the differences between individual seeds as well as seed lots by progressively becoming homogeneous and there by responds positively to dormancy breaking methods, Finch-Saavage- *et al.,* (2002).

4. **Ensure continued physiological development:** Several method were advocated time and again to improve physiological seed performance such as seed priming, Heydecker and Coolbear, (1977); repeated wetting and drying of seed, Hegarty, (1978); Short term imbibition for allowing repair of ageing, Wlaters (1998), Powell et al., (2000); and manipulation of imbibition, Halmer, (2004). All these methods involve moist seed treatment after drying. For example the priming techniques, limit the water sufficient only to progress metabolism but insufficient enough for radicle extension and completion of germination, duly leaving seeds desiccation tolerant. Osmotic priming of Heydecker *et al.,* (1975); ABA Priming of Finch-savage and Mc Quistan (1991); Solid matrix priming of Taylor *et al.,* (1988); Aerated osmoticum of Bujalski and Nienow (1991); Drum priming of Rowse and McKee, (1999), Membrane priming of Rowse *et al.,* (2001), In all these cases water is applied in small quantity to shorten the lag phase on re-imbibition so as to have uniform and more rapid germination. The mechanism of priming though not fully under stood, partly result in repair, Powell et al., (2000) or cell wall elasticity suggesting re-modifying gene expression, Karssen et al., (1989). Typically priming treatment is used not only to overcome specific dormancy issues, but also carried out in dried, high value vegetable/ornamental flower seeds with heterogeneous seed development (containing under developed seeds). Since poor quality aged seeds fail to respond to priming compared good quality fresh seeds hence it is not economical to use the technique to reverse the deterioration caused by ageing. Imbibitional damage is usually associated with planting in soils with low temperature and hence can be alleviated with controlled hydration rates, Taylor et al., (1992).

5. **Pre-germinated seed treatments:** In order to further exploit seed performance, opportunities such as fluid drilling of Gray, (1981) or dry the seed before desiccation tolerance is lost , Finch –Savage (1989); reintroduction of desiccation tolerance Bruggink and van der Toorn,(1997) etc., facilitates selection of 100% viable seeds or most vigorous.

6. **Synthetic seeds**: Somatic embryogenesis offers great potential for the production of superior performing seeds by direct genetic manipulation of embryo. As on date the technology is not fully developed or economical.

7. **Conventional Breeding**: A diversity of genetic backgrounds may have the capacity to be vigorous. QTL analysis is one approach that not only cracks the genetic potential but also realizes the performance of plant in the field. The potential genetic improvements that affect seed performance include hard-seediness in beans, mechanical damage in navy beans, (Copeland and Mc Donald,(2001), modification of seed coats to alter water uptake and minimize the negative impact of imbibitional damage on seed vigour, Powell and Matthews (2012).

8. **Genetic manipulation**: Transgenic approaches can be used to improve seed vigour. Targeted alteration of a positive ABA signalling factor, or negatively acting GA signalling factor results in the more rapid germination of seeds within stress conditions. Seed vigour at deeper planting depths has been reported, Amram et al. (2015).

 ## 10.9 Summary

The processess of ageing, imbibition and their varied interactions leads to loss of seed vigour and has direct relevance in four broad areas, viz., plant breeding; seed production and quality control; marketing including quality enhancement; storage and handling including exposrt and import. The reliable vigour information generated can be utilized by plant breeders and seed growers for planning and executing production and by commercial crop growers; store house and marketing managers; exporters and importers; and wholesalers and retailors as a predictive tool.

Polygenic control of several traits facilitates plant breeders to improve seed vigour traits directly and indirectly into varieties they plan to develop. It is the genetic component of the seed that determines the rate of emergence, speed of growth of the phenotype and ultimately its yield. Though different varieties of same crop exhibit variation for seed vigour and is known to influence yield directly but was never considered as a breeding objective. But both breeders and agronomists used vigour parameters indirectly through measuting seedling vigour rating, capacity to establish rapidly and gorw vigorously. Seedling vigour is associated with seed size and parameters of germination. Seedling vigour index, seedling emergence index and seedling establishment index are most desirable vigour traits. Split coleoptile, husk coverage, hypocotyl length, hardseededness, seed coat etching, seed coat thickness, and maturity parameters exert direct influence on seed vigour.

Application of vigour tests during seed production facilitates identification of crop growth stages where loss in vigour is occuring. The massive amount of data generated during the course of identification of strains when passed on to agronomists and certification agencies together with appropriate pre and post harvest procdures determins the successful quality seed production and supply with maximum vigour. For ensuring quality, Standards together with tolerances have been identified and prescribed for ensuring quality.

This also facilitates identification of critical limits for further improvement of seed quality through enhancement techniques. Information on initial seed quality is essential for determining the economic worthiness of priming treatment. Printing of vigour data on seed lables is not mandatory but still several public/ private seed companies engaged in seed production obviously adopt inhouse quality control measures so as to take up timely corrective measures. Different seed companies in tune with their internal policies, conduct more than one vigour test and combine the results. Since it is not possible to store the seeds for a long time even in storage because it adds up only to maintenance costs. So, the managers of storage houses, export/import house, whole sale/ retail store can decide on which lot can withstand the rigours of transport and pushed in to market on priority basis without unimpairment of quality. Now where, the seed is marketed based on the vigour test results. Even the multinational companies are offering the seed for sale in the name of "High Quality" seed without making a mention of the type of the tests conducted or results achieved. A farmer has nothing to enquire when high quality is offered for sale. He has to ensure his judgement based on germination percentage only, which is a poor predictor of field emergence under suboptimal conditions.

 ## 10.10 Exercise

I. Answer the following questions.

1. List out the major areas and players engaged in using of vigour test reslts?

2. How the information generated by breeders are useful to agronomists and Seed Certification officials?

3. Breeding stocks exhibit considerable amount of variability. But vigour was never a breeding objective. Why?

4. Though seed vigour is known to influence yield directly, the breeders and agronomists have banked only on indirect selection methods. Why?

5. In which way application of vigour tests during seed production helps seed producers?

6. List out pre and post harvest procedurs required for delivering high quality seed?

7. Initial seed quality information is an absolute necessary for seed companies. Why?

8. Why public/ private seed companies engaged in seed production are using some form of vigour tests as inhouse quality conrol measures?

9. How vigour information is usefl to storehouse managers, exporters, wholesale traders and retail seed traders?

10. What is your reaction as a consumer when high quality seed is offered for sale?

11. Guess what will happen to a vigour tested seed placed in a stressful situation?

12. List out the tolerances adopted in seed prodution fields to minimize seed vigour?

13. Cultivars with improved seed performance canbe developed by using seed vigour tests in regualr breeding programmes. How?

14. List different stages of production where seed producers use vigour data?

15. Explain how vigour information is helpul in international seed marketing?

16. Summarise the limitations of seed vigour testing?

17. Suggest precautionary measures to be taken while conducting vigour tests?

II. Fill in the blanks with suitable words.

1. The most underlying reason for vigour testing of seed lots is determination of ----------.

2. Vigour data is useful to seed producers during different stages of seed production to monitor ----------.

3. Vigour information is useful to inventory management and ---------- decissions.

4. Standard germination percentage is a poor predictor of ----------.

5. No where, the seed is promoted or marketed on the basis of ---------- results.

6. The twinprocesses governing the applications of seed vigour are ---------- and ----------.

7. Multinational companies are offerring the seed for sale in the name of ---------- seeds.

8. The quantative trait of soybean contributed much to seed vigour is length of ----------.

9. Seed Coat etching indicator of seed vigour is associated with large seeded ----------.

10. ---------- is associated with weathering of legumes even after maturation.

11. Seedsmen require vigour information because they knew that loss in ---------- preceeds loss in ----------.

12. Agronomists and seed certification staff are expected to adopt more stringent ---------- limits in seed production fields than grain production fields.

13. ---------- the most commonly used method for inhouse quality control in most of the seed companies

14. A high vigour seed is capable of meeting the requirements of ---------- and ---------- of germination.

15. Storage potential of a seed lot is rlated to ---------- of the seed lot at the time of entry in to store house.

Answers

1. True value,
2. Seed quality.
3. Carryover,
4. Field emergence,
5. Vigour test,
6. ageing and imbibition,
7. High – quality,
8. hypocotyl,
9. legumes,
10. Hardseededness,
11. Vigour, germination.
12. Tolerances,
13. Cold test,
14. Rapid and uniform,
15. Vigour status.

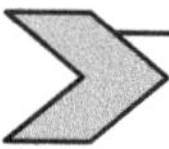 ## 10.11 References

Akbar, M., (1989). Genetic control of Embryo size and Implications of that trait on seed vigour in rice. IRRI, Saturday seminar, June 17, 1989, IRRI, Los Banos, Philippines.

Amram A, Fadida-Myers A, Golan G, Nashef K, Ben-David R, Peleg Z. 2015. Effect of GA-sensitivity on wheat early vigor and yield components under deep sowing. *Frontiers in Plant Science* 6.

AOSA (1983). Seed Vigour testing Hand Book. Contribution No. 32 to the Hand Book of Seed Testing. (ed.) Clark, B.E.., McDonald, M.B., and Joo, P.K., pp 88., Lincoln. NE. AOSA.

Barriga, C., (1961). Effects of mechanical abuse of Navy bean seed at various moisture levels. Agronomy Journal. 53: 250-251.

Bewley JD, Bradford KJ, Hilhorst HWM, Nonogaki H. 2013. *Seeds: Physiology of development, germination and dormancy*. New York, Springer.

Bruggink T, van der Toorn P. 1997. Induction of desiccation tolerance in germinated *Impatiens* seeds enables their practical use. In Ellis RH, Black M, Murdock AJ, Hong TD (eds) *Basic and Applied Aspects of Seed Biology. Pp 461-467. Kluwer, Dordrecht.*

Bujalski W, Nienow AW. 1991. Large scale osmotic prioming of onion seeds: A comparison of different strategies for oxygenation. *Scientia Horticulturae* 46, 13-24.

Chen M, MacGregor DR, Dave A, Florance H, Moore K, Paszkiewicz K, Smirnoff N, Graham IA, Penfield S. 2014. Maternal temperature history activates Flowering Locus T in fruits to control progeny dormancy according to time of year. *Proceedings of the National Academy of Sciences of the United States of America* 111, 18787-18792.

Chiang GC, Bartsch M, Barua D, Nakabayashi K, Debieu M, Kronholm I, Koornneef M, Soppe WJ, Donohue K, De Meaux J. 2011. DOG1 expression is predicted by the seed- maturation environment and contributes to geographical variation in germination in Arabidopsis thaliana. *Molecular Ecology* 20, 3336-3349.

Copeland LO, McDonald MB. 2001. *Principles of seed science and technology*. 4th edition, Massachusetts: Kluwer Academic Publishers.

Debeaujon I, Leon-Kloosterziel KM, Koornneef M. 2000. Influence of the testa on seed dormancy, germination, and longevity in Arabidopsis. Plant Physiology 122, 403-414.

Dekkers BJW, and Bentsink L. 2015. Regulation of seed dormancy by abscisic acid and Delay of Germination. 1. Seed Science Research 25, 82-98.

Dornbos, D. L., (1995). Seed Vigour. In AS Basra (ed) Seed Quality basic mechanisms and agricultural implications. Pp 45-80. Food Products Press. New York.

Ellis, R.H. and Pieta Filho, C., (1992). The development of seed quality in spring and winter cultivars of barley and wheat. Seed Science research 2: pp 9-15.

Ferguson Spears, J., (1995). Introduction to seed vigour testing. In H A Vande Venter (ed) Seed Vigour Testing seminar. Pp 1-9. ISTA. Zurich.

Finch-Savage WE, McQuistan CI. 1991. Abscisic acid: an alternative priming medium for tomato seeds. *Seed Science and Technology* 19, 537-544.

Finch-Savage, W. E. 1989. Seed treatment. *UK Patent* no. 2177488B

Finch-Savage,F W.E, Clay, H.A, and Dent, K.C. (2002) Seed maturity affects the uniformity of cherry (Prunus avium L.) seed response to dormancy-breaking treatments. *Seed Science and Technology* 30, 483-497.

Geneve, R.L (2008). Vigor testing in small-seeded horticultural crops. Acta Hort. 782: 77-82. Gray D. 1981. Fluid drilling of vegetable seeds. *Horticultural Reviews* 3, 1-27.

Halmer P. 2004. Methods to improve seed performance in the field. In Benech-Arnold RL, Sánchez RA, eds *Handbook of Seed physiology: Applications to Agriculture*, New York: Haworth Press 125-166.

Hampton, J.G., and Coolbear, P. (1990). Potential versus actual seed performances- Can vigour testing provide an answer? Seed Science and Technology 18: 215-228.

Hegarty TW. 1978. The physiology of seed hydration and dehydration, and the relation between water stress and control of germination: A review. *Plant Cell and Environment* 1, 101-119.

Heydecker W, Coolbear P. 1977. Seed treatments for improved performance – survey and attempted prognosis. *Seed Science and Technology* 5, 353-425.

Heydecker W, Higgins J, Turner YJ. 1975. Invigoration of seeds? *Seed Science and Technology* 3, 881-888.

Jianhua, Z., Mc.Donald, M.B. (1996). The saturated salt accelerated aging tests for small- seeded crops. Seed Sci. Technol. 25: 123-131.

Karlovich, P.T. (1998). Flower seed testing and reporting needs of the professional grower. Seed Technol. 20: 131-13.

Karssen CM, Haigh A, van der Toorn P, Weges R. 1989. Physiological mechanisms involved in seed priming. In Taylorson RB ed. *Recent advances in the development and germination of seeds.* New York Plenum Press pp. 269-280.

Kendall SL, Hellwege A, Marriot P, Whalley C, Graham IA, Penfield S. 2011. Induction of Dormancy in Arabidopsis Summer Annuals Requires Parallel Regulation of DOG1 and Hormone Metabolism by Low Temperature and CBF Transcription Factors. *Plant Cell* 23, 2568-2580.

Lee, J. M. and M. Oda, (2003). "Grafting of herbaceous vegetables and ornamental crops", *Hort. Rev.* 28: pp. 61-124.

Lee, J. M., (2003). "Advances in vegetable grafting", *Chronica Horticulturae* 43(2): pp. 13-19. MacGregor DR, Kendall SL, Florance H, Fedi F, Moore K, Paszkiewicz K, Smirnoff N,

Penfield S. 2015. Seed production temperature regulation of primary dormancy occurs through control of seed coat phenylpropanoid metabolism. *New Phytologist* 205, 642-652.

Matthews, S., Khajeh-Hosseini, M. (2006). Mean germination time as an indicator of emergence performance in soil of seed lots of maize (Zea mays) Seed Sci. Technol. 34:339-347

McDonald, M.B. *(*1975*).* A review and evaluation of seed vigor tests. *Proceedings of the Association of Official Seed Analysts.* 65*:*108-139.

McDonald, M.B., Kwong, F.Y. *(*2005*).* Flower seeds: Biology and technology *(*CABI*)* Publishing, Cambridge, MA.

Okelola, F.S., (2005). Variation and relationship between seed vigour and seed yield of west african rice (Oryza sativaL). Dissertation, Dept. plant breeding and Seed Technology, University of Agriculture, Abeo- Kuta, Nigeria, pp 89.

Okelola, F.S., Adebisi, M.A., Kehinde, O.B., Ajala, M.O., (2007). Genotypic and phenotypic variability for seed vigour traits and seed yield in west african rice. J. American Sci. 3(3): 34-41.

Pandey, P.K., R.D. Goyal, V. Parakash, R.P. Katiyar and C.B. Singh, (1990). Association between laboratory vigour tests and field emergence in cucurbits. Seed Res., 18: 40-43.

Potts, H.C., Duangpatra, J.D., Hairston, W.G., Delouche, DC, (1978). Some influence of Hardseededness on soybean quality. Crop Science 18: 221-224.

Powell AA, Matthews S. 2012. Seed Aging/Repair Hypothesis Leads to New Testing Methods. *Seed Technology* 34, 15-25

Powell, A.A., Yule, L.J., Jing, H.C., Groot, SPC., Bino, R.Y., and Pitchard, HW, (2000). The influence of aerated hydration seed treatment on seed longevity as assessed by the viability equations. J. Expt. Bot. 51: 2031-2043.

PradeepKumar, Shivani Rana, Parveen Sharma and Viplove Neg: (2015). Vegetable grafting, a boon to vegetable growers to combat biotic and abiotic stresses. Himachal Journal of Agri. Research. 41(1): pp- 1-5.

Revard, C.L and Louws, F.J. (2008). Grafting to manage soil borne diseases in Heirloom Tomato production. Hort Science. 43:2104-2111.

Rowse HR, McKee JMT, Finch-Savage WE. 2001. Membrane priming – a method for small samples of high-value seeds. *Seed Science and Technology* 29, 587-597. Shenoy, V.V., Dadlani, M., and Seshu, D.V., (1990) Association of laboratory assessed parameters with field emergence in rice: The nonaoi acid stress as a seed vigour test. Seed Research, 18: 60- 69.

Rowse HR, McKee JMT. 1999. Seed priming. World Patent No. 9608132.

Sataka, Y., Ohara, T and Sugiyama, M. (2008). The history of melon and cucumber grafting in Japan. ActaHorti. 767:217-228.

Smith, P.G., and Mollett, A.H., (1964). Germinating and sprouting response of the tomato at low temperatures. Jour. American Soc.Hot. Sci. 84: 480-484.

Styer, R.C. and Koranski, D.S. (1997). Plug and transplant production. Ball Publishing, Batavia, IL.

Sun L, Miao Z, Cai C, Zhang D, Zhao M, Wu Y, Zhang X, Swarm SA, Zhou L, Zhang ZJ, Nelson RL, Ma J. 2015. GmHs1-1, encoding a calcineurin-like protein, controls hard- seededness in soybean. *Nature Genetics* 47, 939-943.

Suszka B, Muller C and Bonnet-Masimbert M.1996. *Seeds of forest broad leaves: from sowing to harvest.* (English Translation), Paris: INRA Editions.

Taylor AG, Klein DE, Whitlow TH. 1988. SMP: Solid matrix priming of seeds. *Scientia Horticulturae* 37, 1-11.

Taylor AG, Prusinski J, Hill HJ, Dickson MD. 1992. Influence of seed hydration on seedling performance. *HortTechnology.* 2, 336-344.

Venema, J.H., Dijak, B.E., Bax Jm., Hasset, P.R and Elzenga, J.T.M. (2008). Grafting tomato o to the rootstock of a light altitude accession of Solanum habrochaites improves sub optimal temperature tolerance. Environment and Experimental Botany. 63: 359-367.

Walters, C. (1998). Understanding the mechanisms and kinetics of seed ageing. *Seed Science Research* 8: 223-244.

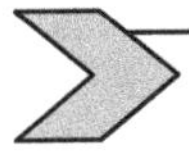

Annexure 1

Acronyms

AASCO	:	Association of American Seed Control Officials
AOSA	:	Association of Official Seed Analysts
AOSCA	:	Association of Official Seed Certifying Agencies
ASSINSEL	:	International Association for the Protection of Plant Breeder's Rights
ASTA	:	American Seed Traders Association
CGIAR	:	Consultative Group on International Agricultural Research
CSC	:	Central Seed Committee
CSCB	:	Central Seed Certification Board
CSTL	:	Central Seed Testing laboratory
CVRC	:	Central variety Release Committee
EMC	:	Equilibrium Moisture content
ERH	:	Equilibrium Relative Humidity
FAO	:	Food and Agriculture Organization
FIS	:	Federation International Seed Trade
GADA	:	Glutamic acid Decarboxylase Test
GATT	:	General Agreement on Tariffs and Trade
ISAC	:	International Seed Analysis Certificate
ISSS	:	International Society for Seed Science
ISST	:	Indian Society for Seed technology
ISTA	:	International Seed Testing Association
NSC	:	National Seed Corporation
NSDC	:	National Seed Development Council
NSP	:	National Seed Project, National Seed Programme
NSSL	:	National Seed Storage Laboratory

OECD	:	Organization for Economic Cooperation and Development
OEEC	:	Organization for European Economic Committee
PBR	:	Plant Breeder's Rights
SCST	:	Society for Commercial Seed Technologists
SFCI	:	State Farms Corporation of India
SIDP	:	Seed improvement and Development Programme
SSDC	:	State Seeds Development Corporation
SSCA	:	State Seed Certification Agency
SVRC	:	State Variety release Committee
TRIPS	:	Trade Related Intellectual Property Rights
WTO	:	World Trade Organisation

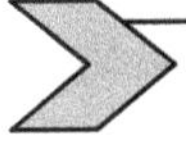

Annexure 2

(Strength does not come from physical capacity, it comes from indeomitable will)

Glossary

Important terminology associated with Seed Vigour and Seed Vigour Testing

"A" Line: Cytoplasmic male sterile seed parent used u production of commercial hybrids.

Abnormal Seedlings: Abnormal seedlings are those seedlings that do not show the potential to develop in to a normal plant even when grown under good quality soil, favourable conditions of moisture, light and temperature and are of the following three types.

(i) **Damaged seedlings:** Seedlings with any of the essential structures missing or so badly and irreparably damaged such that balanced development cannot be expected.

(ii) **Deformed and unbalanced seedlings**: Seedlings with a weak development or physiological disturbances or in which essential structures are deformed or out of proportion.

(iii) **Decayed seedlings:** Seedlings with any of their essential structures so decayed, diseased as a result of primary infection that normal development is prevented.

Abortive: Barren, Defective

Absorption: Intake of water molecules from the ambient air by a seed.

Aeration: Allowing air to move through seeds for maintenance of seed germination or viability.

Accelerated Ageing: Treating the seed under abnormal conditions of temperature and moisture so as to make it deteriorate fast.

Accelerated Ageing Test: A test commonly used for rapid determination of both seed vigour and storage potential, by exposing the seeds to high humidity (80-100%) and high temperature (40- 45°C) for specific period of time. The percent germination indicates the relative storability of seed lots.

Admixture: Anything added to the seed other than the kind or variety specified.

Adsorption: Accumulation of a thin layer of water or gases on the surface of another substance.

Ageing: Gradual accumulation of deleterious changes over a period of time.

Air- tight storage: Storage of seed wherein external air and water vapour are not allowed to enter.

After Ripening: Refers to changes taking place in a seed during the period of rest after harvest that is necessary for germination.

Albuminous seed: A seed with little or no endosperm.

Alternate temperatures: Treating the seeds with high and low temperatures alternately.

Amino acid: Organic acids containing one or more amino groups, at least with one carboxyl group and sulphur and is linked together by peptide bonds to form protein molecules.

Analytical purity: The proportion of intact seeds of a cultivar in a seed lot.

Anthesis: The period of flowering during which florets are open, anthers are extended, pollen sacs burst to release pollen.

Artificial ageing: Involves exposing of seeds to unfavourable conditions.

Basic Seed: A class of seed intended for the production of certified seed.

"B" Line: Fertile counterpart of A – line.

Biological Yield: The total yield of a plant material.

Black layer: The abscission layer formed on the hilum region at physiological maturity of seed.

Brick gravel test: It is a type of seedling emergence test to determine the vigour of seeds.

Bulk: Assembly of similar seeds in to a large volume.

Clone: A group of individuals with common ancestry propagated vegetatively.

Container: A box, bottle, or other things in which seeds are placed or packed.

Cold test: Involves testing of seeds for germination in moist soil at 5-10 °C for one week and then at warmer temperatures.

Cool germination test: A test devised to measure the effect of germination and seedling growth rate of cotton at 18 °C.

Coleoptile: Protectives heath covering the shoot apex of the embryo in monocotyledonous plant.

Constant temperature: A specific temperature used in germination testing which should not vary more than 1°C.

Cultivar: A variety, strain, race or hybrid that has originated and persisted under cultivation or developed for cultivation through selection or hybridization.

Cotyledon: Theembryonic first leaf of a germinating seed, often stores the food material.

Cytoplasmic Male sterility: CMS: A type of male sterility conditioned by cytoplasm rather than by nuclear genes and transmitted only through female parent.

Damaged seed: Broken part of a seed smaller than half of its original size.

Damaged seedling: seedlings with no cotyledon, or constricted seedlings or seedlings with split cracks lesions on the essential structures.

Dead seed: Seeds, which at the end of the test period are neither hard nor fresh and have not germinated or produced seedlings.

Decay: Breakdown of tissues associated with microorganisms.

Defective seedlings: Seedlings that are deformed. Decayed or damaged to such an extent that normal **development** is prevented.

Dehumidification: It is a process by which water vapour from air is removed.

Direct Vigour Tests: Those tests that stimulate potential unfavourable field conditions on a laboratory scale. e.g. AA test, Cold test, Brick gravel test, Paper piercing test etc.,

DNA: A chemical present mostly in the nucleus of a cell that carries the hereditary or genetic information generation after generation.

Dockage: Percentage impurities present in seed sample.

Dormancy: An inherent state of seed that prevents it from germinating despite the presence of favourable conditions.

Dormant seed: The seed with dormancy is termed as dormant seed.

Embryo sac: the eight celled female angiosperm gametophyte.

Emergence Index (EI) = TiNi/S, where, Ti= number of days after sowing, Ni = number of seeds germinated on day I and S= total number of seeds planted. Scot *et al.,* (1984) initial emergence can be calculated as a percentage of the number of seeds planted.

EMC: Equilibrium Moisture content: The moisture content of seed when it is equilibrium with the ambient atmosphere.

ERH: Equilibrium Relative Humidity: A condition in which the air and seed are in same temperature where in the relative humidity of ambient atmosphere is equivalent to the seed with a **predetermined** moisture content.

Endosperm: A tissue that is required for providing nutrition to the developing embryo.

Endospermic seed: Seeds bestowed with special storage tissue at maturity is called endospermic seed.

Epicotyl: The stem of seedling or embryo between the cotyledons and the first leaf.

Epigeal germination: A type of germination in which hypocotyl elongates and carry the shoots and cotyledons to the above ground portion.

Field Emergence Ability: It is the major aspect of seed quality that concerns growers, and high germination serves as a prerequisite for seeds to be sown.

First germination count: Percentage of normal seedlings obtained at first count of germination. A measure of rapidity of germination conducted midway of the final count.

Fresh Ungerminated seed: The seeds that absorb water but do not germinate and remain fresh in a germination test.

Foreign seed: Undesirable portion of seed sample consisting of weed and other crop seeds.

Fumigation: It is a method of controlling storage pests by using a volatile chemical in an airtight chamber.

Genetic male Sterility: GMS: A type of male sterility conditioned by nuclear genes and transmitted by the male or female parent.

Germination: The ability of a seed to produce a normal seedling under favourable conditions. So, germination of a seed in laboratory test is nothing but the emergence and development of the seedling under favourable conditions to a stage where the aspects of its essential structures indicates whether or not it is able to develop further in to a satisfactory plant. Usually, it is measured by percentage, i.e. the number of seeds in a seed lot that are capable of and expected to germinate and grow in to normal healthy plants. The germination percentage is a reproducible index of seed performance, based on which its planting value is assessed and trade in seeds can operate at international level.

Germination capacity: The germination capacity of a seed lot is expressed in terms of percentage of seedlings developing normally under favourable laboratory conditions.

Germination Test: A test conducted under controlled and standardized laboratory conditions to determine the planting value of a seed lot.

Germination Value: Peak value X total number of normal seedlings/ days of final count.

Genetic Purity: Plants and seed conforming to the characters of a variety described by plant breeder at the time of release/ Notification.

GOT: Grow Out Test: Sowing the seeds in controlled plots to determine the genuineness of seed.

Hard seeds: Seeds that have not absorbed water and remained hard till the end of the test period due to the impermeability of the seed coat.

Husk less seed: or dehusked seed: Seeds which are completely devoid of husk as in rice.

Hybrid vigour: It is also known as heterosis and is the increase in vigour in respect of growth, or yield of a hybrid (derived by crossing two genetically unrelated plants) over their parents.

Imbibition: It is a physical process of water up take by air-dry seeds, usually depends on the matrix potential generated by long chain molecules in the cells of an embryo, when it gains access to free water.

Inert matter: Non-living material such as sand, small stones, dried leaf /stem portions of plants.

Immature seeds: Not fully developed seeds.

Indirect Vigour tests: These tests measure the physiological attributes of seeds associated with germination like speed of germination, seedling growth rate, and vigour index.

Isolation: The act of keeping seed crops away from the source of physical and genetical contaminant.

Lab Germination Percentage: It indicates those seeds that have developed the essential. structures needed for development of a normal seedling under favorable conditions.

Normal seedlings: The seedlings, in the presence of favourable conditions of light, temperature and moisture, grow and develop in to mature plants when sown in healthy soil and must conform to one of the following categories.

    (i)   Intact seedlings: Seedlings with their essential structures well developed, complete in proportion and healthy.

(ii) **Seedlings with slight defects:** Seedlings showing certain defects in their essential structures provided they show an otherwise satisfactory and balanced development comparable to that of intact seedlings of the same test.

(iii) **Seedlings with secondary infection:** Seedlings, which have been affected by pathogen sources other than, parent seed.

Official Sample: Sample of seed drawn by law enforcement authorities to make sure whether or not the seed offered for sale is meeting requirements under seeds Act.

Off type: Plant deviating significantly from the characters described by the breeder in respect of any observable trait.

Other crop seeds: Seeds of plants grown as crops as described by Indian Minimum seed Certification Standards.

Other Distinguishable variety seeds: Seeds of other varieties of the same kind identified on the basis of readily apparent differences in the stable seed morphological characteristics.

Peak Germination Value: It is a value assigned for describing germination rate that is used to indicate seed vigour of tree seeds.

Peak Value: It is the cumulative number of normal seedlings/ days on final count day.

Planting ratio: The recommended ratio in which male and female parental rows are to be planted to make a crossing block in hybrid seed production.

Pelleted seeds: Seeds coated to uniformly to enhance their size to facilitate mechanical planting.

Preconditioning: Preparation of seed for processing; usually materials larger or smaller in size than seed size are removed.

Previous crop: Crop of the same kin/ same or different certification stage or different crop grown in the crop season immediately preceding the seed crop.

"R" line: An inbred line when crossed with sterile 'A' line restores the fertility of hybrid.

Radicle: A portion of embryonic axis that develops in to primary root.

Retarded root: A root usually with an intact tip much too short and weak to be in balance with the other structures of the seedling.

Rogue: An undesirable off type plant present in the seed crop.

Roguing: The act of removing off type plants from seed crop.

Quality seed: Seed with high genetic purity, freedom from diseases, pests weed seed, inert matter and other crop seed and had high germination capacity are called as quality seeds.

Quiescence: Inability of an otherwise germinable seeds to germinate under unfavourable conditions

Sample: A portion of seed from which seed quality attributes can be estimated.

Seed: All propagules, *i.e.* all parts of a plant capable of developing in to another such new plant.

Seed Parent: Female parent from which seed is harvested during hybrid seed production.

Seed lot: A physically identifiable specified quantity of seed against which an analysis certificate can be issued.

Seed rate (kg/ha) = (Target plants/m2 × 1000-seed weight in g × 100)/(Germination % × Potential Field Establishment%). The % potential field establishment chosen for the calculation depends on the condition of the seedbed. As a guide rates of 80% are generally used for good seedbeds and 60% for poor expected conditions.

Seedling Vigour: The rapid growth of seedlings during cotyledon and early true leaf stages.

Seedling Vigour Index = (The mean primary radicle length + the mean of first shoot length) x final germination percentage.

Seed Viability and Vigour: Seed longevity is the relationship between viability and vigour. If one is familiar with the general relationship between viability and vigour, from germination data, it is possible to infer partially regarding the vigour of a variety. In the first stage, If germination is approximately 80 per cent or above, the seed is both vigorous and viable. In the second stage, deterioration progresses very rapidly and in the third stage, deterioration slows down and approximately 20 per cent or below, and all seeds will die slowly.

Seed Vigour: The ability of a seed to germinate and grow rapidly so as to establish a normal seedling. Good seed vigour means rapid uniform emergence and development of normal seedlings under a wide range of field conditions.

Seed Vigour: It is another aspect of physiological quality of seeds, defined as those properties, which determine the potential for rapid, uniform emergence and development of normal seedlings under a wide range of conditions.

Seed Vigour Tests: These tests rank seed lots according to their physiological quality and determine the conditions under which the seed lot can be successfully planted. These tests can be used to decide whether to keep a particular lot in storage for a longer period of time and also helps in identification of possible causes of poor performance of a particular seed lot in the field.

Stratification: Exposing and imbibing seed at low temperatures (5-10°C) or warm conditions (35-40°C) for few days prior to germination for breaking dormancy.

Stubby roots: The characteristic of a seedling root. Often short, club shaped, with intact root tip.

Stunted root: The characteristic of a seedling root, wherein root tip is either defective or missing   irrespective of the length of the root.

Test Weight: 1000 seeds are weighed to produce he weight of 1000 seeds in grams. It is used to calculate sowing rates accurately.

Test weight: Weight of 1000 grains or seeds measured in grams.

Tolerance: It is the maximum permissible difference between the results of two test samples drawn from same seed lot.

Ungerminated seeds: Those seeds which have not germinated by the end of the test period when tested under the prescribed conditions. These are of the following three types as per ISTA Regulations.

   (i)   **Hard Seeds:** Seeds, which remains hard at the end of the test period, because they have not absorbed water.

   (ii)   **Fresh seeds:** Seeds, other than hard seeds, which have failed to germinate under the conditions of the germination test, but which remain clean and firm and have the potential to develop in to a normal seedling.

   (iii)   **Dead Seeds**: Seeds, which at the end of the test period are neither hard nor fresh nor have produced any part of a seedling.

Variety: A sub division of a kind with specific set of characters by which it can be differentiated from other plants of same kind.

Viability: It is the probability that a fertilized egg will survive and develop in to an adult plant and is often applied to plant germination experiments with comparisons across phenotypic classes under standard environmental conditions.

Viability: It is simply the ability of living, growing and developing in to a new plant.

Viability equation: An equation to predict the viability of stored seeds based on behaviour of seeds to storage environment.

Viable Seed: It is one, which is live and potentially capable of germination.

Vigour classification: It was carried out during the germination test. The seedlings were split into; normal, strong and weak, abnormal and dead. Only the normal and strong seedlings were considered (Nakagawa, 1999).

Vigorous seed lot: A vigorous seed lot is one that is potentially able to perform well even under environmental conditions that are not optimal for the species.

Vigour: The sum total of all those attributes of seed that favours rapid and uniform stand establishment in the field.

Vigor tests: These tests are designed to reveal subtle signs of damage, and provide a more reliable indicator of field/greenhouse performance than viability (germination) tests.

Viviparous: A condition in which germinated seed remains attached to the mother plant.

Volunteer plant: Unwanted plants growing from residual seeds of previous crops.

Wing: A dry inert appendage of seed or fruit meant for dispersal.

Winnowing: A process of removal of light and inert material from seed lot by flowing away.

Withering: Impact exerted by fluctuations in environmental conditions on seed quality.

Working sample: A portion of submitted sample actually used in the laboratory for analysis.

X-ray: An electromagnetic radiation with a shorter wavelength than those of visible or UV light.

Zygote: A fertilized egg, resulting from the fusion of two gametes.

Annexure - 2

(i)   Speed of germination

Count the number of seeds germinated every day from the first day and cumulative index is made by the following formula N = n1/1+n2/2+ -------------------------nx/x

Where,

N1 to nx are the number of seeds germinated /day 1 to x.; N= vigour

High value of N is an indication of high seed vigour.

(ii)   Seedling growth rate (Copeland,(1976)

It is determined by dividing the mean increase in length from each previous measure by the number of days the seedling had been in germinator, According to Copeland,(1976), sum of each count at the end of the test period is expressed as seedling growth rate and is calculated by the following formula.

SL1/F1 + SL1-Sl2/F2+ --------------------- Sln-Sl(n-1)/ FN

Where,

Sl1= mean seedling length at first count,

Sl2 = mean seedling length at second count,

SL1-Sl2 = mean increase in length in second count,

F1 = days to first count

  Fn = days to final count.

(iii) Vigour index length Percent (Abdul Baki and Anderson,1973)

VI(%)= Germination x seedling length on final count day

Vigour Index mass(%) = Germination x seedling dry weight on final count day

(iv)   Seed Metabolic Efficiency (SME) Proposed by Gangadhararao and Sinha, 1993)

SME = SHW+ RTW + RESP

RESP + SDW – (SHW + RTW + RSW)

Where,

SDW= dry weight of seed before germination,   SHW= dry weight of shoot,

RTW= dry weight of root, RSW= dry weight of seed after germination.

(v)   Mobilization efficiency (ME) = increase in dry weight of seedling/ decrease in weight of cotyledons x100

(vi)   *Mean Time of Germination (MTG)*: This is an index of germination rate and Acceleration can be computed by the following equation.

MTG = □(nd)/ □n

(vii) Daily Germination Speed were obtained by the equation MDG=FGP/d

Where,   FGP is the Final Germination Percentage, and d is the number of days to reach to maximum final germination (Hunter et al., 1984).

(viii)   Final Germination Percentage =Number of germinated seeds / total number of seeds planted.

(ix)   Germination velocity = Number of germinated seeds/ day of count (Magurie, 1962).

(x)   Mean time to germination = ($\square$ni × di) /N, where n is the number of seeds germinated at day I , d the incubation period in days and N is the total number of seeds germinated in the treatment (Brenchley and Probert, 1998).

(xi)   Seed Vigour = Length of root+ length of shoot x germination %/ 100 (Abdul baki and Anderson, 1973).

Index

Symbols

α-amylase content 135

100 seed weight 133

65% cold test 258

90% germination 258

A

ABA or osmotic stress 53

Abiotic stresses 17, 134, 182

Abnormal embryos 150

Abnormal seedlings 6, 10, 42, 63, 92, 132, 138, 142, 182, 207, 287

Accelerated Ageing test 22, 84, 108, 215, 233, 236, 254, 287

Accumulation of ethanol 205

Aggregate sizes 114

Agricultural systems 1

Agriculture 1, 22, 77, 105, 110, 186, 251

Agro-climatic condition 2, 24, 276

Agronomic management 3, 163

Agronomic traits 173,

Aldehyde protein 201

Allelochemicals 112

Allelopathic chemicals 115

Altering plant population 166

Alternating temperatures 242

Ambient temperature 128, 169

Amino acid composition 131, 132, 147

Angiosperm seeds 48

Anoxia 13

Antioxidant defense 212

Aquatic plants 50

Arabidopsis 37, 53, 95, 195

Artificial longevity 184

Ascorbic acid 134, 206, 212

ATP 53, 68, 142, 192, 201

ATP synthesis 54, 68, 142, 201

Auto oxidation 200

Autotrophic growth 13

B

Biochemical vigour 37, 95

Biological yield 11, 165

Biomass 15, 167, 173

Boron deficient 138

Breaking dormancy 62

Breeder seed 5, 187

Brick Gravel Test 83, 90

Broken seeds 139, 181

C

Capital investment 273

Catabolic process 183, 215

Cell division 48, 58

Cell function 106, 190

Central Seed Testing Laboratory 23

Chilling injury 19, 198

Chilling sensitive 187

Chromosome bridges 195

Chromosome frequency 190

Classify seeds 110

Co-adaptive genotype 39, 129

Cold test 22, 83, 108, 116, 233, 254

Comparative testing 235

Competitive market 118, 240

Complete death 142

Complete germination 57, 69, 90, 113

Conductivity test 20, 84, 232, 253

Conductivity test results 234

Cost benefit ratio 245
Cotyledons 9, 59, 138, 202,
Coupling 209
Critical stages 12, 235
Critical threshold level 63, 110
Crop establishment 12, 37, 97, 109, 163, 272
Crop productivity 3
Crop stand 8, 41, 134, 170
Crop varieties 23, 169
Crop yield 16, 82, 94, 149, 164, 166, 171
Crops and varieties 2
Cryopreservation 186
Cultivar 3, 15, 50, 61, 91, 105, 145, 274
Cultivation practices 11, 166

D

Decreased yields 42
De-esterification 191
Dehydration stage 207
Deletions 195
Density grading 136, 278
Desertification 181
Desiccated state 206
Desiccation tolerance 84, 141, 187, 205
Developmental stage 169, 205, 277
Differential storage 134, 151
Direct seeded rice 13
Discrepancy 84, 232
Diversified market 22
DNA 7, 53, 69, 138, 150, 192, 204
DNA repair mechanism 207
DNAase nicking 209
Domesticated species 2
Dominance Hypothesis 36, 95
Dormancy
    2, 21, 37, 57, 92, 112, 132, 244, 277
Dormancy breaking 63, 278
Dormancy character 132, 260
Dormancy characteristics 132
Dormant seeds 115, 244
Driving force 37, 83, 112, 192

Drought stress 138, 165
Dry matter 14, 82, 131, 164
Dry matured seed 107
Dry seed tissues 191
Dry seeds 2, 49, 83, 142, 187, 199
Dry weight 2, 82, 151, 203
During storage
    3, 35, 105, 107, 129, 163, 197, 229, 270
Dwarf 42

E

Ecotype 50, 61
Effects of vigour 44, 82, 91
Electrical conductivity 83, 133, 234, 253
Embryo 9, 51, 57, 112, 147, 189
Embryonic axis 1, 52, 141, 192
Endo-reduplication 58
Enhanced RQ 6, 182
Environmental conditions 3, 38, 68, 77, 105,
    130, 163, 190, 233, 255, 266,
Environmental factors 38, 88, 137, 194
Environmental stresses 36, 60, 163, 190, 241
Enzymes 6, 40, 69, 112, 130, 182, 196, 208
Essential structures 39, 62, 66, 129
Evaporative cooling 114
Excessive pressure 113
Experimental Sample Testing (EST) 235
Exponential 171
Exporters and importers 266, 279
Extensive root tissues 170
Extrinsic variation 129

F

F_1 hybrid 43, 109
F10 recombinant inbred lines 95
Facilitate aeration 278
Farm saved seeds 3
Farmers 3, 12, 23, 68, 107, 230, 266, 272
Fast drying 205
Fertility status of soil 149
Field emergence
    3, 19, 38, 80, 105, 129, 227, 254, 275,

Fine mapping 36
First formed seeds 132
First Referee Test Committee 235
Flowering stage 138
Forage crops 20, 171
Foundation seed 5, 187
Free radicals 199, 209, 211
Fungal pathogens 140, 186

G

GADA test
Gene redundancy
Genetic conservation
Genetic makeup
Genetic resources
Genetic variation
Genotype x environment interaction
Germination
Germination capacity
Germination data
Germination percentage
Germination phase
Germination tests
Germination values
Gibberellic acid
Glassy state
Grafted seedlings
Grafting
Grain production purpose
Grass seeds
Growth parameters

H

Harvesting stage 133
Heat denaturation 48
Hemicellulose 184
High germinating seed
79, 109, 228, 252, 268
High germination results 228
High lysine maize 136
High performance lipid chromatography 211

High quality 14, 82, 106, 170, 183, 215
High Quality seed 14, 106, 171, 271, 280
High seed vigour 8, 119, 251
High soil temperatures 114
High solute leakage 144
High-density seed 136, 269
Hybrid vigour 137, 290
Hydration 7, 52, 212
Hydro Time Model 51
Hypersensitivity 202
Hypocotyls growth 267
Hypoxia 13, 212

I

Ideal conditions 89, 115, 227, 258
Ideal plant population 272
Identification of cultivars 95
Imbibed seed 6, 54, 182, 193
Improper equipment 252
Improved protein and oil content 151
Indeterminate growth 17
Individual seed analysis 244
Initial moisture content 258
Integrity of cell membrane 40, 129, 194
International Seed Testing Association
7, 83, 257, 285
International Seed Trade 241, 274
Intrinsic variation 129
Irreversible changes 193
ISTA Proficiency 237
ISTA Rules 67, 108, 238

L

Label attached 231
Laboratory germination 22, 60
Laboratory germination test 22, 62
Lag phase 55, 195
Large seeded legumes 117, 141, 267
Large seeds 11, 114, 131, 165
Late maturation phase 192
Leaf area index 16, 169

Leakage of solutes 52, 236
Leguminous seeds 9
Lethal damage 215
Life cycle 68, 190
Lipid peroxidation 84, 134, 208
Lipid peroxide 209
Lipids hydrolyzing enzymes 208
Lipoxigenase enzyme 212
Lipoxigenases 200
Liquid-crystalline phase 207
Longevity 5, 48, 150
Long-term conservation 64
Loss of seed viability 199, 214
Low moisture content 9, 25, 134
Low moisture levels 202, 212
Low vigour seeds
      16, 38, 42, 82, 150, 164, 172, 271
Lysosomes 193

M

Marker assisted Selection 96
Marketing 23, 107, 120
Maturity stage 15, 81, 133
Maximum yields 20, 166
Mechanical damage 11, 92, 137
Mechanized harvesting 38, 68
Membrane priming 278
Messenger RNA 54, 190
Metabolic reactions 41, 61
Methionine 54
Microbial concentrations 230
Microorganisms 25, 110, 256
Millard products 192
Millard reaction 192
Minimum tillage 82
Mitochondria 36, 193
Modern greenhouses 273
Molecular evidence 96
Monocot species 59, 114
mRNA and DNA 53
Multiple interval mapping 97

N

NADH and NADPH 53
Natural ageing 183, 198, 202
Natural robustness 38, 68, 88
Negative correlation 204, 234
New seed policy 23
Nitrogen fertilizers 132, 146
Non dormant 2, 51, 63
Non dormant seeds 51, 63
Non linear correlation 119
Non-productive plants 149
Non-uniform conditions 40
Non-viable seeds 194, 200
No-till production 82
Numerical index 119
Nutrient filling duration 167

O

Off coloured seeds 133
Oligo- saccharides 212
Orthodox seeds 51, 60, 134, 185
Over Dominance 36, 95
Oxidative process 50
Oxidative stress 192, 199
Oxygenated fatty acids 142, 208

P

Paper towel 67
Partial folding 197
Pathogens 5, 14, 91, 140, 233
Percentage seedling emergence 15, 253
Phenotype 190, 266
Phospholipase 196, 204
Physical quality 61
Physiological ageing 40
Physiological and physical 253
Physiological quality
      4, 45, 167, 228, 275, 292
Pigmented cultivars 145
Pigmented testa 19, 136
Pine seed germinability 208

Plant breeding 18, 257, 267
Plant genetics 169
Plant population 21, 39, 82
Plant stand establishment 20, 136, 168
Polymerization of proteins 210
Poor seed quality 134
Poor seedbed preparation 15
Poor seeding conditions 265
Positive advantages 171
Positive relationship 136, 171
Post fungicides 254
Post harvest handling 8, 24, 62, 265
Post planting 276
Post-harvest processing 2
Primary branches 133
Processing and marketing 23, 256
Proficiency Test Committee 237
Proficiency Testing 237
Profitable crop production 18, 170
Progressive deterioration 67, 117
Progressive shrinkage 205
Prolonged harvesting 143
Proof containers 187
Protein metabolism 7
Puddled rice system 14

Q

QTLs 36, 151
Qualitative character 47
Quality assurance 4, 23, 120
Quality control 22, 83, 140, 240
Quality evaluation 229
Quality seed 268
Quantification of results 230
Quantitative character 94

R

Radical attack 195, 209
Radical emergence 56, 58
Radicle and plumule 56
Radicle emergence 57, 236

Radicle fresh weight 167
Radicle protrusion 69, 243
Raffinose series 53, 203
Range of vigour 243
Rate of decline 137
Rate of growth 38, 190
Real field conditions 79, 165, 253
Recalcitrant seeds 185
Reduction in leakage 141
Referee testing 235
Re-imbibition 278
Relative humidity 6, 128, 289
Repair enzymes 48, 195
Repair system 193
Reproducible 4, 62, 228
Reproductive maturity 82, 170
Research and Development 240
Respiration 53, 192
Retard crystallization 191
Ribosomal (mRNA) 54
Rights of farmers 4
Ripening period 61
RNA polymerases 55
Robust seedling 67, 113
ROS 55, 195
Rules for Seed Testing 89, 111

S

Scarification 21
Schieff's base 193
Secondary dormancy 61
Secondary wall 194
Seed 1, 4, 17
Seed ageing 5, 25, 48, 117, 143
Seed Analyst's 252
Seed attributes 87, 89, 90
Seed bag 228, 277
Seed bank 20, 39
Seed biology 3
Seed borne fungi 142, 188
Seed Certification 4, 8, 129

Seed Certification Agencies 23, 270
Seed coat 9, 19, 51, 136
Seed coat cracking 19, 137
Seed coat imposed dormancy 195
Seed coat pigments 277
Seed coat thickness 267, 279
Seed companies 107, 120, 256, 269
Seed development 9, 18, 133, 138, 165, 278
Seed dormancy 17, 244
Seed dressing chemicals 109
Seed drying 48, 91, 205
Seed filling period 138
Seed Growers 24, 80, 271
Seed health testing 237
Seed industry 4, 22
Seed longevity 150
Seed lots 3, 15, 42, 77, 143, 147, 227, 257
Seed maturation 19, 38, 58
Seed moisture 18, 113, 186
Seed multiplication ratio 5, 9
Seed physiological resistance 90
Seed physiologist 3, 50, 106
Seed priming 13
Seed producer 8, 15, 25, 227, 240
Seed production areas 163
Seed production programmes 164, 265
Seed properties 39, 88
Seed quality 2
Seed quality evaluation 229, 231
Seed replacement (SRR) 5
Seed singulation 146
Seed sowing rates 82, 168
Seed storage 14, 25, 96, 120, 271
Seed Technologists 5, 41, 106, 115
Seed testing 3, 23, 77, 92, 232
Seed testing laboratory 3, 23, 229, 256
Seed trade 22, 65, 241
Seed vigour chapter 236
Seed vigour index 242, 256, 260
Seed vigour test 22, 77, 90, 117

Seedbed preparation 12
Seedling 11
Seedling abnormalities 42, 92, 150
Seedling development 7, 19, 56, 138
Seedling establishment 3, 8, 13, 170
Seedling vigour 11, 48, 167
Seeds Act 1966 4
Seeds for testing 65, 242
Sharpness of test 231
Shrunken seeds 165
Sigmoid survival curve 46
Single tests 232
Slow emergence 16, 21, 173
Small to medium 185
Soil microorganisms 233, 256
Soil physical conditions 80
Soil tilth 169, 173
Soil-borne diseases 21
Solar radiation 114, 166
Sorting seeds 112
Sowing conditions 39, 119, 129, 268, 273
Sowing season 232, 265
Specific crops 231, 246
Specific microorganisms 256
Specific symptoms 183, 215
Speed of germination
     16, 40, 64, 83, 91, 108, 147, 254,
Split or missing coleoptiles 150
Stable genotypes 164
Stable index of seed 228
Stage of survival 260
Standard germination test 22, 61, 110, 245
Standardization of treatments 231
Standardization of vigour 229, 239
Storage and bagging 240
Store seed lots 270
Storehouse 270
Stresses 6, 17, 36, 108, 165, 238
Sub optimal conditions 38
Subnormal heat 198

Suboptimal conditions 4, 10, 24, 109, 148
Suboptimal soil conditions 107
Sugar- protein 192
Superior parent 36
Supermarket chains 97, 271
Survival curve 50, 83, 260
Synchronous emergence 81, 105, 150

T

Telomerase 55, 56
Temperature 114, 147, 189
Test methods 92, 238
Test parameters 228
Testing protocols 38, 84
Tetrazolium test 84, 233, 254
The Seeds Bill 2004 4
The vigour test committee 234, 254
Threshing conditions 139
Threshold model 49
Tocopherol content 195
Tolerant 127, 185
Tolerate ageing 254
Tolerate stresses 255
Total free amino acids 200
Total lipids 210
Toxic metabolites 115, 196
Transcryptomic analysis 57
Transition elements 200
Triebkraft 37, 68, 83, 90
Turgor pressure 52, 112

U

Uniform 232
Uniform emergence 90, 168, 272, 292
Uniform test 229
Utilization of germination 118

V

Viability and vigour 18, 45, 82, 142, 168, 292
Viability curve 45, 68
Vigour 1, 7
Vigour curve 45, 68
Vigour scale 93
Vigour test 35, 77
Vigour Test Committee 83, 89, 238
Vigour testing 66, 83, 90
Vigour testing protocols 38, 242
Vigour tests 35, 79

W

Warm season crops 274
Weathering 2, 137, 253
Weeds 3, 16, 115, 172
Well-equipped laboratories 259
Wholesalers 230, 266, 279
Wild type seeds 203
WTO standards 4

Y

Yield comparisons 251
"Y" months 252